FUNDAMENTALS OF MODERN BOTANY

FIG. 1.—(*Frontispiece.*) General View of a Wallflower plant (*Cheiranthus Cheiri*), about ¼ natural size. *g–g*, level of ground.

FUNDAMENTALS
OF
MODERN BOTANY

By

DUNKINFIELD HENRY SCOTT

Re-Edited by

F.T. BROOKS, M.A., F.R.S.

WITH 117 ILLUSTRATIONS
AND A GLOSSARY

Vol. I

ANMOL PUBLICATIONS PVT. LTD.

NEW DELHI - 110 002 (INDIA)

ANMOL PUBLICATIONS PVT. LTD.
4374/4B, Ansari Road, Daryaganj
New Delhi - 110 002
Ph.: 23261597, 23278000
Visit us at: www.anmolpublications.com

Fundamentals of Modern Botany

ISBN 81-261-1470-3 (Set)

PRINTED IN INDIA

Published by J.L. Kumar for Anmol Publications Pvt. Ltd., New Delhi - 110 002 and Printed at Mehra Offset Press, Delhi.

PREFACE TO THE THIRTEENTH EDITION

THE text has again been revised and considerably amplified in the present edition. The Types selected have been compared more fully with other kinds of Flowering Plants. In the Chapter on "Outlines of Physiology" additional experiments of a simple kind have been incorporated. I am again indebted to my colleagues, Mr. G. E. Briggs and Dr. E. J. Maskell, for much help.

F. T. BROOKS.

PREFACE TO THE TWELFTH EDITION

THE text has been thoroughly revised in this edition. The principal changes concern Chapter III., "Outlines of Physiology." The results of recent investigations on the ascent of sap, on the translocation of carbohydrates and nitrogenous substances, and on the influence of auxin in promoting curvatures of plant organs in response to external stimuli, have been incorporated. I am much indebted to my colleagues, Mr. G. E. Briggs, F.R.S., and Dr. E. J. Maskell, for assistance generously given.

F. T. BROOKS.

PREFACE TO THE ELEVENTH EDITION

Structural Botany has now had a career of over thirty years, Part I. having first appeared in 1894, and Part II. in 1896. I have repeatedly revised the book for successive editions, but felt that the time had come when a more drastic treatment was desirable. With this end in view, I thought it wise to call in a colleague with recent experience in general botanical teaching. I was so fortunate as to secure the co-operation of my friend, Mr. F. T. Brooks, University Lecturer in Botany at Cambridge. The publishers sanctioned my proposal, and Mr. Brooks has thoroughly revised the present volume and brought it up to date.

The chief changes which he has made are in Chapter III., "Outlines of Physiology." Mr. Brooks has not only much improved the chapter as originally limited, but he has further added two entirely new sections, on Growth and on Movement of Plant Organs. Thus the physiological portion of the book, however brief, now gives a general survey of the mode of life of a typical flowering plant.

The authors are much indebted to Miss E. R. Saunders, Dr. F. F. Blackman, F.R.S., and Mr. G. E. Briggs for help kindly given.

D. H. SCOTT.

PREFACE TO THE FIRST EDITION

THIS book is intended as a first guide to the study of the structure of plants. Botany is now taught in schools of all kinds, and wherever Botany is taught it has become customary to expect some knowledge of the construction of plants, and of the function of their organs. All that I have aimed at in this book is to secure that such knowledge, when first acquired, shall be correct as far as it goes.

My purpose has been to write an Introduction to *Structural Botany*, not a manual of Botany in general. It is absolutely necessary that schoolboys and girls, if they are to learn this science at all, should also gain a knowledge of plants in the field. For this part of the work a guide is necessary, and some such book as Professor Oliver's *Lessons in Elementary Botany* is indispensable.

The type-system has been adopted, as far as practicable, because it seems better to gain as thorough a knowledge as possible of a few plants, rather than to acquire mere scraps of information about a larger number. The types have been specially investigated for the purpose of this book, and many of the figures are original.[1] For the rest, the authority is cited. An effort has been made to point out those structural characters which are of wide importance, as distinguished from those which are peculiar to the type, or its nearer allies.

The subject-matter is not always easy, and some parts of the book will not be followed without close attention. There is really no reason why Natural Science should be regarded in schools as a specially easy subject. If Science is to be taken seriously, it rather seems desirable that those who

[1] Those signed " R. S." have been drawn by Mrs. D. H. Scott.

study it should have to use their brains as much as in learning Euclid, Algebra, or Grammar.

If any real knowledge of the subject is to be gained, practical work is essential. It is expected that the teacher should have sufficient training to be able to demonstrate to his class most of the structural features described in the book. We are now well supplied with laboratory guides, such as the works on Practical Botany by Professors Bower and Strasburger.

If the present volume is found useful, it is proposed to add a second, treating in like manner of Cryptogamic types.

I much hope that not only pupils in schools, but also private students, may derive some help from this book, and that it may awaken in some readers a genuine interest in the study of living things.

D. H. SCOTT.

CONTENTS

CHAPTER II

TYPE II

CHAPTER III

CHAPTER IV

TYPE III

THE SPRUCE FIR (*Picea excelsa*, Link)—

INTRODUCTION

In beginning the study of plants we cannot perhaps do better than fix our attention on any common kind of plant with which we may happen to be familiar, and consider what are the most striking points in its construction. In this way we may hope, starting from the general knowledge which we all possess as the result of everyday observation, to pass on to that more accurate and systematic knowledge which is called Science.

It matters little with what plant we begin. The Wall-flower is well known to everybody, so we will take that as our example.

The chief parts of which such a plant consists are easily distinguished (see Fig. 1). There is a branched colourless root, which is fixed firmly in the ground ; from this rises the upright stem bearing the flat leaves. The stem is branched, each branch arising from the angle between a leaf and the main stem ; the branches also bear leaves, and resemble in all respects the stem on which they grow. Later on the flowers appear, borne on the upper part of the stem and its branches, above the leaves. These flowers, when they wither, give place to the fruits or seed-vessels, and in these the seeds themselves are ripened, which, when sown, will produce a new generation of wallflowers for next year.

Now a plant, like all living things, is made up of *organs* ; thus a leaf, a stem, a root, or a flower is not merely a part of the plant, but it is a part which does some definite work for the good of the whole. The highest or most perfect plants are those in which the division of labour is most complete, in which the principle of setting apart a distinct organ for each distinct kind of work is most thoroughly carried out. The Wallflower is an example of a very highly organised plant. Later on we shall make the acquaintance of plants in which there is little or no division of labour—in which, that is, the organs are not distinct, or at least not evidently so.

The work which an organ has to do is called its *function.* In the Wallflower, as in the higher plants generally, the *root* has two manifest functions : it has to hold the plant firmly in the ground, and also to take up food from the ground. The *leaves* have for their chief office the absorption of food from the air ; we shall soon find that a green plant obtains quite half its food from this source. The *stem* has to conduct the food, which the root and the leaves absorb, to other parts of the plant, and it also has to support the leaves in such a position that they may best be able to do their work ;[1] and thirdly, the stem has also to bear the *flowers.* The flowers themselves have quite a different duty to perform. Their business is to produce the *seeds,* and so to provide a supply of new wallflowers for the future. We see, then, that the organs which we have mentioned are of two kinds. On the one hand, the root, stem, and leaves do work for the benefit of the particular plant to which they belong ; these we call the *vegetative organs.* On the other hand, the flowers are concerned in the production of fresh plants ; they are therefore called the *reproductive organs.*

One important division of Botany, then, is concerned with such questions as those which we have just roughly indicated. We aim at finding out how a plant lives, what different kinds of work it has to do in order to live and to

[1] It is often convenient to speak of the stem and leaves together as *the shoot.*

produce others of its kind, and by means of what organs the work is done ; also what is the structure of these organs by which they are enabled to perform their office. All these questions belong to *Physiology* in its widest sense. Physiology is concerned with the question what a plant *does*, and what its various organs do. The answer to such questions must be obtained by *experiment*. The study of structure, or *Anatomy*, is from this point of view a necessary auxiliary to physiology.

But we can also look at plants from a different point of view. Suppose, for example, that instead of considering the Wallflower alone, we compare it with some other plant, such as a Potato. A potato plant differs in many points from the Wallflower, but we will confine ourselves to one of these differences. It not only consists of the stem bearing its green leaves and ultimately producing the white or mauve flowers, and of the root with its fibrous branches, but it also forms the potatoes themselves (or *tubers*, as they are called), which we eat. Now these potatoes, as every one knows, grow, like the root, underground, borne on subterranean branches. Both they and the branches which bear them are colourless, and from their general appearance and position every one who has not learnt Botany at once classes them as roots. But is this their true nature ?

If we closely examine the branches on which the potatoes grow, we shall find that they bear small scaly leaves (Fig. 2, *sc*). Similar little leaves (*ec*) will be found on the potato itself, adjoining the " eyes," and the " eyes " themselves are buds (*bd*), which grow out into new plants when the potatoes are sown. Leaves and buds, however, are characteristics of stems ; the former are never, the latter rarely, found on undoubted roots. Besides this, a more thorough examination would show that the internal structure and the mode of growth of the potatoes and of the underground branches which produce them are those of stems and not of roots. Further, we may sometimes find on the parts of the potato plant above the ground bodies which are intermediate in form and structure between potato tubers and

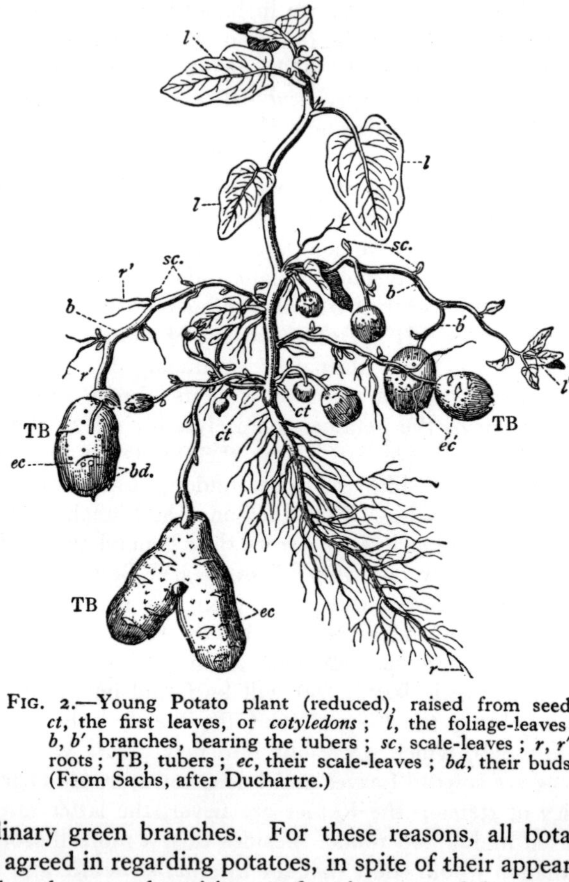

FIG. 2.—Young Potato plant (reduced), raised from seed. *ct*, the first leaves, or *cotyledons* ; *l*, the foliage-leaves ; *b*, *b'*, branches, bearing the tubers ; *sc*, scale-leaves ; *r*, *r'*, roots ; TB, tubers ; *ec*, their scale-leaves ; *bd*, their buds. (From Sachs, after Duchartre.)

ordinary green branches. For these reasons, all botanists are agreed in regarding potatoes, in spite of their appearance and underground position, as forming part of the shoot and not of the root. In other words, we come to the conclusion that certain shoots of the potato plant have become completely modified, so as to lose all the more obvious characters of shoots, and to bear a superficial resemblance to roots. These changes are the external manifestation of a complete alteration of function. The potato tubers have given up the usual functions of shoots, and become adapted to serve as storehouses of food (chiefly starch) for the young plants which will grow from the " eyes " next year.

If, then, we compare the potato plant with the Wallflower, we find that, while in the latter the shoot consists entirely of above-ground branches bearing green leaves, in the potato plant certain parts of the shoot are changed into underground colourless organs, bearing scale-leaves, and ending in tuberous swellings, which are storehouses of food.

But we must not suppose that all underground organs which serve this purpose are shoots. Turnips, radishes, and dahlia tubers are also underground parts of plants, and in function are similar to potato tubers, but here the structure and mode of growth show that we have to do with modified parts of the root (Figs. 3 and 4).

The comparison of various plants thus shows us that quite different parts may be changed, so as to serve the same function, and that the same part may be modified to perform quite different functions. We shall meet with plenty of examples of this later on, but the instances just given will suffice to introduce us to the second great division of the science of Botany, namely, the *Morphology* of plants. This is based upon the comparative method of study as applied to plants and their organs.

To put the distinction shortly, while, as we have seen, physiology asks the question what each part of a plant *does*, morphology inquires what it *is*. The word " morphology " means the " study of form," but this literal meaning, as so often happens with scientific terms, gives a very imperfect idea of what is really implied. External form, at any rate, is quite an insignificant part of the subject of morphology, which is much more concerned with internal structure, relative position of parts, and the changes which they undergo during development. By the accurate comparison of plants in all these respects, morphology aims at finding out the relationships between them—that is, it endeavours to construct a natural classification of the vegetable kingdom. Classification is often treated as a separate part of the science, called Systematic or Taxonomic Botany, and this is a convenient division, but Systematic Botany can only be satisfac-

FIG. 3.—Conical root of Bryony. Reduced. (After Balfour.)

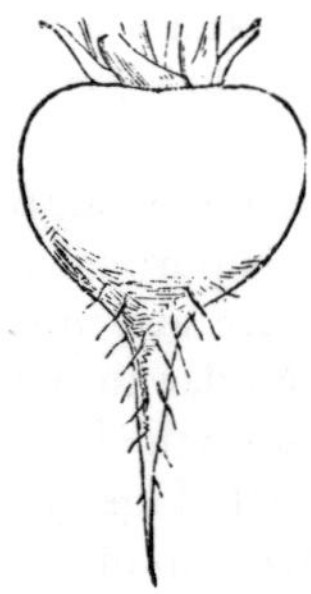

FIG. 4.—Fleshy root of Turnip. Reduced. (After Balfour.)

torily based on the comparative study of plants—that is, on morphology.

While it is easy to distinguish sharply between the physiology and morphology of plants, it is more important to remember that neither can be pursued to any advantage without the other. Physiology without morphology would teach us much about the life of individual plants, but could give us no idea of the vegetable kingdom as a whole, or of the relationship between the innumerable species of which it is composed. Morphology without physiology, on the other hand, would be just as barren, for the complex modifications of the organs of plants would be wholly unintelligible without reference to the functions to which they are adapted. Only by examining plants from both points of view can we attain to any knowledge of them which deserves the name of science.

The method pursued in this book is to take in succession a series of types representing important groups of plants, to examine each of them as fully as our space permits, with reference to both structure and function, and to compare them together. It will often be necessary to supplement the study of our main types by that of other plants which resemble them on the whole, but better illustrate some particular point, for Nature does not provide us with perfect types ready made.

In this way we may hope to gain some real knowledge of a few plants, which may serve as a firm foundation for more extended study afterwards. In choosing the order of our types, we will begin with the more highly organised plants, in which the division of labour among the organs is most complete. We do this, partly because the higher plants are most familiar to us in everyday life, and partly because we shall gain from them the clearest conceptions of organs and their functions.

Besides morphology and physiology, there are some more divisions of Botany which we must mention. One of these is the study of *fossil* plants—that is, of the vegetable remains which have come down to us preserved in the earth from ages long past. This is called the *Palæontology* of plants, or sometimes *Palæobotany*—that is, the science of ancient plants. From one point of view this is simply a part of morphology ; we compare the fossil plants among themselves and with recent ones, and draw conclusions, so far as we are able, as to their relationships. We cannot study their physiology, for they are dead ; we may, however, draw some inferences as to their mode of life from their structure. So far, palæontology is concerned with the same questions as recent Botany, but we also have to learn the succession of the strata in which the fossils are found, and so to trace their relative geological age, or, in other words, their distribution in time. Inquiries of this kind are peculiar to palæontology, and are of the greatest possible importance,

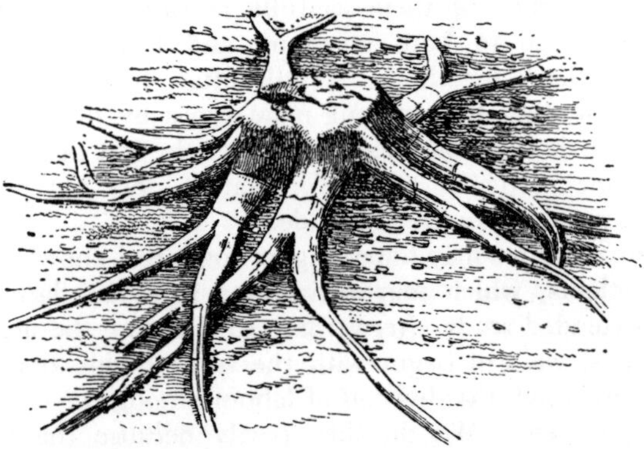

FIG. 5.—Stump and roots of a Fossil Tree (*Stigmaria ficoides*) from the carboniferous sandstone. The diameter of the stump is 4 ft. 4 in., and the spread of the roots 28–30 ft. (After Williamson.)

because they enable us to a certain extent to trace the past history of the different kinds of plants. To take only one example : the enormously rich vegetation which is so well preserved in the coal-measures consisted of forms of plants which have long since become extinct. The classes to which many of the great trees of that period belonged (such as the *Stigmaria* figured) are now only represented by a few small herbaceous plants (such as *Selaginella*).[1] On the other hand, most of the groups which are now of the

[1] See Part II., *Flowerless Plants*, p. 1.

greatest importance had not then appeared, and their first representatives are found in much later rocks. This part of the science, however, cannot be further touched on here. We shall only be able to concern ourselves with plants now living,

Then we have *Geographical Botany*, from which we learn the parts of the world in which the different kinds of plants grow wild, or, in other words, their distribution in space. This teaches us, for example, the very uniform character of alpine vegetation in many parts of the world ; the peculiarities of the Flora[1] of oceanic islands, deserts, and so on. This also leads to very important conclusions, especially when we study the geological and climatic changes which have influenced the present distribution of plants. We shall call attention to the distribution of our various types, but this part of Botany cannot be pursued far without a much wider knowledge of plants than is aimed at in this book.

Another branch of Botany which is being pursued very actively at the present time is called *Ecology* (*οἶχος*, "a house"), and is concerned with the study of the plant *at home*—that is, in relation to its surroundings. Ecology aims, for example, at finding out why some plants grow in woods, others in marshes, others on sand-dunes ; how plants are affected by one another or by animal neighbours ; the causes of their succession according to the seasons ; and so on. Ecology has a great deal to do with Geographical Botany, because it helps to explain the distribution of plants, and with Physiology, because one must know how a plant lives in order to understand its relation to its surroundings.

In plant nomenclature each plant receives a Latin name consisting of two words, the first being the name of the *genus* and the second that of the *species*. Thus the scientific

[1] By the *Flora* of a country we mean all the plants which grow wild there, collectively.

name of a particular plant is the same throughout the world, whereas its popular name varies from country to country and even in different parts of the same country. The scientific name of the Wallflower is *Cheiranthus Cheiri*, L., the letter L. being an abbreviation of the name of the author (Linnæus) who first gave it this name and described it precisely. The genus *Cheiranthus* comprises several different species which have certain characteristics in common. Some genera contain many species, others only one. This binomial nomenclature of plants, as it is called, was stabilised by Linnæus, a great Swedish naturalist, in the eighteenth century. Genera which have several identical characters, but otherwise differing, are said to belong to the same *Family*. Thus the Wallflower (*Cheiranthus Cheiri*) and the Shepherd's Purse (*Capsella Bursa-pastoris*) belong to the family Cruciferæ as their flowers are constructed on essentially the same plan. In the classification of plants greatest weight is attached to the structure and arrangement of the reproductive parts as these are less liable to change under different environmental conditions than are the vegetative organs.

A plant species like the Wallflower consists of many approximately identical individual plants which can breed freely together and in which the seeds give rise to the same kind of plant again on germination. There are, however, a number of different *varieties* of Wallflowers, *e.g.* some have yellow flowers and some have dark red flowers. Each variety usually breeds true to its own particular character unless cross-pollinated by another variety.

After these introductory considerations, we will now go on at once to our first type.

CHAPTER I

TYPE I

THE WALLFLOWER (*Cheiranthus Cheiri*, L.)

I. *EXTERNAL CHARACTERS*

A. Vegetative Organs

a. The Shoot

The Wallflower, a native of Southern and Central Europe, is occasionally found naturalised in England, and derives its popular name from growing commonly on old walls. It is a *perennial*—that is to say, it lives for a number of years. It does not die down in winter, nor does it lose its leaves in autumn. The leaves, however, gradually drop off from the older parts of the stem.

The stem is erect and branched (see Fig. 1), and its lower part is covered by scales of pale brown bark. This part is hard and woody. The upper portions, both of the main stem and branches, are softer and coloured green, and are rather hairy. The surface of the stem is ridged lengthways, the ridges being more conspicuous on the younger parts. The ridges are generally five in number ; each of them runs vertically along the whole length of the stem or branch, and passes through the places where the leaves are inserted. The ridges are most prominent just below the insertion of the leaves.

Both in the main stem and its branches it is evident that the lower parts are the older. For some distance above the ground there are no leaves ; they have dropped off, and we only find the scars which they have left. Then we come to leaves which are beginning to wither; a little higher up they are green and fresh, and of full size. Higher up still we find them smaller and more delicate, while at the top of each branch the leaves are only just forming, and we can trace them up to the extreme tip, until they are so small that we can no longer distinguish them. We see, then, that the formation of new organs goes on from below upwards, or, in other words, the growth takes place at the apex of the branch. The extreme end of the branch, where the young leaves are just making their appearance is called the *growing-point*. The growing-point with the young leaves which surround it is called the *leaf-bud*.

The branches do not arise indiscriminately on any part of the stem, but every branch grows from the angle between leaf and stem, just above the insertion of the leaf. This angle is called the *axil*, and the branches are said to be *axillary*. The main stem continues its growth indefinitely in the same direction, giving off its branches laterally.

The arrangement of the leaves on the stem requires rather close attention to make out. We see at once that no two leaves arise at the same level. Every leaf is placed a little above or below its next neighbours. The leaves, therefore, are said to be *alternate*.

We find further that they are arranged in a spiral. An imaginary line connecting the bases of the successive leaves together would wind upwards round the stem, just like the thread of a screw; here, however, the spiral ascends from the right up to the left, like a left-handed screw, in the reverse direction to an ordinary right-handed one.[1] Now, if we count the leaves carefully, beginning from below, and *not* counting the leaf taken as our starting-point, we shall find that the *fifth* leaf comes exactly above the one we started with, and in order to reach this fifth leaf we have had to follow the spiral exactly twice round the stem.

[1] In the diagram (Fig. 6) the spiral ascends from left to right

The ridges make it much easier to see which leaves come in the same straight line; for, as we saw just now, each ridge runs straight from one leaf to that vertically above or below it. This arrangement of the leaves may be expressed by the fraction $\frac{2}{5}$, in which the denominator indicates the number of the leaf which comes directly above the starting-point, while the numerator tells us how many times the spiral line travels round the stem before the fifth leaf is

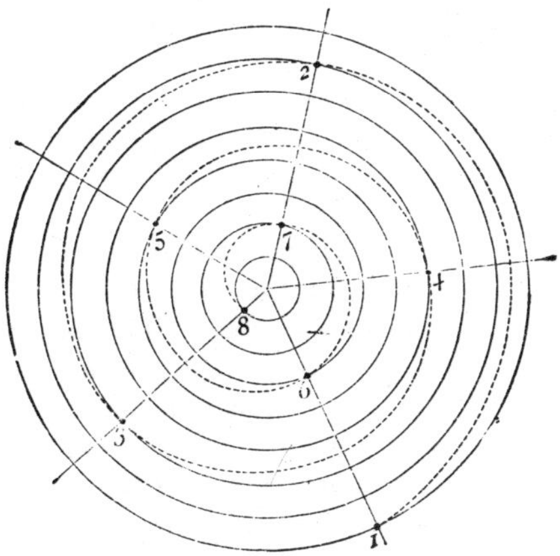

FIG. 6.—Diagram to illustrate $\frac{2}{5}$ Phyllotaxis. The dotted line represents the leaf-spiral, which is seen in horizontal plan, the centre corresponding to the apex. The concentric circles represent the successive nodes. (After Van Tieghem.)

reached. This fraction $\frac{2}{5}$ also expresses the *divergence* or angle between two successive leaves, which is equal to $\frac{2}{5}$ of the circumference of the stem. The way in which leaves are arranged on the stem is called *phyllotaxis*, and varies very much in different kinds of plants.

The whole length of the stem is divided into *nodes* and *internodes*. The part where a leaf is borne is called a *node*, the spaces between the leaves are the *internodes*.

Each leaf is attached to the stem by a narrow base. It has no distinct stalk, but the blade of the leaf is very narrow near the base, and broadens out quite gradually for about two-thirds of the whole length, and then narrows again rather more rapidly to the pointed end. The leaf is technically called *sessile*, because the blade is seated on the stem directly, without the intervention of a leaf-stalk or petiole, such as we see in the leaf of a Vine or Geranium. The narrow part of the blade, however, only differs from a leaf-stalk in being slightly winged. The part of the leaf where it joins the stem is called the *leaf-base* ; this remains in connection with the stem after the rest of the leaf has dropped off.

FIG. 7.—Leaf of Wallflower, natural size, showing the chief veins. (R. S.)

The shape of this leaf is expressed in botanical language by the words *lanceolate*, *acute*, and *entire*. The word *lanceolate* refers to the general outline, which is like that of a lance-head ; the word *acute* indicates the pointed apex ; while the term *entire* means that the edge is not toothed or divided in any way.

The upper surface of the leaf is dark green and very slightly hairy. The under surface is of a lighter green, and has many more hairs. The hairs are closely applied to the surface, and cannot be seen well without a lens, though they can easily be felt.

The leaf is traversed from end to end by a stout midrib, or principal *vein*, which is more prominent on the lower surface, and gradually tapers towards the apex. From the midrib branch-veins are given off at irregular intervals, which turn upwards so as to run nearly parallel to the midrib. These main branches join on to each other, and themselves give off innumerable lesser branches, which permeate every part of the leaf, and are united among themselves into a fine network. The finest branches of all, however, end blindly within the meshes of the network. The smaller veins can only be traced with the help of a lens (see Fig. 24). This type of venation is said to be *reticulate* and is characteristic of the large Class of the Dicotyledons, to which the Wallflower belongs.

We have now described the chief external characters of the stem and leaves. To complete our sketch of the vegetative organs, we will next consider the root.

b. The Root

If we pull up a Wallflower and wash away the earth which clings to the root, we find that this, like the stem, has a main axis, which grows straight downwards, in the same line with the erect axis of the stem. A main root of this kind is called a *tap-root*. Of course the root meets with more resistance to its growth than a stem does, and sometimes it is forced to diverge more or less from its vertically downward direction (see Fig. 1). From the tap-root a number of branches are given off. These branches do not grow straight down, but take an obliquely descending course, forming an angle a little less than a right angle with the tap-root. From these principal branches other much smaller ones arise, and these also ramify repeatedly. The finer branches of the root do not take a fixed course, but penetrate the soil in all directions equally. If the plant has been very carefully taken up, we shall find that particles of earth stick firmly to the very youngest branches of the root a little above their extreme tips. This is because these parts of the root are clothed with very fine *root-hairs* (see Fig. 29), which have withered away from the older parts. These *root-hairs* are barely visible to the naked eye, and must by no means be confused with the *rootlets* or fibres already described, to which they bear no resemblance in structure, as we shall find when we come to the microscopic work. Each growing-point of the root and its branches is covered by a *root-cap* (see pp. 67 and 74). We notice as conspicuous differences between the root and the shoot, that the former bears no leaves and is not green in any part.

We shall find a young seedling more convenient for studying the root than an older plant. On such a seedling it is possible to make out that the branches of the tap-root are not arranged irregularly, but form four vertical rows.

B. Reproductive Organs

a. The Flowers

The branches which bear the flowers do not at first differ in any way from ordinary vegetative branches. They arise

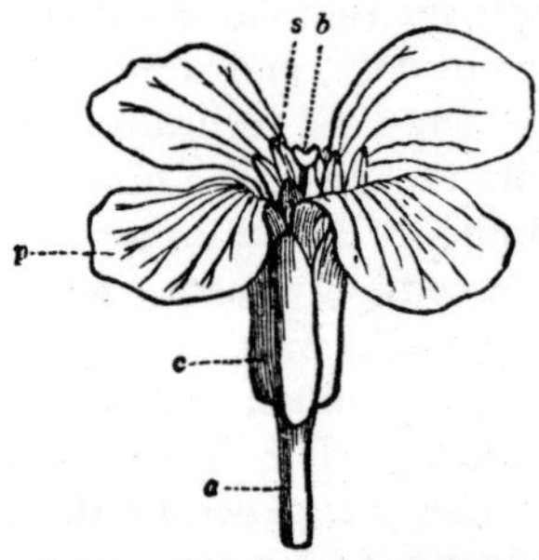

Fig. 8.—Flower of Wallflower. *a*, flower-stalk ; *c*, calyx ; *p*, corolla ; *s*, stamens ; *b*, stigma. (After Balfour.)

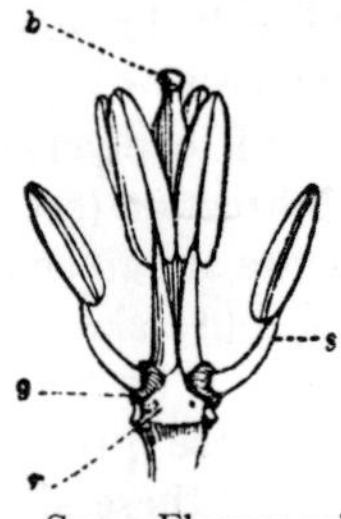

Fig. 9.—Same Flower, with calyx and corolla removed. *s*, stamens ; *b*, stigma at top of pistil ; *r*, receptacle ; *g*, honey-glands at base of short stamens. (After Balfour.)

in the axils of leaves, and for some time bear leaves themselves in the usual way. But soon the production of leaves ceases, and from that time onwards the branch bears flowers only. Each flower may itself be regarded as a modified branch (borne on the main branch), the leaves of which differ greatly in arrangement, form, structure, and colour from the vegetative leaves. The flowers, unlike the ordinary branches, do not arise in the axils of leaves, for there are no leaves at all on the part of the plant which bears them.

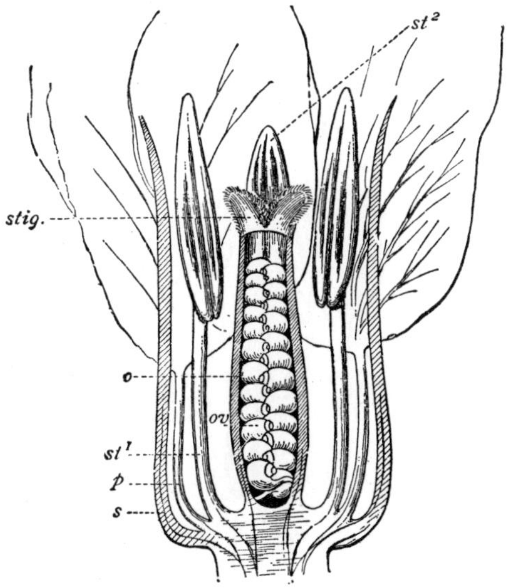

FIG. 10.—Diagrammatic view of the flower of Wallflower (enlarged) in longitudinal section, cut in median plane of outer sepals. *s*, sepal ; *p*, petal ; st^1, two of the long stamens ; st^2, one of the two short stamens ; *o*, ovary; *ov*, ovules ; *stig*, stigma. (R. S.)

This is characteristic of the Wallflower and its relations. In most plants the flowers, like other branches, are borne in the axils of leaves.

The entire group of flowers borne upon the stem is called the *inflorescence*. In the Wallflower the flowers are borne laterally on a main axis, and are stalked. The main axis terminates in a bud which is continually forming new flowers as long as growth continues. Thus the oldest flowers are at the bottom, and the youngest at the top. An inflorescence of this kind is said to be *indefinite*. An in-

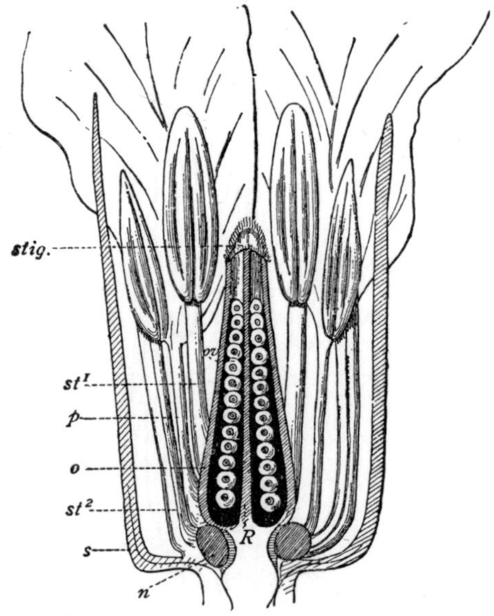

FIG. 11.—Diagrammatic view of the flower of Wallflower (enlarged) in longitudinal section, passing through median plane of inner sepals. *s*, sepal; *p*, petal; st^1, long stamens; st^2, short stamens; *o*, ovary; *ov*, ovules; *stig*, stigma; *n*, nectaries at base of short stamens; *R*, septum between the two carpels. (R. S.)

definite inflorescence with a single elongated main axis and stalked flowers is a *raceme*.

Now we will consider the parts of a single flower in detail. The first point that strikes us is that, while the stalk (*pedicel*) is quite bare, all the leaves are crowded together at the end of the branch. The floral leaves are not arranged spirally, but two or more grow at the same level. The enlarged end of the stalk, on which the floral organs grow, is the *receptacle*, or *thalamus*.

Beginning at the outside of the flower, we first come to four small narrow leaves of a purplish or greenish colour (see Fig. 8, *c*). These leaves are not all inserted at the same

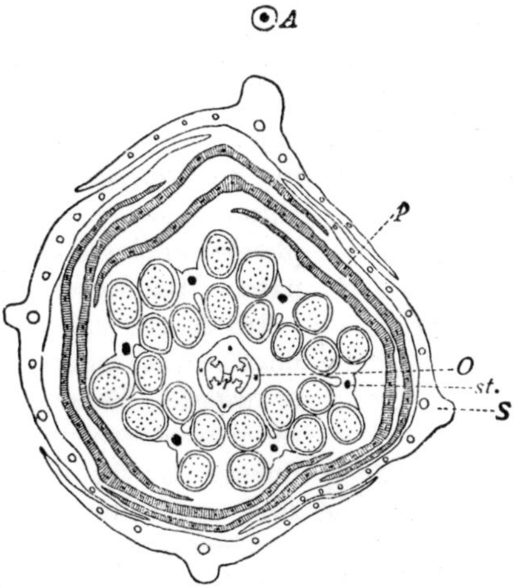

FIG. 12.—Transverse section through a flower-bud. *A*, position of axis of inflorescence ; *S*, sepals (four in number) : *p*, petals (shaded) (four in number) ; *st*, stamens (six in number, each with four pollen-sacs) ; *O*, ovary (of two carpels). Compare the diagram, Fig. 13. Magnified about 12 times. (R. S.)

level. They are in two pairs, placed at right angles to each other, and one pair is inserted rather farther in than the other pair (see Figs. 12 and 13). This is seen very clearly in the bud (Fig. 12), though less easily in the open flower. Each of these four leaves is called a *sepal*, and collectively they form the *calyx* ; as the sepals are not joined together, the calyx is said to be *polysepalous*.

The diagram, Fig. 13, represents as it were the ground plan of a flower, and shows the relative position of all its parts. In order to fix this, it is necessary to know how the flower is placed with reference to the axis of the inflorescence. This axis is indicated by a dot in the diagram. It will be seen that the two outer sepals are so placed that one lies next the axis (*posterior*), and the other remote from it (*anterior*). The two inner sepals are placed at the sides (*laterally*), alternating with the outer two. The two inner sepals are bulged out at the base; we shall see the reason for this directly. In the bud before it opens, the calyx completely encloses all the other parts of the flower.

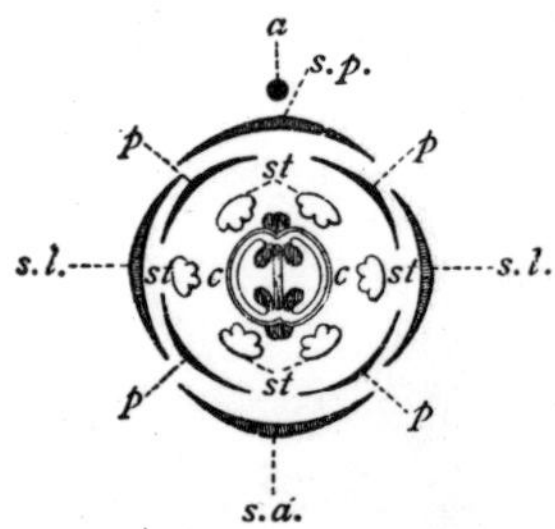

Fig. 13.—Floral diagram, or ground plan of the flower of *Cheiranthus*. *a*, axis of inflorescence; *s.p.*, posterior sepal; *s.a.*, anterior sepal; *s.l.*, lateral sepals; *p*, the four petals; *st*, the six stamens (the four long ones are shown connected in pairs); *c*, the two united carpels, forming the pistil.

Inside the calyx we find four floral leaves forming a single ring, or *whorl*. These are the *petals*, the most conspicuous part of the flower. In the wild Wallflower they are yellow; in cultivated specimens they are often more or less brown. Each petal is broad and rounded above, and tapers into a narrow stalk below (see Figs. 10 and 11). The four petals form collectively the *corolla*, which is here said to be *polypetalous*, because the petals are not connected together. The petals are so placed as to alternate with the four sepals—that is to say, the middle of each petal comes exactly opposite the space between two sepals (see Figs. 12 and 13).

Both the sepals and the petals are veined, but not in such a complicated way as the vegetative leaves.

So far the organs of the flower have borne an obvious

resemblance to ordinary leaves. The next whorl is very different, and consists of bodies which, so far as their outward form is concerned, are not at all leaf-like. Each consists of a longish stalk, bearing at the top a little boat-shaped case of a yellowish colour. These are the *stamens* (see Figs. 8, 9, 10, and 11). There are six of them, four of which are longer than the other two. The two shorter stamens stand on either side of the flower, and each of them is placed just between two of the petals, and just opposite the two inner lateral sepals. The four longer stamens are placed rather farther in, so that all six stamens might be regarded as forming *two* whorls. The longer ones are in two pairs, each pair standing between two petals, and just opposite the two outer (posterior and anterior) sepals. (See Fig. 12 and diagram, Fig. 13). The stamens collectively are called the *andrœcium.*

The stalk of the stamen is called the *filament*; the body at the top of the filament is the *anther.* The filament is continued up through the middle of the anther, and is here called the *connective*, because it connects together the two halves (*lobes*) of the anther. When ripe, the anther bursts to discharge a yellow powdery substance called *pollen.*

Just inside the bases of the two shorter stamens are two large green honey-glands (*nectaries*). The filaments are bent outwards to make room for them, and this causes the bulging of the adjacent lateral sepals, which we noticed in examining the calyx (see Figs. 9 and 11).

The middle of the flower is occupied by the *pistil* or *gynœcium.* This is a nearly cylindrical hollow body, green in colour, tapering a little towards the top, and surmounted by a forked outgrowth. The lower thicker part, which occupies more than three-quarters of the whole length, is the *ovary*, and the forked outgrowth at the top of all is the *stigma*, while the short tapering portion which connects the two is the *style* (see Figs. 9, 10, and 11).

The hollow ovary is divided lengthways into two compartments by a membrane called the *septum*. This membrane marks the boundary between the two[1] *carpels*, the name given to the parts of which the pistil is composed. We see, then, that the two carpels are so placed as to lie opposite the two inner lateral sepals. Within the ovary are the young seeds, or *ovules*, as they are called at this early stage. They grow on the walls of the ovary adjoining the septum. As, however, we are only dealing with external characters at present, we will postpone any further examination of the interior of the ovary and of the ovules (see p. 90).

The ovules only develop into the seeds after fertilisation, which is brought about by the pollen. In order that the ovules may become ripe seeds, which can grow into young plants, it is necessary that the pollen should come into contact with the stigma of the pistil. The whole process of fertilisation will be described later on, when we have made ourselves acquainted with the internal structure of the floral organs (see p. 105).

b. The Fruit

The *fruit* is the ripened pistil, and contains the *seeds*, which are the fertilised ovules. It continues to show the same general structure as the pistil, but grows much larger, and undergoes some important changes, a description of which will be found on p. 118. During the development of the pistil into the fruit all the other floral organs—calyx, corolla, and stamens—drop off.

[1] E. R. Saunders, who has investigated the morphology of the ovary in detail, considers that there are four carpels in the ovary of the wallflower, two of which are sterile and form the valves, the other two being fertile and forming the septum (see Fig. 50).

II. *INTERNAL STRUCTURE*

A. The Cells

If we examine under the microscope a thin slice, or section as it is called, from any part of an ordinary plant, say, for example, a section cut straight across the stem of a Wallflower, we find that its substance is not uniform, but is divided up by a network of walls into innumerable minute chambers of varying size and shape (see Fig. 20). If we make another section at right angles to the first, say a section lengthways through the stem, we find the same structure, except that in this view the chambers appear to have somewhat different shapes (see Fig. 22). Combining the two views together, we come to the conclusion that the whole substance of the plant is composed of a system of closed compartments. This is the reason why, if a juicy fruit like the melon is cut into pieces, all the juice does not run out, but only a little is lost. The juice is contained in these little closed chambers, which hold it fast, and only those which are actually opened by the knife are emptied.

Now these chambers, of which the whole substance of almost all plants consists, are called *cells*. They are nearly always so small that they cannot be distinguished at all except under the microscope. Few reach a diameter of $\frac{1}{100}$ of an inch, while many measure only $\frac{1}{1000}$ of an inch, or less. The fact that plants have a cellular structure is far from being a modern discovery. Cells were discovered in plants by Robert Hooke [1] in 1667, and within the succeeding fifteen years the internal structure of many plants was worked out by Nehemiah Grew in England, and Marcello

[1] See his *Micrographia ; or, Some Physiological Descriptions of Minute Bodies made by Magnifying Glasses, with Observations and Enquiries thereupon* (London, 1667), especially chapter xviii., "Of the Schematisme or Texture of Cork, and of the Cells and Pores of some other such frothy Bodies."

Malpighi in Italy. The early anatomists were struck by the similarity of many vegetable tissues to the honeycomb of a bee, and they applied the word " cell " to the chambers in both cases in the same sense.

From the time of Grew and Malpighi onwards until well within the nineteenth century—for a period, that is, of quite a century and a half—those botanists who troubled themselves at all about internal structure attended chiefly to the walls of the cells. These are much the most conspicuous features in the section through any fully formed organ, when seen under the microscope. The network of walls enclosing the cavities strikes the eye at once. We do not at first particularly notice the contents of the cavities, but the research of the last century has shown that it is the cell-contents, after all, which are the most important part of the cell ; it is the contents which form the wall, not *vice versa*, and in some plants the cell-contents escape, and are then quite able to grow and increase without any wall at all, while the cell-wall without its contents is a dead shell, incapable of growth or activity of any kind.

a. General Structure of the Cell

We will now shortly consider the structure of a single cell, and we will begin by describing one from quite a young part of a plant, such, for example, as the bud at the growing end of a branch.

A young cell (see Fig. 14, *A*), like one of those of which a growing-point is formed, is surrounded on the outside by a delicate but firm and elastic *membrane*, the *cell-wall*, by which it is separated from its neighbours. The wall consists chiefly of a substance called *cellulose*, which is chemically very similar to starch, and is composed of the same chemical elements (carbon, hydrogen, and oxygen) in the same proportions.[1]

[1] The formula is $(C_6H_{10}O_5)_n$.

The strength and elasticity of all parts of the plant are ultimately due to the cell-walls, which serve as a firm supporting framework for the whole structure. In such young parts as we are now considering, however, the cell-walls are still very thin, and for this reason growing-points are very delicate, and require external protection, which in the case of stems is afforded by the surrounding leaves.

The interior of the cell is almost completely filled by a soft, viscid, finely granular, colloidal substance, the *protoplasm*. The outer part of the protoplasm, where it borders on the cell-wall, is clearer—that is, more free from granules,—than the rest. The protoplasm consists of a mixture of various substances, called by chemists *proteins*, which are more or less similar in composition to albumen or white of egg. These are very complicated bodies, not yet fully investigated by chemists ; they are composed of the elements, carbon, hydrogen, oxygen, nitrogen, and sometimes also sulphur and phosphorus. Besides the proteins, active protoplasm always contains a large proportion of water. Physically it resembles a very thick fluid, but it has powers of spontaneous movement and continuous changes of form such as no other fluid in nature possesses.[1] Protoplasm is found in all cells without exception in which growth is going on, or in which food is assimilated, or any new structure formed, or any kind of spontaneous movement carried out. It is, in fact, the seat of all those processes (whether in plants or animals) which distinguish living organisms from lifeless matter. The word "protoplasm," meaning that which is *first formed*, was invented by the Bohemian physiologist Purkinje in 1840, and was used independently by the German botanist von Mohl in 1846, to express the fact that this substance is the primary part of

[1] No part of the Wallflower plant is convenient for observing these movements of protoplasm, so some special illustrations of them from other plants will be given below (see p. 36).

the cell, while the cell-wall is a secondary product derived from it.

Embedded in the protoplasm is a roundish denser body, which has a more coarsely granular appearance than the rest of the protoplasm. This body is the *nucleus*. A nucleus is certainly present in all living cells of the higher plants, and also in the great majority of those of the lower plants.

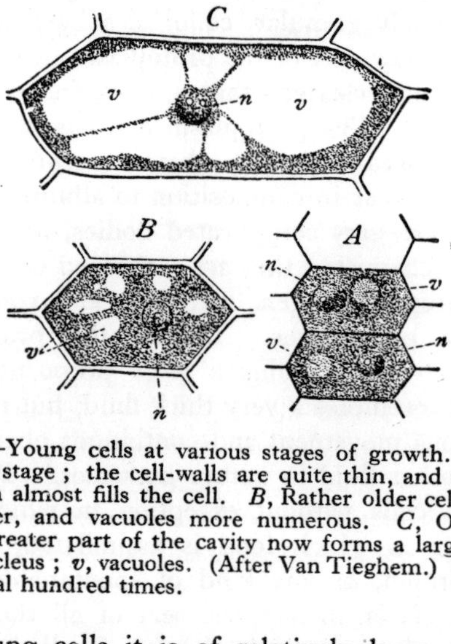

FIG. 14.—Young cells at various stages of growth. *A*, Very early stage ; the cell-walls are quite thin, and the protoplasm almost fills the cell. *B*, Rather older cell, cell-wall thicker, and vacuoles more numerous. *C*, Older still ; the greater part of the cavity now forms a large vacuole. *n*, nucleus ; *v*, vacuoles. (After Van Tieghem.) Magnified several hundred times.

In very young cells it is of relatively large size, and its diameter may be more than half that of the entire cell. Its complicated structure cannot be fully described here, but it may be mentioned that in many cases the apparent granules are found, under a sufficiently high power of the microscope, to represent extremely minute threads or fibrils. The nucleus also contains one or more comparatively large rounded masses distinct from the fibrils, and called the *nucleoli* (see Fig. 17). The nucleus has a definite membrane.

The substance of the nucleus consists of various chemical bodies which are similar to, but not altogether identical with, those of which the rest of the protoplasm is built up; a complex compound containing phosphorus is constantly present in the nucleus. The protoplasm excluding the nucleus is called *cytoplasm*.

Most cells contain only one nucleus, but in very long cells a number of nuclei are often found.

The cytoplasm, nucleus, and cell-wall are the most constant constituents of a vegetable cell. The cytoplasm, as we have seen, is absolutely constant in all living cells without exception; the nucleus is equally general in the higher plants, though in some of the lower ones its presence has not been proved; the cell-wall is formed sooner or later in all cells of the higher plants, but is very frequently absent, at any rate for long periods, from certain cells of the simpler plants.[1]

If we examine rather older cells, such as those which we find at a little distance below the growing-point, we shall see that the whole cell is larger, and that its cavity is no longer so nearly filled by the cytoplasm (Fig. 14, *B*). Clear spaces, which had already appeared in the cytoplasm at an earlier stage (*v* in the figures), have now increased in number; they are filled with water containing various organic and inorganic substances (salts, sugars, acids, etc.) in solution, and called the *cell-sap*. The spaces filled with cell-sap are known as the *vacuoles*, because they look as if they were empty.

The nucleus increases little in size, and now bears a much smaller proportion to the whole cell than it did at first. The cell-wall will have become somewhat thicker (see Fig. 14, *B* and *C*).

A full-grown but still living cell, such as we might find in the older parts of the Wallflower, has its cytoplasm

[1] See Part II., *Flowerless Plants*.

reduced to a thin layer, lining the inside of the wall. The whole interior of the cell appears empty—that is, it is occupied by one large vacuole containing cell-sap. This has been formed by the smaller vacuoles running together. The layer of cytoplasm lining the cell-wall is often called the *primordial utricle*. The nucleus has undergone little change, but is now placed close against the wall, embedded in the cytoplasmic layer which lines it. The cell-wall itself has been further thickened by the deposit of new cellulose from the cytoplasm.

This is the structure of most mature living cells of plants. The protoplasm does not keep pace with the growth of the whole cell ; it may even actually diminish in amount, owing to its being used up to make new cell-wall and for other purposes, quicker than it is renewed by taking in food. Some cells lose their protoplasm altogether ; then they are dead, and incapable of any further growth, though they may still be of use to the plant as passages for sap or as a mechanical support. We shall find plenty of examples of altered cells of this kind in various organs of such plants as the Wallflower, more especially in the wood.

b. Special Contents of Cells

Many other structures are found in cells besides cytoplasm, nucleus, and cell-wall. Some of these we will not trouble about until we come to them in describing the internal structure of our types. One or two, however, are so important that they must be mentioned at once.

a. Chlorophyll-Granules or Chloroplasts

If we examine, under the microscope, a living cell from the green succulent parts of the leaf or stem, we shall find, embedded in the cytoplasm, a number of round bodies of a bright green colour (see Figs. 19 and 26). If a leaf is killed by immersion in boiling water and then transferred to alcohol, we shall find that it becomes colourless, the colouring matter having been dissolved by the alcohol, which becomes of a dark green tint in consequence. The round bodies themselves remain, however, unaltered in size and shape, though now colourless. The green colouring matter is *chlorophyll*, the round granules which contain it are the *chlorophyll-granules* or *chloroplasts*. Cytoplasmic granules of this kind are called *plastids*. The solution obtained by extracting a green leaf with alcohol contains actually a number of distinct pigments the exact chemical composition of which is known. Two of these pigments are green (*chlorophyll a* and *chlorophyll b*), and two are yellow (*xanthophyll* and *carotin*). Carbon assimilation (photosynthesis) cannot take place without the green pigments. It will be convenient in this book, however, to use the term chlorophyll to denote all these pigments collectively. The chlorophyll-granules consist of cytoplasm, rather denser and firmer than the ordinary cytoplasm of the cell, and saturated by the green colouring matter. All green parts of plants owe their colour to the presence of chloroplasts in their cells.

We have said above that one of the two main functions of the shoot is to take up food from the air. The great food-substance which green plants obtain from the air is *carbon*. About half the solid substance of plants consists of this element, which enters into all the chemical compounds of which living things are built up. The whole of this carbon is obtained by ordinary green plants from the carbon dioxide (CO_2) which is contained in small quantities in the air. In ordinary air the amount of this gas is equal to about three parts in ten thousand. The carbon dioxide is decomposed, the oxygen given off again, and the carbon retained, to help in forming the various organic substances of the plant. Now this decomposition of carbon dioxide is entirely performed by the chlorophyll-granules, and they can only do this work under the influence of *light*. We see, then, that the chlorophyll-granules are of the greatest possible importance to the plant, for they are the organs by which alone, with the help of the sun's rays, it can obtain its carbonaceous food from the air. Both parts of the granule are necessary for the process ; the chlorophyll itself is useless without the cytoplasm of the granule, and the cytoplasm is quite unable to do the work unless it contains chlorophyll. The process of nutrition just described is known as *assimilation*, or, more accurately, as *carbon-assimilation* or *photosynthesis*, for assimilation is sometimes used in a wider sense. To this process the formation of nearly all naturally occurring organic substances is ultimately due, and carbon-assimilation by green plants may thus be fairly regarded as the most fundamentally important of all physiological processes. The assimilation of carbon is not really so simple a matter as might appear from the rough sketch just given ; we shall return to the subject later on, when we come to the special consideration of physiological questions (see Chapter III.).

β. Starch Granules

Another important body very generally, though not always, found in the cell-contents is *starch*. Starch contains the three elements, carbon, hydrogen, and oxygen in the proportion $C_6H_{10}O_5$, but the molecular weight is much larger than is indicated by this empirical formula. It occurs in the form of granules, which are easily identified under the microscope on treating with iodine, which gives them a deep blue colour. The larger granules show *stratification* (see Fig. 15, *A* and *C*)—that is to say, the substance of the granule is made up of a number of successive layers, which have been deposited one after the other, so that the innermost layers are the oldest and the outermost the youngest. The oldest part of the granule round which all the layers have been formed is called the *hilum*. It does not necessarily lie at the centre, for sometimes the layers are deposited very unequally, so that in an old granule the hilum is in a very eccentric position. This is always the case in potato starch.

In a green plant starch is formed in two quite different ways. First it is formed in the chlorophyll-granules under the influence of sunlight. It is easy to demonstrate the presence of small starch granules in the chloroplasts of a Wallflower leaf, after exposure to light. These starch granules are the product of carbon assimilation—that is to say, they are formed by the chloroplasts in sunlight as the result, though not the direct result, of the decomposition of carbon dioxide and water, and the rearrangement of their elements.

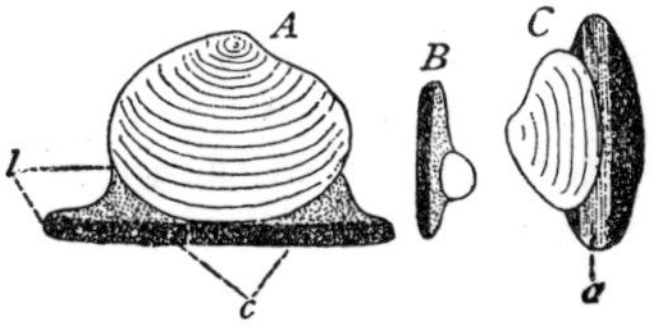

FIG. 15.—Starch granules, and leucoplastids. From an orchid (*Phajus*). *A*, Large granule, showing stratification; *l*, the leucoplastid; *c*, protein crystalloid; the hilum is at the opposite side of the granule, near the letter *A*. *B*, Very young granule, still much smaller than the leucoplast. *C*, Intermediate stage. (After Strasburger.) Magnified 540 times.

But starch is also formed in the deep-seated tissues of the stem and the root, *i.e.* in parts which receive little or no light, and in which there is no chlorophyll. This starch cannot be the product of assimilation, for it is formed under conditions which render assimilation impossible.

What happens is this: the starch formed in the chlorophyll-granules is converted into sugar, and in this soluble form passes down into the stem and root. When it has reached the cells in which starch is to be deposited, the sugar is taken up by certain cytoplasmic bodies, which are essentially similar to chloroplasts without the chlorophyll. These cytoplasmic bodies (called *leucoplasts*)[1] use up the sugar to re-form starch granules (see Fig. 15, *l*).

It is evident that the function of the leucoplasts is dependent on that of the chloroplasts, for unless the latter carried on the work of assimilation there would be no supplies of sugar available for the use of the leucoplasts. Starch is a very important reserve food-substance, and is very commonly found stored up in seeds, tubers, and fleshy roots, and, in fact, in organs of almost every kind. It is especially important inasmuch as it affords the material for new cell-walls, but it is used in many other ways also. Starch belongs to the great group of organic bodies called *carbohydrates*, which all consist of the three elements, carbon, hydrogen, and oxygen, the two latter elements being present in the same proportion as in water, H_2O. Sometimes we find carbohydrates stored up in the form of some kind of sugar, as in many fruits and in Beetroot. These and some other forms of reserve carbohydrates are always dissolved in the cell-sap.

γ. *Oil*

In many plants we find that the place of carbohydrates is taken by various kinds of oil. Oils are especially common as reserve food-substances in seeds, *e.g.* Castor-oil seed, Hemp-seed, Linseed, etc. Oil contains the same elements (C, H, O) as carbohydrates, but in different proportions, the proportion of oxygen being very much less. The oil occurs in the form of small drops embedded in the cytoplasm.

[1] The word *plastid* is a general term, including *leucoplasts*, chlorophyll-granules, or *chloroplasts*, and also cytoplasmic bodies which contain other colouring matters (*chromoplasts*). The three kinds of plastids are convertible into one another; they usually arise by the division of pre-existing plastids.

δ. *Protein Granules*

These form a very important part of the contents of certain cells, especially in seeds and other organs in which food is stored up for the use of the plant. They are chemically similar to the substances of which the protoplasm is composed, but are themselves lifeless, and serve simply as food. An example is shown in Fig. 16, which represents the complex protein granules (aleurone-grains) of the Castor-oil seed.

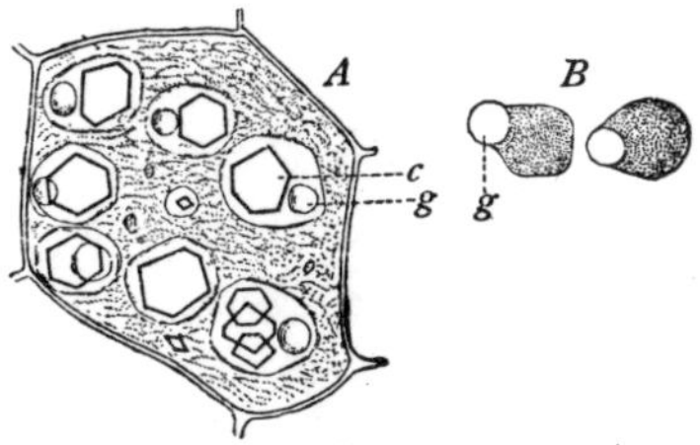

Fig. 16.—*A*, Cell from the endosperm of the Castor-oil seed, examined in water; note the large protein granules, each of which contains one or more crystalloids (*c*), also composed of protein, and *globoid* (*g*), of calcium-magnesium phosphate. *B*, Isolated granules, examined in olive-oil. (After Strasburger.) Magnified 540 times.

c. Cell-Formation

Cells increase in number entirely by division. Every cell which exists owes its origin to the division into two or more parts of some cell which existed before it. At present we will only describe the kind of cell-formation which goes on in the vegetative parts of plants such as the Wallflower, and indeed in all plants except the lowest. Wherever new organs are in course of development we find cell-division going on, as, for example, at all growing-points whether of stem or root. We will suppose that the cell which is about to divide has a single nucleus to start with, as is generally the case. The first obvious changes which take place are in the nucleus. The changes gone through are of a very complicated character, but their ultimate effect is that the nucleus becomes divided into two exactly equal and similar parts, the *daughter-nuclei*. So precise is the accuracy of the division, that every single minute fibril (or *chromosome*) of which the framework of the parent nucleus consists is split lengthways into two identical halves, one of which goes to each of the daughter-nuclei. The latter remain for a time connected by delicate threads of cytoplasm.[1] The new cell-wall is formed at right angles to these threads and midway between the two daughter-nuclei. It is formed by the protoplasm itself. Generally the new cell-wall does not stretch all the way across the cell, but has to be completed

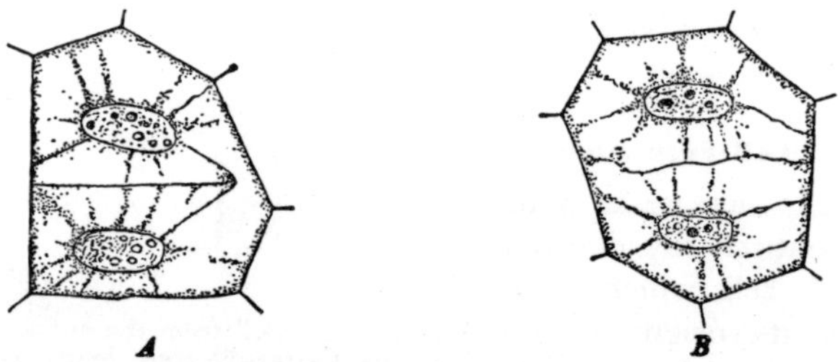

FIG. 17.—Dividing cells from the young seed of *Fritillaria*. *A*, The two daughter-nuclei have been formed, but the new cell-wall between them does not yet reach across the cell. *B*, Later stage ; the new cell-wall now completely divides the mother-cell into two parts. (After Strasburger.) Magnified 240 times.

at the sides later on (see Fig. 17). The nucleoli disappear during the division, and their substance passes out into the cytoplasm. New nucleoli make their appearance in each of the daughter-nuclei, which soon assume the size and structure of the nucleus before division. We see, then, that the essential points in vegetative cell-formation are the division of the nucleus into two exactly similar halves, and the formation of a cell-wall between them, separating the cytoplasm of the cell into two distinct parts.

[1] The details of nuclear division (*mitosis*), a subject of the highest possible importance, must be studied in more advanced text-books.

d. Continuity of the Protoplasm

We have hitherto been considering the structure of the individual cells of which a plant is built up ; it has been shown, however, that in all vegetable tissues the protoplasm of each cell is in communication with that of its neighbours by means of excessively fine cytoplasmic strands, which extend through the cell-wall from one cell to another. The

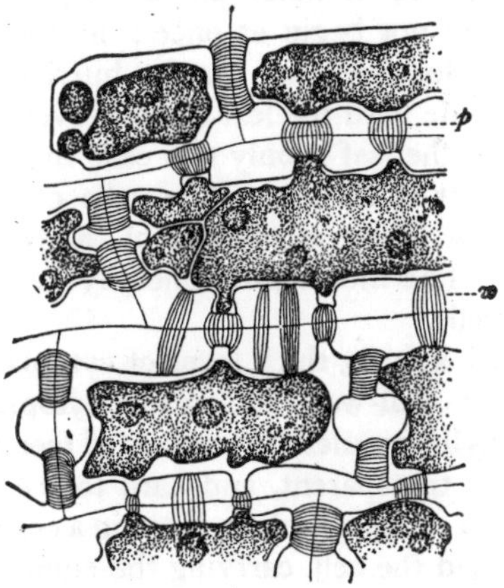

FIG. 18.—Cells from the endosperm of the seed of a Palm (*Bentinckia*), showing the fine strands of cytoplasm connecting the cells. Some (*w*) pass through the full thickness of the wall, others (*p*) through the thin places or pits. (After Gardiner.) Magnified 550.

perforations through which the connecting cytoplasmic strands pass are extremely minute, but are often very numerous (see Fig. 18). The importance of the discovery lies in the fact that we now know that the living matter in the plant is continuous, and not absolutely severed into isolated portions by dead cell-wall. The plant, in fact, is, as a whole, a living protoplasmic body, supported, but not interrupted, by the skeleton of cell-walls.

e. Protoplasmic Movements

A good example of protoplasmic movement is afforded by the cells of the leaf of the American Waterweed (*Elodea canadensis*), a transatlantic alien, which is commonly found in our slow-moving waterways. In order to see the movements of the protoplasm, a young leaf is removed from the stem and examined in water under the microscope. The movements may not begin at once ; in cold weather they are especially sluggish, but can be hurried on by gently warming the slide with the leaf upon it over hot water.

The blade of the leaf is only two cells thick, except at the midrib, where it is thicker. The cells of the midrib are long and narrow, those of the rest of the blade are shorter and broader. The movement generally begins first in the long narrow cells.

Each cell has a fairly thick layer of cytoplasm (*primordial utricle*) lining its cell-wall, and in this cytoplasm numerous large chlorophyll-granules are embedded (see Fig. 19). The nucleus is very transparent, and hard to see. (It is shown dark in Fig. 19.) The cytoplasm flows in a continuous stream round and round the cell, carrying the chlorophyll-granules with it. Owing to the presence of these granules, the movement is very easy to follow. If we fix our eye on one of the granules which is being swept along by the cytoplasmic stream, we shall see it travel steadily along one of the side-walls until it reaches a corner of the cell. Here there may be a momentary hesitation, for at these points the current is often choked by the accumulation of chlorophyll-granules. Soon, however, the movement goes on again ; the granule is carried along the short end-wall of the cell, the next corner is reached and turned, and the journey continued along the other side-wall, and so on until the circuit is complete. The revolving cytoplasm here forms one single stream, which flows round the whole cell. The current, as we have seen, follows the lateral and terminal walls of the cell. What, then, is the behaviour of that part of the cytoplasm which is in contact with the other two walls, those, namely, which form as it were the roof and floor of the cell ? This cytoplasm also takes part in the general movement of rotation,

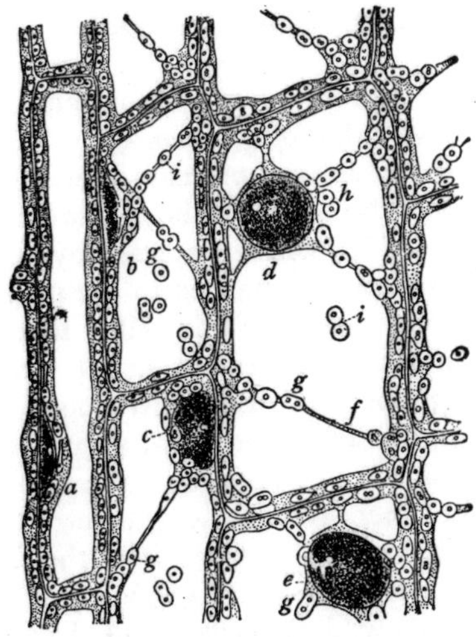

FIG. 19.—A few cells from the leaf of the American Waterweed. *a*, *b*, *c*, *d*, *e*, nuclei of the various cells ; *f*, a strand of cytoplasm crossing the cell-cavity ; *g*, *h*, *i*, chlorophyll-granules, some of which are dividing. The small bodies inside the chlorophyll-granules are starch. (After Kny.) Magnified several hundred times. The narrow cell, to the left, would show *rotation* ; the broader cells, to the right, *circulation*.

and revolves in the same direction as the rest. But it has a shorter distance to cover, and the nearer we approach to the middle of the top and bottom walls, the shorter is the circuit, and consequently the slower is the speed. Now, just in the middle of each of these two walls there is a band of cytoplasm which does not move at all, but forms a quiet eddy in the stream. This still part is called the *indifferent band*, and it is always present where rotation of the cytoplasm goes on.

But even in the other parts of the cell it is not really the whole cytoplasm which rotates. The *ectoplasm*—that is, the thin outer layer in immediate contact with the cell-wall—remains stationary, while the revolving *endoplasm* (containing the chlorophyll-granules) travels over its inner surface.

This will serve to represent an important type of protoplasmic movement—that of *rotation*.

Another form of movement may often be observed in the broader cells of the lamina of the *Elodea* leaf (see Fig. 19). Here the vacuole of the cell is traversed by strands of cytoplasm, most of which are attached to the nucleus, which is suspended near the middle of the cell. In these cells the cytoplasm is in active movement in all directions. The strands crossing the cell show active currents by which chlorophyll-granules are often swept along. In the thicker strands there may be two opposite currents flowing simultaneously. The inner layer of the cytoplasm lining the cell-wall is also in movement, but not as in the first case all in one direction, for here there are varying streams running in different directions in different parts of the layer. Neither the arrangement nor the direction of movement of the protoplasm is constant. Sometimes a cytoplasmic strand is snapped in two, and both halves drawn back into the main mass. Sometimes a new strand is put out, which extends itself across the vacuole and joins on to the cytoplasm on the opposite side. Moreover, in any part of the cytoplasm the movement may stop altogether for a time, and then start again, perhaps in the opposite direction to that which it pursued at first.

Cytoplasmic movement such as this, in which numerous distinct and varying currents are flowing at the same time in the same cell, is called *circulation*. In *Elodea* circulation tends to pass over into rotation; all the cytoplasmic strands become withdrawn into the primordial utricle, which then settles down into a simple movement of revolution. In circulation as in rotation the ectoplasm remains at rest.

In one form or other it is probable that the protoplasm in every active cell is in movement, and this power of spontaneous motion is the constant characteristic of all living matter. But it is often difficult to examine a cell under sufficiently natural conditions for its protoplasmic movements to be observed. When we come to consider the simpler plants, we shall meet with several other kinds of movement executed by the protoplasm.

B. The Tissues

We have now been able to form some idea of the essential points in the structure of a living plant cell. Of such cells and their modifications all plants consist, if we except some of the lower plants with which we are not concerned at present. We shall find, however, that the modifications undergone by many of the cells, of which the higher plants are built up, involve very considerable changes of structure.

The different kinds of cells are not distributed uniformly all through the substance of a plant. They have a definite arrangement, and the various kinds of cells are associated together so as to form more or less sharply defined layers or strands. Such an association of similar cells is termed a *tissue*. Certain kinds of tissue, again, are generally associated together, and such a combination of tissues is called a *tissue-system*.

We are now in a position to undertake the study of the internal structure of the Wallflower. We will begin with the stem, and will make ourselves acquainted with the tissues of which it is composed.

a. Structure of the Stem

The stem of the Wallflower is traversed lengthways by a number of strands of rather tough, stringy tissue. It is possible to dissect away the soft substance in which they are embedded, and so lay bare the strands themselves. If a stem be allowed to rot, it is this stringy part which will longest resist decay. Most people will have noticed at some time or other an old rotten cabbage stalk, in which all the soft parts have perished and only a network of hard, woody, interlacing strings is left behind. These strands are the *vascular bundles*. They constitute one of the systems of tissue of which the higher plants are composed. Their great function is to serve as conducting channels through which water and the various food-materials pass. In most cases they are also of importance as forming a supporting skeleton for the plant.

In the Wallflower the tissue of the young stem, with the exception of the vascular bundles, is soft and succulent. A part of the soft tissue lies to the inside of the vascular bundles, and between them ; the remainder is situated to the outside of the bundles. The bundles collectively, together with all the tissue that lies between and within them, form the *central cylinder* or *stele* ; the tissue lying to the outside of the vascular bundles is called the *cortex*. The surface of the stem is covered by a definite skin, which we can easily strip off with the aid of a pair of forceps. We find that it is a very thin, colourless membrane, which, in spite of its delicacy, is fairly tough and elastic. This skin, the *epidermis*, is only found intact on the younger parts of the stem. The older portions are covered by a much thicker and harder skin, the *bark* ; all the external " protective " tissues are called by the general name of *dermal* tissue, which thus includes both epidermis and bark.

We will now examine, with the help of the microscope, a

transverse section cut through the stem of the Wallflower (see Fig. 20). This shows us the cut ends of the vascular bundles, and we see that they are arranged in a ring. The space inside the ring of bundles is occupied by a large-celled tissue, the *pith* or *medulla*. A similar tissue extends between

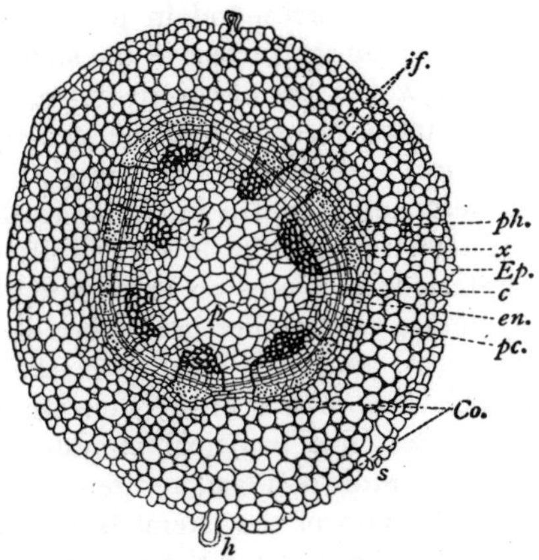

FIG. 20.—Transverse section through the second internode of the stem of a seedling Wallflower, with seven vascular bundles. *Ep*, epidermis ; *h*, hair ; *s*, stoma ; *Co*, cortex ; *en*, endodermis, or innermost layer of cortex ; *pc*, pericycle, or outermost layer of cylinder ; *ph*, phloëm of a vascular bundle ; *x*, xylem of vascular bundle ; *c*. cambium ; *if*, medullary ray, or interfascicular tissue ; *p*, pith. Magnified 55.(R. S.)

the bundles as far as the cortex which surrounds them. These bands of *interfascicular* tissue (*if* in Fig. 20) are called *medullary rays*. Vascular bundles, medullary rays, and pith collectively form the *central cylinder* or *stele*. This is surrounded on the outside by the large-celled cortex, many cells in width (*Co*), and this again by the single layer of epidermis (*Ep*).

a. The Vascular Bundles

We will now consider the vascular bundles more in detail, and first we will determine their *course* as traced longitudinally. Each bundle runs out into a leaf. This fact is most easily proved by tracing the course of the five largest bundles, which correspond in position to the five projecting ridges seen on the surface of the stem. It is most convenient to follow the bundle from above downwards. It enters the stem from the midrib of the leaf, runs inwards for a short distance, and then turns straight downwards, continuing its descending course in the stem through five internodes. It then reaches the leaf vertically below that from which it started, and joins on to its bundle. Besides the principal bundle, two smaller lateral bundles enter the stem from each leaf, and these behave in the same way. Hence in any transverse section of the stem we cut through about fifteen bundles, five principal strands, each representing the middle bundle of a leaf, and ten smaller strands, forming the continuation of the lateral leaf-bundles. The number is not always exact, for the smaller bundles may branch or unite together on their downward course. It is quite easy to understand the arrangement if we bear in mind that each ridge on the outside of the stem corresponds to a principal bundle, which is accompanied by its two smaller side bundles. At the nodes, *i.e.* at the insertion of each leaf, the various bundles are connected together by cross branches.

The description given applies, with slight variations in detail, to the whole stem of the Wallflower, with the exception of the first few internodes of the seedling. These have a simpler arrangement of the leaves, and consequently a smaller number of bundles. The section (Fig. 20) is taken from one of these first-formed and simpler internodes.

Vascular bundles of this kind, which are continuous from

the stem into the leaf, are called *common* bundles, because they are common to both organs. That part of a common bundle which passes through the stem is called a *leaf-trace*, because its position is dependent on that of the leaf with which it is connected. We see, then, that the whole bundle-

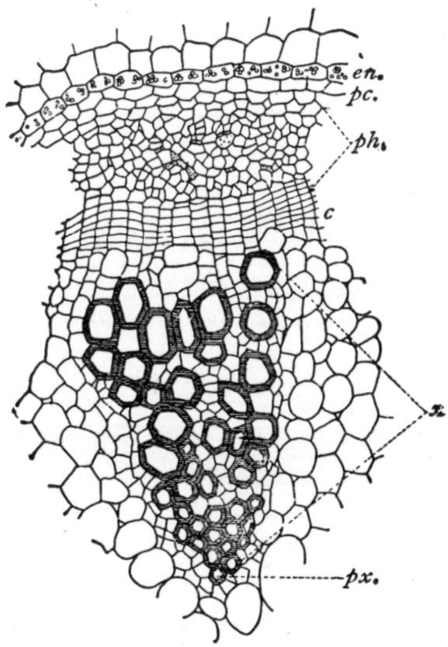

FIG. 21.—Transverse section through a vascular bundle of a young stem of Wallflower. *en*, endodermis, containing starch granules ; *pc*, pericycle ; *ph*, phloëm ; *c*, cambium ; *x*, xylem or wood ; *px*, protoxylem, or first-formed wood. Portions of pith, interfascicular tissue, and cortex are also shown. Magnified about 210. (R. S.)

system of the stem is built up of these leaf-trace bundles, the upper ends of which run out into the leaves themselves. As the various bundles are united to one another at the nodes, it is equally easy for the water and food-substances which they conduct to pass out into one of the leaves or to continue their course straight up the stem, while, on the other hand, those food-substances which are formed in the leaves have ready access through the stem-bundles to all parts of the plant. In fact, the arrangement of the bundles is adapted to facilitate communication in every direction between leaf and stem. The connection with the root will be described later on (p. 67).

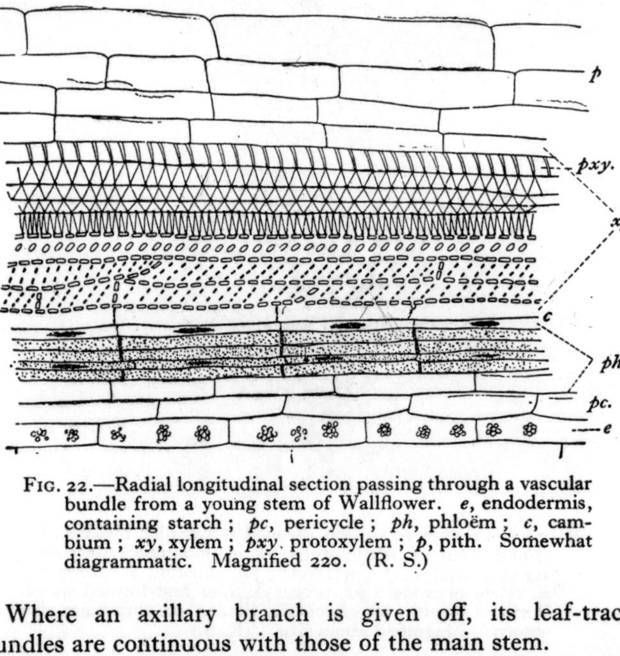

FIG. 22.—Radial longitudinal section passing through a vascular bundle from a young stem of Wallflower. *e*, endodermis, containing starch ; *pc*, pericycle ; *ph*, phloëm ; *c*, cambium ; *xy*, xylem ; *pxy*. protoxylem ; *p*, pith. Somewhat diagrammatic. Magnified 220. (R. S.)

Where an axillary branch is given off, its leaf-trace bundles are continuous with those of the main stem.

Having described the longitudinal course of the vascular bundles, we have next to examine their internal structure. We will fix our attention on a single leaf-trace bundle, such as that which is shown in transverse section in Fig. 21.

The bundle is made up of two quite different kinds of tissue. The inner half of the bundle—that, namely, which is turned towards the centre of the stem—consists chiefly of rather large cells, the walls of which are somewhat thickened, and are hard and woody. Many of these cells are devoid of living contents, containing water only. This inner half of the bundle is the *xylem* or *wood* (*x* in Figs. 20 and 21).

The outer part of the bundle, lying towards the cortex, consists of much smaller cells, with thin walls of pure cellulose, and with abundant protoplasmic contents. This part of the bundle is called the *phloëm* or *bast* (*ph* in Figs. 20 and 21).

Between xylem and phloëm is a band of very regular cells, which appear oblong in transverse section, and have very thin walls, showing that active division is going on (see Fig. 21, *c*). This layer is the *cambium*, which in plants of the Class to which the Wallflower belongs is constantly undergoing cell-division, and adding new cells to the xylem internally, and to the phloëm externally (see pp. 52 and 76).

The xylem in the Wallflower consists of three kinds of elements, which are best studied in longitudinal section (see Fig. 22).

The Xylem or Wood

1. The Vessels

These are long, continuous tubes, which extend through the plant for considerable distances without any interruption. When mature, they appear empty, containing no protoplasm, but only water and air. The vessels have very peculiar markings on their walls, which are due to their being unequally thickened, some parts of the wall growing much in thickness, while other parts remain quite thin. In the vessels nearest the pith the thickening generally takes the form of a spiral thread, which winds along the inside of the wall. The parts of the wall between the coils of the spiral are quite thin. The structure of one of these *spiral vessels* is exactly like that of an indiarubber hose with a stiff spiral wire coiled inside it to prevent its collapsing. In the innermost vessels the spiral is always loosely coiled. This is because the innermost vessels are the first to be formed, and their walls become thickened before the tissues have stopped growing in length, so that as growth goes on the spiral gets drawn out.

Sometimes, instead of the continuous spiral thread, we find isolated transverse rings round the vessel, which is then called *annular*. This form also admits of subsequent growth in length, for as the vessel becomes longer the rings become pulled farther apart. This form is not so very common in the Wallflower. These innermost vessels constitute the *protoxylem*, so called because this part of the wood is differentiated first, and is therefore the oldest (see Fig. 22, *pxy*). As the protoxylem is the innermost part of the xylem it is said to be *endarch*.

The vessels a little farther towards the outside are still spirally thickened, but here the coils of the spiral are closer together, for the thickening has been developed a little later, and so has not got so much pulled out by subsequent growth.

Next we come to vessels in which the thickening on the walls forms a network (*reticulated vessels*), the meshes of which are occupied by a thin membrane. This form is not capable of longitudinal extension, and is only differentiated after growth of the stem in length is complete.

Lastly, in the outer and later-formed parts of the xylem, we find vessels in which the wall generally is thickened, but a number of little oval spaces are left thin. These vessels are said to be *pitted*, the thin parts of the wall being the pits. The pits serve for the passage of liquids from one vessel to another ; the various kinds of thickening have the object of stiffening the vessel and enabling it to resist pressure.

We have seen that the vessels are open, continuous tubes. They are therefore quite different from ordinary cells such as we described above. A vessel is not a single cell, but it is formed from a number of cells. Each vessel consisted at first of a whole row of superposed cells, each with its own cytoplasm and nucleus. Then the cross walls separating these cells from one another were dissolved away, and so the whole row came to form one open passage. The cross walls, however, do not disappear entirely. A ring is left at the edge, and in the Wallflower we can easily see these rings, even in the mature vessels, marking the limits between the original cells which have fused to form a vessel (see Fig. 36, p. 81). These rings are found in all vessels, and must not be confused with the rings of thickening, which are found in the annular vessels of the protoxylem only.

We have already seen that the cell-wall is formed by the protoplasm. Where, as in vessels, certain parts of the cell-wall are specially thickened, these thickening masses also are deposited by the protoplasm, which at last becomes entirely used up in the process. Thus the vessels when fully formed are *dead* structures ; they contain no living matter, and are incapable of further growth. They serve simply as open channels through which the sap or transpiration stream can pass.

2. The Fibrous Cells

The xylem of the Wallflower also contains fibrous cells, which are scattered about among the vessels, and are more common in the later-formed parts of the wood. They are very long, thick-walled cells, with sharply pointed ends. The wall is thickened almost all over, but there are a few small narrow pits where the wall is left thin. These cells, unlike the vessels, keep their living contents for some time, and do not fuse with one another. The chief function of the fibrous cells is a mechanical one ; they serve to give rigidity to the stem. At the same time their protoplasm retains its activity, and consequently these cells are able to form and store up starch for future use. Tissue composed of cells of this shape, with long pointed ends overlapping one another, is called *prosenchyma*, and is commonly found where mechanical strength is required (see Fig. 36).

3. The Xylem-Parenchyma

Besides the vessels and fibrous cells, the xylem also contains cells with square ends. They are of considerable length, though very much shorter than the fibrous cells, and have rather thick walls with small pits. They always contain cytoplasm and a nucleus, and often starch granules as well, and form the chief tissue in the xylem in which starch is stored up. They do not undergo fusion. These square-ended cells constitute the woody or xylem parenchyma, the word *parenchyma* being used for any cellular tissue with cells of a rounded, square, or oblong form. Cells of this kind are always found in contact with the vessels (see Fig. 36.)

The woody character of the cell-walls of the xylem is due to the presence of a substance called *lignin*, a chemical modification of cellulose, which greatly alters its properties. Lignified cell-walls are much harder and stiffer than those of pure cellulose. Such walls are characteristic of the xylem, which is often a supporting as well as a conducting tissue, and then forms part of the skeleton of the plant. But we may find cells with thick, lignified walls outside the vascular bundle altogether, or in its phloëm portion, though this is not the case in the Wallflower. The xylem, therefore, does not of necessity form the whole skeleton of the plant, and in some cases is a conducting tissue only.

In the vessels of the xylem the hard, thick, lignified, stiffening bands have a special function, for the water in the cavities of the vessels is often under considerable tension, which tends to pull in the walls. Vessels in this state would very readily yield to the pressure of the surrounding tissues, and so collapse, if their walls were not specially strengthened.

It will have been noted that the primary xylem of a vascular bundle differentiates centrifugally, *i.e.* outwards from the protoxylem. Thus the thick-walled vessels, which form one of the main supporting elements of the young stem, develop towards the periphery of the stem. This is in agreement with a well-known mechanical principle that in erect columns, as *e.g.* the stem of a wallflower, which have to withstand the lateral stress of the wind, it is best to concentrate the greatest strength towards the periphery of the column.

The Phloëm or Bast

In the Wallflower the phloëm consists entirely of cells with soft walls of pure cellulose. The most important cells are of three kinds (see Fig. 23).

1. The Sieve-Tubes

These, like the xylem vessels, are made up of long cells arranged end to end. There are, however, important differences between sieve-tubes and vessels. In the sieve-tubes the living contents persist for some time. Each cell has its nucleus, cytoplasm, and vacuole. Also the end-walls are not dissolved away but are perforated instead with a number of small pores. Thus, in transverse section the end-walls appear like small sieves (Fig. 23, *A*). In longitudinal section the pores are seen as fine channels in the wall. These pores are filled with cytoplasm, continuous with the cytoplasm lining the walls of the sieve-tubes. The vacuole is usually rich in sugars and nitrogenous compounds. All young and active sieve-tubes have a nucleus, but it is absent from old cells in which the protoplasm breaks down into a slimy mass. Where sieve-tubes are in contact with each other side by side, their lateral walls are also perforated, not all over, but in certain parts. Through the perforations, whether of the transverse or of the longitudinal walls, the contents of the sieve-tubes are continuous. We must remember that this perforation of the cell-walls and open communication between the cytoplasm of neighbouring cells is not peculiar to sieve-tubes. We have

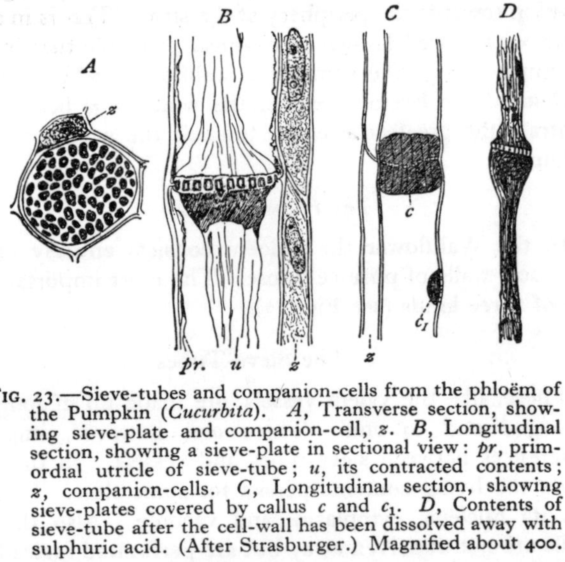

FIG. 23.—Sieve-tubes and companion-cells from the phloëm of the Pumpkin (*Cucurbita*). *A*, Transverse section, showing sieve-plate and companion-cell, *z*. *B*, Longitudinal section, showing a sieve-plate in sectional view: *pr*, primordial utricle of sieve-tube; *u*, its contracted contents; *z*, companion-cells. *C*, Longitudinal section, showing sieve-plates covered by callus *c* and c_1. *D*, Contents of sieve-tube after the cell-wall has been dissolved away with sulphuric acid. (After Strasburger.) Magnified about 400.

already learnt that living cells generally have their protoplasm continuous through minute perforations in their walls. The perforations in the sieve-tubes, however, are often much larger than those of ordinary cells, and continuity of the protoplasm was recognised in them long before it was even suspected to exist between other living cells. That part of the cell-wall of the sieve-tube which is perforated is called the *sieve-plate*. The sieve-plate and its pores may become coated, especially during the winter, with a substance (distinct from cellulose) called *callus*, which may accumulate to such an extent in old sieve-tubes as completely to close the plate (Fig. 23, *C*).

2. The Companion-Cells

The sieve-tubes are accompanied by narrow, longish cells, each of which is densely filled with cytoplasm, and contains a large nucleus, which is permanently retained. These cells are called the companion-cells. In a transverse section the companion-cells look as if they had been cut off from the corners of the sieve-tubes, and that is how they really originate, for sieve-tube and companion-cell are formed by the division of the same mother-cell (see Fig. 23, *A*, *B*, and *C*, *z*). The companion-cells are in cytoplasmic communication with the sieve-tubes by fine pores.

3. The Phloëm-Parenchyma

The rest of the phloëm is made up of rather large parenchymatous cells, which have the typical structure of living cells, each containing a layer of cytoplasm lining the wall and a single nucleus. This tissue undergoes no special modification, and is called simply *phloëm parenchyma*.

The cells at the extreme outer edge of the phloëm are the first to be differentiated, and are therefore called the *protophloëm*. Unfortunately, the phloëm of the Wallflower is not well adapted for showing the minute structure, as all the cells are very small in this part. It has therefore been necessary to introduce Fig. 23, taken from another plant (the Pumpkin). The phloëm, like the xylem, is partly a conducting tissue; its special function is the transport of sugars, nitrogenous compounds, and other materials from the leaves (or storage organs) to parts where growth is going on. This transport by the phloëm involves the activity of its living cells and takes place much more

5

rapidly than diffusion through dead tissues. The channels of this rapid transport are the sieve-tubes.

Between xylem and phloëm there are some layers of cells which do not as yet show the characters of either tissue. On the inner side these layers pass over gradually into xylem, and on the outer side into phloëm. The layer in the middle is itself constantly dividing, except in winter, and thus contributes new elements to xylem and phloëm respectively. The thinnest walls indicate the most recent divisions (see Fig. 21). The divisions take place chiefly in the tangential direction—that is to say, in the direction of the line separating xylem from phloëm, which is parallel to a line tangential to the surface of the stem at a point opposite the bundle. As already mentioned, it is characteristic of the class to which the Wallflower belongs to have this actively dividing layer of *cambium* between the xylem and phloëm. Vascular bundles which have a cambium are known as *open* bundles, because there is no definite limit to their growth in thickness, the formation of new xylem and phloëm from the permanently active cambium going on as long as the plant lives. We shall, however, understand the nature of the cambium better after we have considered the development of tissues in the growing points.

β. *Other Tissues of the Central Cylinder*

The central cylinder or stele consists of the vascular bundles and of parenchyma, which is called *conjunctive* tissue because it unites the bundles together. That part of the conjunctive tissue which lies inside the ring of bundles, and is called the pith, is of very simple structure. It is made up of large, rather elongated, square-ended cells (Fig. 22, *p*), which retain their living contents, and usually form starch granules. The *interfascicular* tissue, constituting the medullary rays, is similar to the pith, but its cells are smaller. It is important, because its cells divide up to form the *interfascicular cambium*, which unites the cambial layers of the separate bundles into a continuous ring of actively multiplying cells (see Figs. 20 and 21). Owing to the presence of this layer of cambium, the stem is able constantly to form new tissues, and thus to grow indefinitely in thickness (see p. 76).

Surrounding the whole ring of vascular bundles on the outside is a layer of thin-walled cells, the *pericycle* (*pc* in Figs. 20, 21, and 22). This layer forms the outer limit of the central cylinder. Beyond this we come to the cortex.

Those cells of the conjunctive tissue which border on the vascular bundles are generally rich in grape-sugar, dissolved in their cell-sap.

γ. *The Cortex*

Adjoining the pericycle on the outside is a single layer of cells, with slightly thickened radial walls, and with numerous starch granules in their contents. This layer is the *endodermis* (*en* in Figs. 20 and 21, *e* in Fig. 22), and forms the inner limit of the cortex. The rest of the cortex consists of ordinary parenchyma, like that of the pith. Its outer layers, however, have rather thick walls, though they consist of pure unlignified cellulose. The outer cells contain chlorophyll-granules, and can therefore take some part in carbon-assimilation, though this function belongs chiefly to the leaves. The thick-walled layers are of some mechanical importance in helping to give greater stiffness to the stem.

Parenchymatous cells generally, whether belonging to the central cylinder or to the cortex, do not as a rule fit closely together, but separate slightly from one another at their corners, so as to leave little spaces between them. These *intercellular spaces*, as they are called, contain air and aqueous vapour, but not water ; they ensure the aeration of the internal living cells.

δ. *The Epidermis*

The epidermis of the stem is one layer in thickness. The epidermal cells are of three different kinds ; the ordinary cells are elongated, with square or sometimes pointed ends, and fit closely together without any intercellular spaces. These cells each contain cytoplasm and a nucleus, but no chlorophyll-granules. The outer layer of the superficial cell-wall of the epidermis forms the cuticle, and does not consist like the other walls of cellulose, but of a substance (*cutin*) resembling cork, which is extremely impervious to water and water vapour, and which is chemically a remarkably stable body, resisting the action even of strong sulphuric acid, and of other substances which easily dissolve cellulose. The inner and lateral walls of the epidermal cells consist simply of cellulose, and the same is the case with the innermost layer of the superficial wall.

The epidermis of the Wallflower bears very large and characteristic *hairs*. Each hair is a single cell. It has the form of a spindle, and is attached at the middle by a short stalk, which is inserted between the ordinary epidermal cells (see Figs. 26, 27, and 43, which show similar hairs on the leaf). These hairs are always so placed that the spindle lies close to the surface of the epidermis, and parallel to the axis of the stem. They have an external cuticle, like other epidermal cells. Their surface is rough, with projecting knobs, which contain calcium carbonate (chalk). Experiments have shown that these rough and sharp hairs are of great importance to the plant, as a protection against the attacks of slugs and snails.

It has already been mentioned that the ordinary epidermal cells have no intercellular spaces between them. There are, however, intercellular spaces in the epidermis, but they only occur at definite points, between specially modified cells. Such a space, with the cells which enclose it, is called a *stoma*. Each stoma consists of a pair of cells, much smaller than those of the epidermis generally, and of quite a different form (see Fig. 20, *s*, also Figs. 26 and 27, *s*, from leaf). These two cells are called the *guard-cells*. Each guard-cell is sausage-shaped and curved, the ends of the cells being firmly joined together, while in the middle they are separated a little, leaving a narrow pore or chink between them, which communicates with the intercellular spaces of the ground tissue. The guard-cells of the stomata differ from other epidermal cells in containing chloroplasts and starch granules. We will defer the more detailed consideration of the structure of the stomata until we come to the leaf, where their more important functions are discharged.

b. Structure of the Leaf

We have already learnt so much of the structure of the leaf as can be made out with the help of a lens only (see Fig. 7). The leaf is built up of the same tissues as we have found in the stem, but with many differences in arrangement and in the details of structure.

The vascular bundles of the leaf are the direct continuation of those in the stem. We have already traced the general course of the veins of the leaf, and this is identical with the course of the vascular bundles, each vein corresponding to a bundle. In the larger veins, however, such as the midrib, the bundles are accompanied by a massive, enveloping zone of parenchyma, differing from that of the rest of the leaf in containing little chlorophyll and plenty of starch. The principal bundle traverses the midrib from end to end, tapering towards the apex of the leaf, and giving off branches on either side. All these branches ramify repeatedly, but always in the same plane, the successive offshoots becoming finer and finer, until we come to the extremely slender ultimate branchlets, which end blindly within the meshes formed by the larger bundles (see Fig. 24).

The two lateral bundles also branch immediately on entering the leaf, and form the bundle-system of the basal part of the lamina.

The larger vascular bundles of the leaf agree in structure with those of a young stem, but the two parts of the bundle are so placed that the xylem is directed upwards, and the phloëm downwards. This is a constant rule for vascular

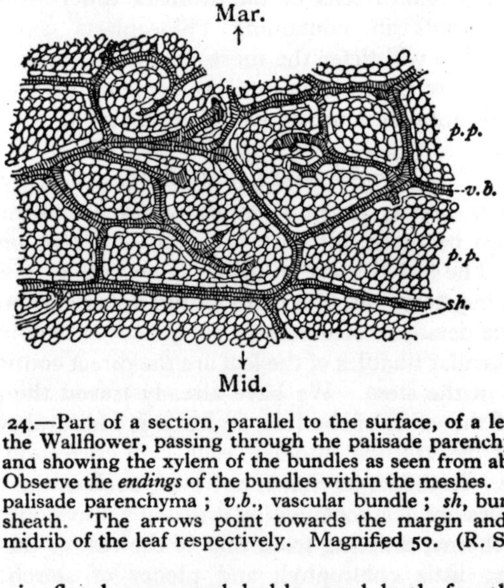

FIG. 24.—Part of a section, parallel to the surface, of a leaf of the Wallflower, passing through the palisade parenchyma, and showing the xylem of the bundles as seen from above. Observe the *endings* of the bundles within the meshes. *p.p.*, palisade parenchyma ; *v.b.*, vascular bundle ; *sh*, bundle-sheath. The arrows point towards the margin and the midrib of the leaf respectively. Magnified 50. (R. S.)

bundles in flat leaves, and holds good both for the main bundles and their branches. We know that in the stem the xylem faces inwards and the phloëm outwards, and as the bundle passes into the leaf without any twisting, the xylem necessarily comes to lie towards the upper, and the phloëm towards the lower surface of the leaf.

The loosely coiled spiral vessels of the protoxylem are found at the extreme upper edge of the wood, while the protophloëm lies at the extreme lower edge of the phloëm. The differentiation of the parts of the bundle follows precisely the same order as in the stem. In the leaf, however, the bundles are *closed*. The life of the leaf being limited, there is no need for a permanently active cambium.

In the finer bundles of the leaf the structure is greatly simplified, the xylem consisting exclusively of spirally thickened elements. In the ultimate branches these elements are of the kind called *tracheids*; they resemble vessels in all respects except that the cells of which they are composed do not fuse with one another, so that they do not form continuous tubes. If we trace the finest bundles to their termination, we find that the phloëm comes to an end before the xylem, so that the extreme end of the bundle consists of spiral tracheids only. All the bundles, including their finest branches, are enclosed in sheaths of closely-fitting parenchymatous cells (see Fig. 24).

We must now say something as to the tissue which surrounds the vascular bundles in the larger veins of the leaf, especially in the midrib (see Fig. 25). The parenchymatous tissue of the midrib contains but little chlorophyll, and closely resembles the cortical tissue of the stem. The cells towards the upper and lower surfaces of the midrib are rather thicker-walled than the rest. The other main veins of the leaf show a somewhat similar structure on a much smaller scale. Together with the midrib they form a mechanical supporting system for the leaf, constituting a stiff framework, between the ribs of which the delicate tissue of the lamina is extended, and thus kept in position. The larger bundles also have a little parenchyma within their sheath.

Now we will consider the structure of the chlorophyll-containing tissue (*mesophyll*), of which the thin, deep-green lamina chiefly consists.

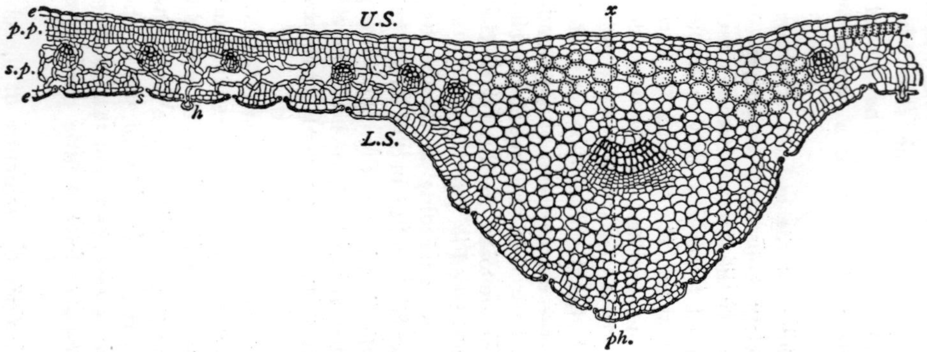

FIG. 25.—Transverse section through part of the leaf of the Wallflower, passing through the midrib. *e*, epidermis ; *h*, hair ; *s*, stoma ; *p.p*, palisade parenchyma ; *s.p.*, spongy parenchyma ; *x*, xylem of largest bundle ; *ph*, phloëm ; *U.S.*, upper surface ; *L.S.*, lower surface, of leaf. Magnified about 35. (R. S.)

On examining a transverse section (see Figs. 25 and 26) we see at once that the parts of this tissue towards the upper and lower surface respectively have a different structure. The upper part of the mesophyll is made up of three or

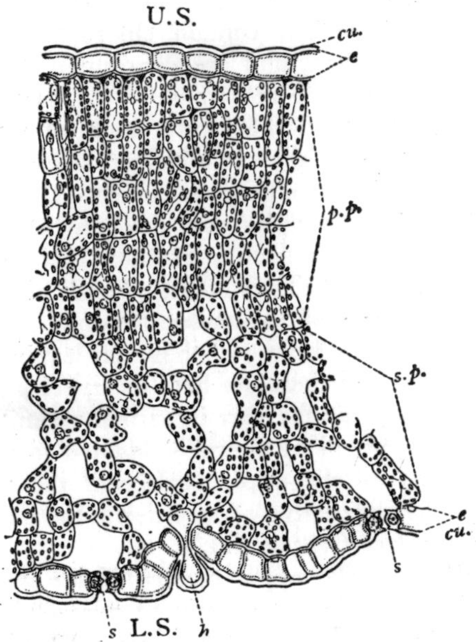

Fig. 26.—Part of a transverse section of a leaf of the Wallflower. *e*, epidermis ; *cu*, cuticle ; *p.p*, palisade parenchyma ; *s.p*, spongy parenchyma ; *s*, stoma ; *h*, hair ; *U.S.*, upper, *L.S.*, lower, surface of leaf. Magnified 165. (R. S.)

more layers of closely-packed cells, which are elongated in the direction at right angles to the surface of the leaf. These cells contain an immense number of chloroplasts embedded in their cytoplasm, and chiefly ranged on the side-walls of the cells (see Fig. 26). From the appearance of its upright and regularly arranged cells this tissue is called the *palisade parenchyma.* Its cells are cylindrical in form, and have narrow intercellular spaces between them.

The lower half of the mesophyll, on the other hand, consists of very loosely arranged irregular cells, with very large air spaces between them. These cells, like those of the palisade parenchyma, contain chloroplasts, but they are less numerous here, and are distributed uniformly in the cytoplasm lining the cell-walls. This tissue is called the *spongy parenchyma.* The lower ends of the palisade cells never end blindly in an intercellular space, but always join on to cells of the spongy parenchyma.

The finer vascular bundles pass through the upper part of the spongy tissue in such a position that their xylem almost reaches the lower ends of the palisade cells.

The differentiation of the mesophyll into palisade and spongy tissue is characteristic of those flat leaves which are horizontal, and which therefore have an upper and a lower surface. Both tissues take part in carbon-assimilation, the palisade-parenchyma being specially adapted to the intense light which falls on the upper surface of the leaf. The spongy parenchyma is actively concerned in the transpiration of water vapour (see Chapter III.).

The *epidermis* of the upper surface of the leaf consists entirely of closely-packed cells, without intercellular spaces. It bears a few long, spindle-shaped hairs, like those of the stem. There are no *stomata* on this upper surface.

The epidermis of the lower surface has more numerous hairs, and possesses an immense number of stomata (see Fig. 27). The general structure of a stoma has already been described. Each stoma opens into a large intercellular space in the spongy tissue immediately within it. This space is called the *air-chamber* (see Fig. 26). The air-chambers, again, are in communication with all the intercellular spaces of the leaf, and through them, with those of the whole plant. The stomata, then, are the pores by which the intercellular passages of the whole plant, containing air and water vapour, open into the external atmosphere.

A section across a stoma shows the two guard-cells with the pore between them (Fig. 26). The transverse section of each guard-cell is roughly a square; the walls are very thick, especially at the two corners towards the pore, which are provided with projecting ridges. We have already learnt

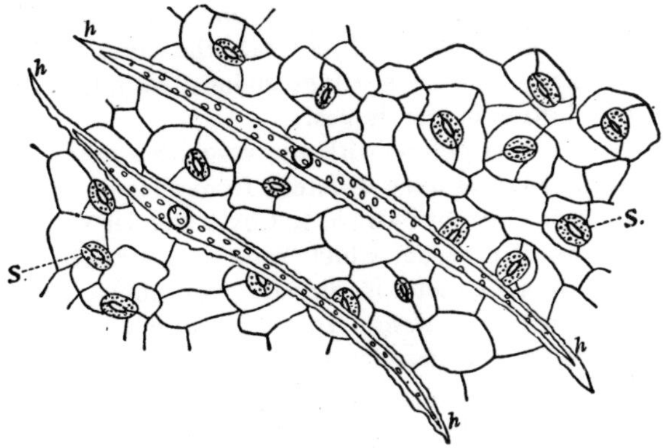

FIG. 27.—Portion of epidermis from the under side of a Wallflower leaf, in surface view. *h*, *h*, hairs; *S*, *S*, stomata. Magnified about 210. (R. S.)

that the guard-cells contain chloroplasts and starch grains, and thus differ from the ordinary epidermal cells.

The stomata have the power of opening and closing. As a general rule, they open under the influence of light and warmth, and close when it is dark or cold. The pore opens when the guard-cells become more curved, so that the space between them is widened. It closes when they straighten, so that their sides are brought into contact. We shall see later on how this opening and closing of the stomata works in with the mode of action of the leaf as a whole (see p. 197).

c. Structure of the Root

We will now examine the structure of the main root or *tap-root*, which forms the direct downward continuation of the stem. It does not differ essentially from its branches, but has the advantage of being rather larger, and more convenient for investigation.

In order to find the primary structure of the root unaltered, we must take a very young one ; for in the Wallflower, and most members of its Class, the cambium begins its activity early, and soon brings about a complete change in the distribution of the tissues.

We will start as before with the vascular system, which at first sight appears very different from that of the stem. The middle of the root is traversed by a slender cylindrical strand of vascular tissue, which runs in a straight course from base to apex, and is directly continuous with the bundle-system of the stem. In a root which has reached the stage shown in transverse section in Fig. 28, the xylem forms a plate of tissue passing through the centre of the vascular cylinder ; the phloëm is grouped in two distinct bundles, one on each side of the xylem-plate. If we had examined a still younger root, we should have found that only the cells at the two ends of the xylem-plate were mature and lignified, the middle part of the plate consisting of thin-walled cells still in course of differentiation. In other words, the protoxylem, or oldest wood, of the root is formed at the outside (it is said to be *exarch*), and differentiation advances from without inwards, *i.e.* centripetally, just in the reverse order to that which is followed in the formation of xylem in the stem. Consequently we should have to look for the first-formed spiral vessels at the extreme ends, as seen in a transverse section, of the xylem-plate of the root. The phloëm differentiates from without inwards, just as in the stem, so we see that, in the root, phloëm and xylem are both formed in the same direction, and not in opposite directions, as in the stem. We find that in this root there is no pith, for the xylem extends quite to the centre. This is characteristic of very many roots, though not of all. There are a few parenchymatous cells between

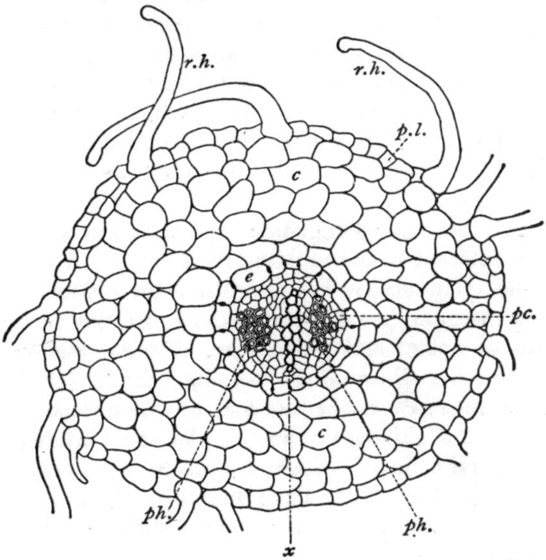

FIG. 28.—Transverse section of a very young root of the Wallflower. *r.h.*, root-hairs ; *p.l.*, piliferous layer ; *c*, cortex ; *e*, endodermis ; *pc*, pericyle ; *x*, xylem-plate ; the protoxylem is at its two ends ; *ph*, *ph*, the two phloëm groups. Magnified 110. (R. S.)

xylem and phloëm ; these cells form the conjunctive tissue of the central cylinder.

A root with the structure just described is called *diarch*, because there are, to begin with, two groups of xylem and two of phloëm, the two xylem groups soon uniting in the middle to form the continuous plate. When there are three groups of each, the root is called *triarch* ; when four, *tetrarch* ; and so on.

The centripetal differentiation of the primary xylem strands of the root, *i.e.* inwards from the protoxylem, is correlated with the conditions under which the root lives. Unlike the stem it is not liable to lateral wind stresses. In the absence of these stresses no advantage is gained by the differentiation of the strengthening elements, *i.e.* the vessels, towards the outside. On the other hand, the first formed elements of the xylem to be differentiated in the roots, *i.e.* the protoxylem, are conveniently situated, towards the periphery, to obtain the water absorbed by the root-hairs from the soil.

We see that the arrangement of the xylem and phloëm is quite different in the root and in the stem. While in the latter the xylem always lies just inside the phloëm, the two together forming a collateral bundle, in the former the bundles of xylem and phloëm are distinct, and alternate with each other all round the vascular cylinder (cf. Fig. 64).

The arrangement in the root may be called *radial*, because the xylem and phloëm groups are placed on alternate radii of the central cylinder, which, as seen in transverse section, is almost circular. The chief points, then, in which the primary vascular system of the root differs from that of the stem are—

(1) The alternating arrangement of the xylem and phloëm groups.

(2 The centripetal differentiation of the xylem.

(3) The greater concentration of the bundles, and consequent smaller diameter of the central cylinder, which contains but little parenchyma between the bundles, and often has no pith.

The vascular cylinder is surrounded by a layer of thin-walled cells, the *pericycle*, which here, as in the stem, forms the outermost layer of the stele or central cylinder. In the root this layer is of special importance, for from it all the branches of the root arise.

The *pericycle*, again, is surrounded by another layer, the *endodermis*, which consists of somewhat elongated cylindrical cells. Here it has the peculiarity that the middle part of its cell-walls, which border on other endodermal cells, is cuticularised. This is not the case with the inner and outer walls, which abut on the pericycle and cortex respectively. The thickened part of the radial walls of the endodermal cells extends as a narrow strip (called the *Casparian strip*) over their entire length. This strip extends also over the transverse walls. We can recognise the endodermis in the root in transverse sections by the dark appearance of its cuticularised radial walls (see Fig. 28, *e*) ; we know that cuticularised walls resist the passage of water and air much more than cellulose walls do, so one effect of this arrangement is that water has free passage through the endodermis from the cortex into the vascular cylinder, but cannot turn aside to pass from one endodermal cell into another. The cuticularised walls of the endodermal cells fit closely together, so that the intercellular spaces of the cortex are completely shut off from those of the cylinder. This structure of the endodermis is sometimes found in the stem also, but not so constantly as in roots. The endodermis is really the innermost layer of the cortex. The rest of the cortical issue consists of living parenchymatous cells, with thin cellulose walls. They do not, of course, contain any chlorophyll, because the roots in the earth receive no light, without which chlorophyll cannot be formed, though the plastids may be present.

The outermost layer of the young root, corresponding in position to the epidermis of the stem, consists of thin-walled living cells, many of which grow out into long hairs (see Fig. 28). These root-hairs are very important organs ; they take up all the water and food which the plant obtains from the soil. Each root-hair is a single cell, with cytoplasm and a nucleus, and a thin cell-wall. The delicacy of the outer cell-walls, and the absence or slight development of cuticle, are points in which the external layer of the root differs from the epidermis of the shoot. The meaning of these differences is, that the epidermis of the stem is a *protective* layer, which among other duties has to hinder the passage of water out of the plant, while the external tissue of the root is *absorptive* (so long, that is, as the root is quite young), and has to admit into the plant water and various food-substances which are dissolved in it. Partly because of this great difference in function, and partly on account of some differences in the development, it is better not to speak of the external layer of roots as an epidermis, but to call it the *piliferous layer*, *i.e.* the layer which produces the root-hairs.

These root-hairs grow between the small particles of soil, and attach their ends to them quite firmly, so that it is impossible to pull up a plant out of the ground without lifting with it innumerable fragments of soil, which remain sticking to the hairs on the young roots so firmly that they cannot be washed off. This adhesion to the soil is due to the conversion of the outer layer of the cell-wall into mucilage. The student must be very careful never to confuse the root-hairs with the finer branches or fibres of the root itself. The latter may appear thin to the naked eye, but they are very thick in comparison with the hairs, and always have the same structure as the main root on a smaller scale, while, as we have seen, the root-hair is a single long cell (see Fig. 29).

The root bears no leaves; its only appendages are—(1) the rootlets, which repeat in essentials the structure of the root which bears them; and (2) the hairs.

The growing-point of the shoot is, as we know, protected by the leaves of the terminal bud, which close in over it.

The growing-point of the root requires protection even more, for it has to make its way between the hard particles of the soil, and hence a special organ is present which performs this sheltering function. This organ is the *root-cap*, a sheath of tissue shaped rather like a thimble, which envelops the delicate growing apex. As the root elongates, the cap is pushed before it, and has to bear the brunt of the friction with the soil (see Figs. 33 and 34, *c*). The outer layers of the root-cap become gradually converted into mucilage, which serves both to diminish the friction and to

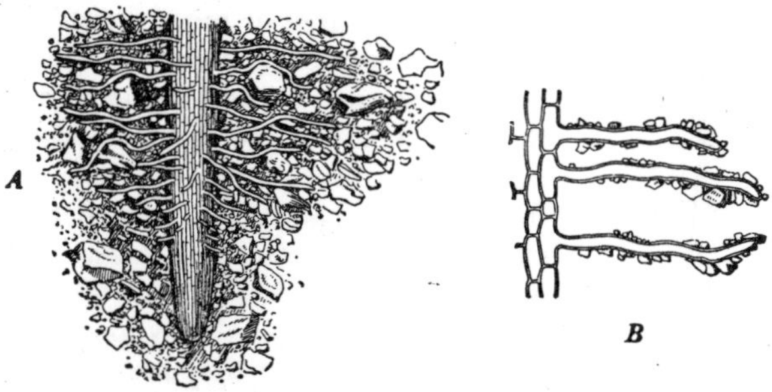

FIG. 29.—*A*, Young root, with root-hairs penetrating the soil. Magnified about 7 times. *B*, Root-hairs more highly magnified, showing their close adhesion to the particles of soil. (After Kerner.)

keep the apex moist, and as this happens, the cap is renewed from within by the growing-point itself, which thus has to provide new tissue for the cap in front of it as well as for the body of the root behind (see Fig. 33, *c*).

The presence of the cap is another very constant character by which roots are distinguished from stems.

d. Transition from Stem to Root

It has already been said that the tissues of the root are continuous with those of the stem. We will now see how the one organ passes into the other. In order to understand this, it will be necessary to study a fairly young seedling, before there has been much growth in thickness.

The first two leaves which the Wallflower seedling forms are placed opposite to one another, as is almost always the case in the Class of which it is a member. These two first leaves are called the *seed-leaves* or *cotyledons*, and differ considerably from the other leaves of the plant. For example, they are nearly round in shape, and have distinct leaf-stalks, their venation is also different from that of ordinary leaves, and the hairs mostly have three or more arms. The part of the stem which comes below the cotyledons, *i.e.* between them and the tap-root is called the *hypocotyledonary stem* or *hypocotyl* (*hyp* in Fig. 30). It is in this region that the transition between stem and root structure takes place. From each cotyledon we can trace two vascular bundles into the hypocotyl. As soon as they enter it they are joined by the bundles of the next leaves above; but with these we need not concern ourselves, for only the bundles of the cotyledons extend down into the root. The hypocotyl, therefore, contains four vascular bundles only, two on each side. Their position is different from that of the ordinary vascular bundles of the stem. In each pair the two protoxylem groups are united, and not directed inwards, but are turned towards the periphery; this is the case even in the cotyledons themselves, before the bundles have entered the hypocotyl. The rest of the xylem forms two distinct groups (*metaxylem* [1]) in

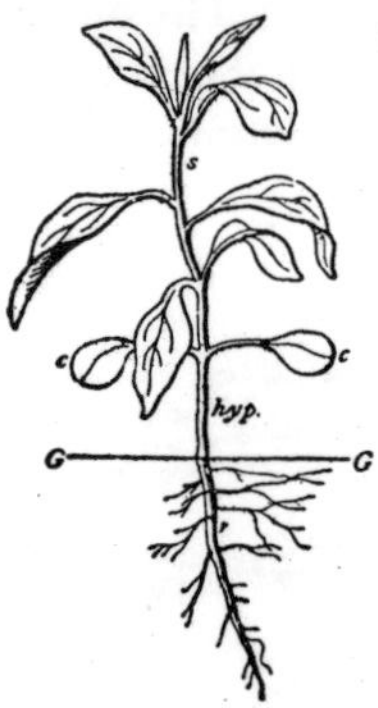

FIG. 30.—Wallflower seedling. *s*, stem; *c*, *c*, cotyledons, or first leaves; *hyp*, hypocotyledonary stem; *G–G*, level of ground; *r*, root. One-half natural size. (R. S.)

[1] The word *metaxylem* is applied to that part of the primary xylem which is differentiated after the protoxylem.

each pair. The phloëm groups are not placed directly opposite the xylem, but lie rather farther to the right and left, so that in each pair of bundles the two groups of phloëm are farther apart from each other than are those of the xylem. Now, the farther we trace the vascular bundles downwards, the nearer to each other do we find the xylem groups of each pair, and as they approach each other their protoxylem keeps its peripheral position, on the right and left (see Fig. 31, *px*).

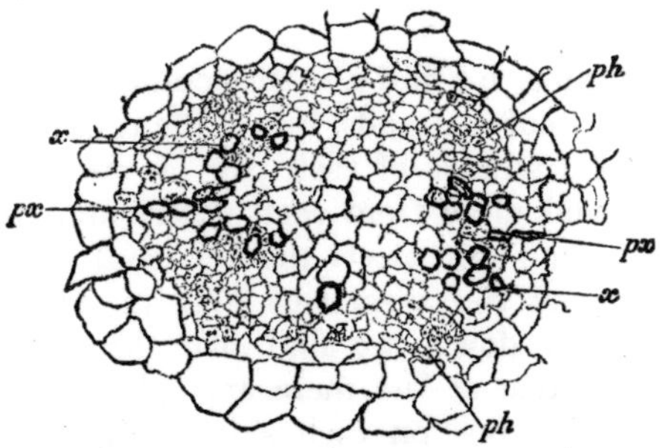

FIG. 31.—Part of transverse section of hypocotyl of Wallflower, showing cylinder. *px*, *px*, protoxylem groups of the four bundles from the cotyledons; *x*, *x*, metaxylem; *ph*, *ph*, phloëm groups of right-hand pair. From the middle of the transitional region between stem and root. Magnified about 200. From a drawing by Dr. Ethel N. Miles Thomas, F.L.S.

Lower down still we find in place of each pair of xylem groups a single one, with its protoxylem still directed straight outwards. The two bundles of each pair are at no time so separate as to possess distinct protoxylem-groups; there is a single, central, and common protoxylem throughout, the position of which, with reference to the rest of the xylem and the phloëm, varies according to the level. Ultimately the whole of the xylem on each side becomes united. From this point onwards, the hypocotyl contains two groups of xylem only, and these groups now converge, and ultimately unite to form the diarch xylem-plate of the root itself. We see, then, that this xylem-plate forms the downward prolongation of the two pairs of xylem strands from the cotyledons. Consequently the xylem-plate lies in the vertical plane which passes through the insertion of the two cotyledons.

In the Wallflower and many other Dicotyledons, the outwardly directed protoxylem is not only a character of the root, but is common also to the hypocotyl and the cotyledons.

In the meantime, we find a corresponding change in the position of the phloëm strands, but in the reverse direction. The two phloëm strands of each pair of vascular bundles diverge more and more from each other, until the right-hand group of one pair meets the left-hand group of the opposite pair, and *vice versa* (see Fig. 31, *ph*). Tracing them still farther down, we find that these two groups become one, so that at last there are only two groups of phloëm altogether, one on each side of the xylem plate. Each phloëm strand of the root, then, is continuous with two phloëm strands of the hypocotyl, one belonging to each cotyledon, and thus we see that in spite of the great difference in the arrangement of the xylem and phloëm strands, the vascular system of the main root is the direct downward continuation of that of the stem, the hypocotyl forming the connection between the two; the parenchyma of the central cylinder undergoes a great reduction in the transitional region, the pith gradually disappearing as the bundles converge. The pericycle forms the external layer of the cylinder throughout, whether in stem or root. The continuity of the endodermis and other cortical tissues between stem and root is also complete. The change from the epidermis of the stem to the piliferous layer of the root is, however, a sudden one.

e. Apical Development

We will now describe rather more fully how the development of the Wallflower plant takes place. We already know that the growing-points, *i.e.* the regions where the formation of new organs and tissues goes on, lie at the extreme ends of the branches of both shoot and root. In every branch the base is the oldest, the apex the youngest part. It is at the apex, then, that we shall find development in progress.

(*a*) *The growing-point of the shoot.*—If we examine a longitudinal section passing exactly through the middle of a growing-point, we find that the end of the stem has a nearly flat or slightly domed surface. If the section has passed through one or more of the very youngest leaves, we shall see that they arise as slight projections on the sides of the growing apex. Lower down, we come to rather more advanced leaves, which at first simply show an increase in size; while farther down still, their tissues begin to be differentiated.

Now the growing-point and the very youngest leaves consist entirely of thin-walled cells rich in cytoplasm, with relatively large nuclei. Such cells have already been described (p. 25). All these cells are in course of active growth and division, and therefore the tissue of the growing-point is called a *meristem* or *dividing tissue*. From this meristem all other tissues are derived. In favourable sections we can distinguish three distinct meristematic strata at the apex; the outermost of these is a single layer of cells, dividing exclusively by walls at right angles to the surface, and giving rise to the epidermis only. This is called the *dermatogen* (Fig. 32, *ep*). It extends not only over the growing-point of the stem, but also over the young leaves, for the dermatogen forms the epidermis of the leaves as well as that of the stem. Next below the dermatogen we find a stratum which may be only one cell thick at the extreme apex, but which lower down divides by cell-walls parallel to the surface as well as by walls at right angles to it, and thus becomes several layers in thickness. This is the *periblem*, or young cortex (Fig. 32, *i*). Lastly, underneath this again, we find a group of meristematic cells which divide in all directions, and form the central cylinder or stele of the stem, including the pericycle, vascular bundles, pith, etc. This last group is called the *plerome* (Fig. 32, *pl*).

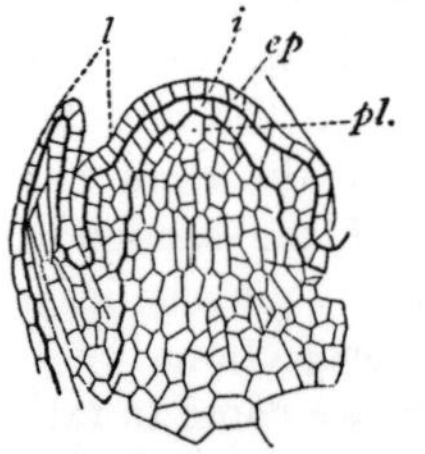

FIG. 32.—Growing-point of a stem in longitudinal section. *l*, young leaves ; *ep*, young epidermis, or dermatogen ; *i*, periblem, or layer from which the cortex is developed ; *pl*, plerome, which develops into the central cylinder. (After Douliot.) Magnified.

The dermatogen is always a distinct layer. The periblem and plerome, however, often unite in a common layer of meristem at the extreme apex.

The vascular bundles begin to develop at some little distance from the apex. The outer part of the plerome divides up more actively than the rest, giving rise to a small-celled tissue. Within this zone certain strands of cells divide more frequently by longitudinal than by transverse walls, so that they come to consist of rather elongated cells. These constitute the *procambial strands*, each of which develops into a vascular bundle, as already described. The innermost elements of each strand thicken their walls and become the protoxylem, while the outermost are converted into the protophloëm. The successive differentiation of the elements then goes on from within outwards in the xylem, and from without inwards in the phloëm. But, as it is an open bundle with which we are concerned, there is always a layer of cells left between xylem and phloëm, which remains capable of dividing, and forms the cambium of the bundle, by means of which it can increase the number of its elements indefinitely.

Only the dermatogen and periblem take part in the formation of the leaves. To form a leaf, the outer layers of the periblem grow out, and divide by walls parallel to the surface, while the dermatogen follows their growth (see Fig. 32, *l*). The latter gives rise to the epidermis of the leaf only. All its other tissues—vascular bundles as well as mesophyll—are formed from the periblem. The plerome of the stem takes no part in the development of the leaf.[1]

The growth of the leaf itself, unlike that of the stem, goes on chiefly at the base, the lower part of the leaf continuing to develop long after the apex is fully formed. We see that in each leaf-trace bundle that part which belongs to the stem is formed from the plerome, while that which belongs to the leaf may arise from the periblem. The part of the node where the vascular bundle bends out towards the leaf is generally the first to be formed. From this point the differentiation of the vascular tissue proceeds upwards into the leaf and downwards into the stem.

When a branch arises in the axil of a leaf, it is formed, like the leaf itself, entirely from the more external tissues. The branch is developed a little later than the leaf, and rather more layers of the periblem take part in forming it. But here again the plerome is said not to be concerned in the process.

(*b*) *The growing-point and branching of the root.*—The root, like the shoot, forms its new tissues at the apex of each branch, but as the structures to be produced are different it is evident that the meristem which produces them must be different also. In roots there are no leaves to be formed, and, on the other hand, the root-cap has to be provided for. Hence arise the chief differences between the growing-points of shoots and roots. It is obvious that we

[1] Most observers agree that this is the case. It has been stated, however, that the cells of the plerome grow out to form the leaf-trace bundle.

must seek here the meristem, not at the extreme end of the root, but inside the root-cap.

In a root like that of the Wallflower the meristem of the apex consists of three layers (see Fig. 33). The innermost forms the plerome, which here, as in the stem, gives rise to the vascular cylinder. The next outer layer is the periblem, which develops into the cortex ; the pericycle is formed from the outermost layer of the plerome, and the endodermis from the innermost layer of the periblem. So far there is no difference from the meristem of the shoot.

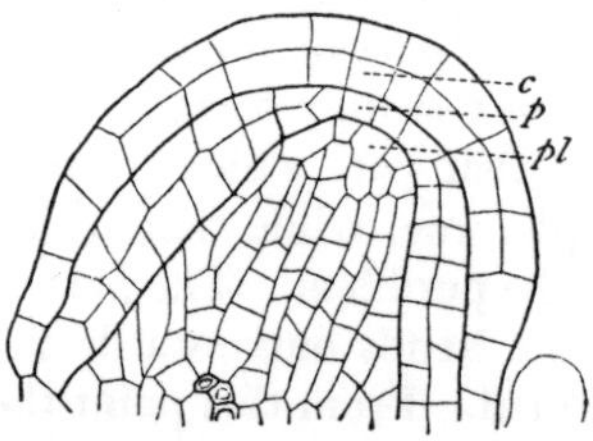

FIG. 33.—Young rootlet of *Sinapis* (allied to Wallflower) in longitudinal section. *c*, calyptrogen ; *p*, periblem ; *pl*, plerome. (After Van Tieghem.) Magnified about 250.

The external part of the meristem gives rise at once to the piliferous layer and to the root-cap. We may call it the *calyptrogen*, or cap-forming layer. Its cells divide not only by walls at right angles to the surface, but also in the direction parallel to it. The cells which are thus cut off towards the outside of the calyptrogen go to form the root-cap, those cut off towards the inside form the piliferous layer, which always remains one cell thick.

As regards the differentiation of the various tissues from the meristem, there are a few points to be noticed, though, on the whole, what has been said of the stem applies here also. The cells of the root-cap remain living for a time, and grow larger. The oldest of them, however, which lie farthest towards the outside, gradually die off, becoming partly converted into a slimy mucilage, which serves to diminish friction between the root-tip and the soil ; they are replaced by new cells formed from the calyptrogen. The other tissue formed from the calyptrogen, namely, the piliferous layer, is remarkable for the great growth in length of those of its cells which form the root-hairs, organs of which we have already learnt the high physiological importance. The development of the cortex calls for no remark.

The plerome here, as in the stem, gives rise to *procambial strands*, which in the root occupy much the greater part of the young vascular cylinder, as there is only a little *conjunctive parenchyma* between them, and no pith is present. In some plants, however, the root possesses a pith. The external part of the conjunctive parenchyma forms a single-layered pericycle, which extends all round the cylinder. In the root each procambial strand gives rise to xylem or to phloëm only.

The *branching of the root* is very characteristic of that organ, and quite different from that of the stem. All the branches arise *endogenously*—that is to say, they are formed from deep-seated tissues, and have to make their way through the external structures in order to reach the surface. In the Wallflower, and indeed in all Flowering Plants, the *pericycle* is the layer from which the rootlets are developed.

The position in which the branch roots or rootlets are formed has a definite relation to the xylem. In the Wallflower and its nearest allies the rootlets usually form four rows, which are so placed that one lies a little to the right, and another a little to the left, of each of the two protoxylem strands of the parent root. Sometimes, in place of the four rows, there are only two, which are then situated exactly opposite the two protoxylem strands. The development of the rootlets begins fairly early, but not so close to the apex as is the case with the appendages of the stem. It begins, however, before the central vessels of the xylem-plate have become lignified. A group of pericyclic cells, situated (in the more usual case) a little on one side of one of the protoxylem strands, begins to elongate radially, the middle cell elongating most. All these cells then divide by tangential walls. The inner cells thus cut off go to form the plerome of the rootlet (see Fig. 34, *pl*). The outer layer divides again (with the exception of the cells lying at the extreme edge of the group), and of the two layers thus formed the more internal gives rise to the periblem (*p*), while the outermost becomes the calyptrogen (*c*), producing the piliferous layer and root-cap. We thus have the three meristematic layers marked off from the first. They continue to grow and divide in the usual manner. The young rootlet grows out through the cortex of the parent root, the cells of which are not mechanically pushed aside, but are actually digested by the rootlet, and their substance absorbed by it as food (see Fig. 34, *d*). Eventually the piliferous layer is reached and absorbed, and then the tip of the rootlet becomes free and penetrates into the ground. We see that the origin of the branches of the root is just the converse of that of the leaves and branches of the stem. In the latter the tissues of the leaf or branch are formed from the dermatogen and periblem ; it is doubtful whether the plerome takes any part in the process. In the root it is the plerome alone which forms the branch. It is clear that the rootlet, owing to its endogenous origin, has the advantage of a more protected position from the first—a protection which it needs, as its delicate tissues would otherwise be exposed to injury by contact with the soil. The *endogenous mode of branching* is one of the most constant characters by which roots may be distinguished from stems.

f. Secondary Growth in Thickness

The stem and root of the Wallflower, like those of most members of the Class to which it belongs, go on increasing in thickness as long as the plant lives. This is due to the activity of the cambium, the origin of which, in the vascular bundles themselves, we have already described. It is impossible to draw a sharp line between procambium and

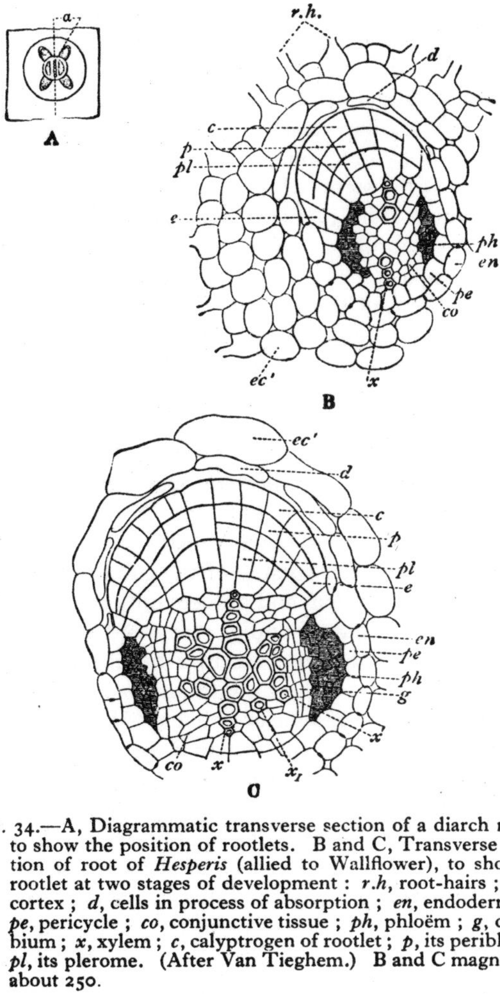

FIG. 34.—A, Diagrammatic transverse section of a diarch root, to show the position of rootlets. B and C, Transverse section of root of *Hesperis* (allied to Wallflower), to show a rootlet at two stages of development : *r.h*, root-hairs ; *ec'*, cortex ; *d*, cells in process of absorption ; *en*, endodermis ; *pe*, pericycle ; *co*, conjunctive tissue ; *ph*, phloëm ; *g*, cambium ; *x*, xylem ; *c*, calyptrogen of rootlet ; *p*, its periblem ; *pl*, its plerome. (After Van Tieghem.) B and C magnified about 250.

cambium ; for the latter, as we have seen, is simply the permanently active middle part of the procambial strand ; but it is to be noticed that while, in the earlier stages of development, the cell-divisions in the procambium take place indiscriminately in all directions, after a time they become very regular, nearly all the new walls being formed in the tangential direction, *i.e.* parallel to a tangent drawn to the surface of the stem, while radial walls are formed at long intervals only. This regular tangential division is characteristic of the cambium as distinguished from the procambium (Fig. 21, *c*, p. 43).

The cambium of the bundle, then, goes on growing and dividing without limit, except during the dormant season of the year ; on its inner side it cuts off cells, which, after one or two further divisions, gradually become converted into vessels, parenchyma, or fibres of the xylem. In the same way the cells cut off on the outer side of the cambium divide once or twice, and then become sieve-tubes, companion-cells, or parenchyma of the phloëm. The amount of xylem formed on the inside is much greater than the amount of phloëm formed on the outside of the cambium. Now, this production of new xylem and phloëm only adds to the size of each individual vascular bundle. In an old Wallflower stem, however, we find that the vascular bundles are no longer isolated, but there is a thick continuous zone of xylem, surrounded by a narrower zone of phloëm. How is this brought about ? It depends on the fact that the cambium does not remain isolated within each bundle, but becomes united to form a complete ring in the central cylinder of the stem.

When the stem is quite young, we find the vascular bundles separated from one another by radial bands of conjunctive parenchyma—the primary medullary rays (see Fig. 20, *if*). There is already an active cambial layer between the xylem and phloëm of each bundle, and soon the infection of division spreads to the neighbouring cells of the ray parenchyma, which also begin to divide up by tangential walls. Thus a layer of cambium is formed across each primary ray, joining the cambium of the bundles on either side, and so the whole central cylinder comes to contain a zone of actively dividing cells, producing new xylem internally and new phloëm

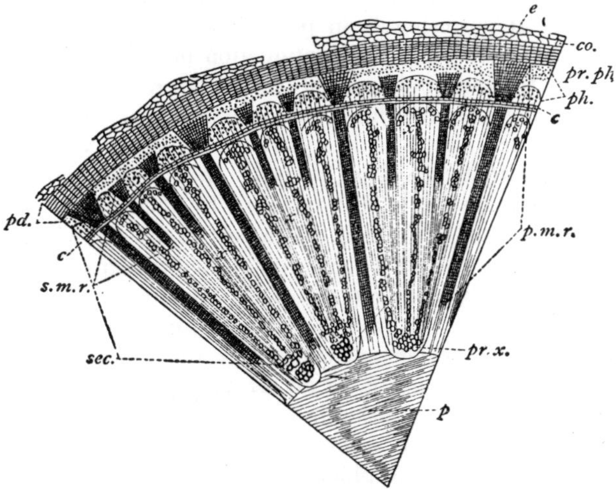

FIG. 35.—Diagrammatic transverse section of part of an old stem of Wallflower, including three of the original bundles. *e*, epidermis ; *co*, cortex ; *pd*, periderm ; *pr.ph*, primary phloëm ; *ph*, phloëm ; *c*, cambium ; *sec*, secondary tissues ; *pr.x*, primary xylem ; *p.m.r*, a principal medullary ray ; *s.m.r*, a secondary medullary ray ; *p*, pith. Magnified about 20. (R. S.)

externally. The cambium within the vascular bundles, derived from the original procambial strand, is called the *fascicular* cambium ; that between the bundles, formed by division of the cells of a primary medullary ray, is the *interfascicular* cambium.

The interfascicular xylem, *i.e.* that formed by the parts of the cambium between the bundles, consists at first chiefly of fibrous cells and parenchyma, but farther to the outside vessels are formed here also. The limits between the original bundles can, however, always be traced, for the cambium produces a radial band of parenchymatous tissue two or more cells thick, which extends right through from pith to pericycle ; this is called a *principal medullary ray* (see Fig. 35). It corresponds in position with a primary ray, and forms, as it were, the continuation of it, but is much narrower, for the cambium only forms the ray parenchyma just opposite the middle of the primary ray. The cells of the ray are short, with square ends ; their walls are thick and pitted (see Fig. 36 *mr*) ; they form starch in summer, and are easily distinguished by their shape from any other cells of the xylem. The part of the ray which is formed on the outside of the cambium broadens out, its cells becoming both larger and more numerous. Between the principal medullary rays there are shorter secondary ones (Fig. 35, *s.m.r*), which are started by the cambium after growth in thickness has gone on for some time, and which therefore do not extend either inwards to the pith or outwards to the pericycle. The secondary as well as the principal rays broaden out in the phloëm. As the xylem constantly increases in thickness, it is evident that all the tissue outside it must become stretched. This stretching is partly provided for by the tangential expansion, followed by radial division, of the cells forming the phloëm-rays, so that the phloëm itself does not get distorted.

The cells of the secondary xylem and phloëm are essentially like those of the primary tissues. The chief forms of xylem-elements are shown in Fig. 36.

The cambial cells are elongated prisms with inclined ends (see Fig. 22, p. 44). Where a medullary ray is being formed, however, the cambium has short, square-ended cells ; and when a new secondary ray is to be started, the cambial cells first divide up by transverse walls, so that in this case the form of the cambial cells is similar to that of the elements which they produce. This, however, does not hold good elsewhere, for the vessels and fibres of the xylem undergo considerable changes during their development from the

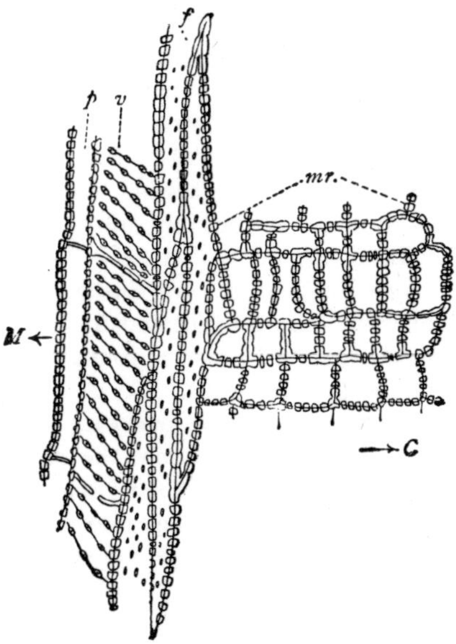

FIG. 36.—Part of a radial section through the secondary wood of Wallflower. *p*, xylem-parenchyma ; *v*, vessel ; *f*, fibrous cell ; *mr*, medullary ray. The arrows *M* and *C* point respectively towards pith and cortex. Magnified 220. (R. S.).

cambium. The vessels have a much greater diameter than the cambial cells, while the fibres are more elongated.

If we think the matter over carefully, we shall see that these changes of dimensions in particular cells cannot go on without some disturbance of the original arrangement of the cells. Take a vessel, for instance : as its diameter increases, it has to find room for itself among the surrounding parenchymatous and other cells which do not grow much in diameter. To do this it has to displace its neighbours, and eventually it comes into contact with many more cells than it touched when first formed from the cambium.

Next consider the great growth in length of the fibrous cells ; each of these when full grown is about twice as long as it was in the cambial condition. Possibly the young fibrous cells become elongated owing to compression by the neighbouring cells. In attaining its full length a fibrous cell has to insert its pointed growing ends between its neighbours, and as they, if they are fibrous cells, are all doing the same thing, the ends of these cells have to grow past each other (see Fig. 36, *f*). All this may happen in a part of the stem which has long ceased to grow in length, so that all the room required must be found by the mutual displacements of the growing cells. To all changes of this kind in which developing cells grow past each other, whether in transverse or longitudinal direction, the name of *sliding growth* is applied. The meaning of the term is, that the walls of these cells slide upon each other as they grow.

The Wallflower is a short-lived perennial. Its stems are always relatively soft because no very great amount of wood is formed; the Wallflower therefore is spoken of as an herbaceous plant. Contrasted with this is the shrubby or arborescent habit of many plants, in which the stems live for many years and in which large amounts of wood are formed, so much in fact in trees that it is used as timber. An Oak tree may live for hundreds of years and its trunk may attain a diameter of two or three feet. Most of the trunk consists of wood, but only the younger part, the *sap-wood*, functions as a water-conducting tissue, the older part, the *heart-wood*, eventually serving only as a mechanical support. Oak wood contains essentially the same elements as occur in the xylem of the Wallflower, but all, including the cells of the medullary rays, become lignified soon after formation from the cambium; consequently the wood is very hard. The conspicuous rings seen in a block or plank of oak wood are due to differences in the character of the xylem cells produced by the cambium in spring and summer respectively. Each ring represents the wood developed from the cambium in a single growing season; the rings are therefore spoken of as *annual rings*, which are characteristic of the wood of dicotyledonous trees that grow in temperate climates. The *silver-grain* of oak wood represents the large medullary rays.

g. Periderm

We know that the young stem is covered by the epidermis; the old stem, however, is coated by a greyish bark, much thicker and harder than the epidermis; this bark splits lengthways here and there; as the stem grows thicker, the cracks widen and expose the softer tissues below, until these too become coated over by new bark.

If we examine the bark of the Wallflower microscopically, we find that its outer layers consist of the epidermis and the whole of the cortex in a dried and withered condition (see Fig. 35, *e* and *co*). The inner part of the bark is formed of closely-packed cells, which are arranged in regular radial rows, and which have their walls corky or *suberised*. These walls have undergone a similar change to the cuticularisation of the outer epidermal cell-wall, and so the corky tissue has become almost impermeable to water or water vapour, just as is the case with the ordinary corks used in bottles, which are made of this tissue, obtained from the Cork Oak (*Quercus suber*). The cortex, which lies outside this corky zone, is thus completely cut off from all supplies of water and food from within, and consequently withers and dies. Now we will see how the bark is formed.

The pericycle, which encloses the whole vascular cylinder,

is at first one cell thick. Soon after secondary thickening has begun, however, its cells divide, so that it becomes three or four cells in thickness. These divisions take place rather irregularly. Then the outermost layer of the pericycle begins to divide with great regularity by tangential walls (see Fig. 37, *pd*), so that a cambial layer is formed, which, like the true cambium, extends all round the stem. This is called the *cork cambium* or *phellogen*. It produces cells, both on its inner and outer side. Those cut off towards the inside retain their living contents, and have rather thick cellulose walls. There are several layers of these cells in the Wallflower; together they constitute the *phelloderm*. The cells cut off on the outside of the phellogen have their walls suberised and their contents disappear. These cells form the *cork*.

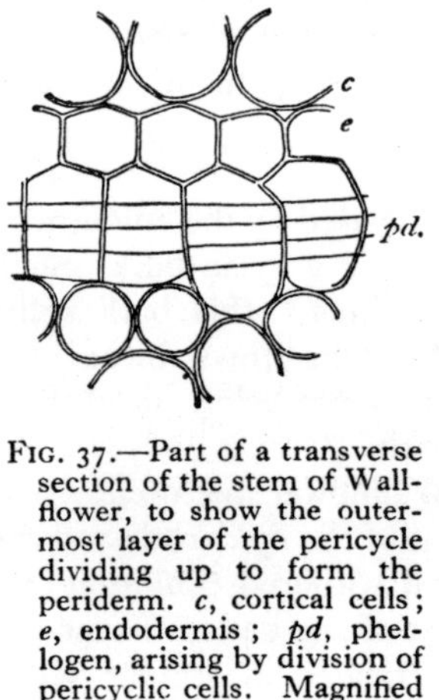

FIG. 37.—Part of a transverse section of the stem of Wallflower, to show the outermost layer of the pericycle dividing up to form the periderm. *c*, cortical cells; *e*, endodermis; *pd*, phellogen, arising by division of pericyclic cells. Magnified 460. (R. S.)

The whole of this secondary tissue, including phelloderm, phellogen, and cork, is called the *periderm*. The word *bark* is applied to everything outside the phellogen, and so here it includes the corky layers, together with the endodermis and the rest of the cortex and the epidermis. The bark consists entirely of dead tissues, for the corky walls allow of no absorption of water or food-materials. It is, however, of great importance as a protection to the living tissues within. Almost all plants with secondary growth in thickness form periderm and bark through the activity of a phellogen, which, however, does not always arise from the pericycle. The bark of our trees and large shrubs arises essentially in the same way as that of the Wallflower, though there are many differences of detail.[1]

In certain trees successive layers of phellogen arise more and more deeply in the outer tissues as time elapses, occasionally even in the outer phloëm. Deeply seated phellogens give rise to a very rugged kind of bark, as in the Oak.

Gaps, called *lenticels*, occur at intervals in cork, where the phellogen produces loose, thin-walled cells instead of impermeable cork cells. These lenticels, which allow of aeration of the internal tissues, are conspicuous features of some kinds of bark, *e.g.* birch, in which they appear as long horizontal lines. The dark streaks seen in bottle corks (from the Cork Oak) are lenticels.

h. Secondary Thickening of the Root

The root, like the stem, grows indefinitely in thickness, owing to the formation of new tissues from a cambium. This secondary growth begins very early in the root, just after the stage shown in transverse section in Fig. 28, so that in the older roots much the greater part of the tissue is of cambial origin. The first formation of the cambium in the root is rather different from that in the stem, owing to the different arrangement of the primary tissues. It has already been mentioned that the two phloëm groups are separated from the xylem by parenchymatous cells belonging to the conjunctive tissue. It is in these cells that the tangential divisions begin. The cambium, consequently, at first forms two bands of actively dividing cells, lying on either side of the xylem-plate and immediately within the two phloëm groups (see p. 77, Fig. 34, *C*, *g*). These cells give rise to new xylem on the inside and new phloëm on the outside in the usual way, but the cambium does not as

[1] In popular language the word " bark " is commonly used to denote all tissues outside the cambium, and thus includes the phloëm. This wider use of the term is adopted in some botanical works.

yet form a complete ring. It is completed in this way. The pericycle, which was at first a single layer of cells, divides and becomes two or more cells in thickness, at least at the places opposite the two protoxylem groups. The innermost of these pericyclic cells, *i.e.* those lying next the protoxylem, now undergo regular tangential divisions, and so form the continuation of the cambium round the ends of the xylem-plate. It is a constant rule in plants of the Class to which the Wallflower belongs, as well as in plants of the Fir kind, which belong to a different Class (cf. Figs. 104 and 106), that the formation of the cambium in the root begins by division of the conjunctive cells inside the phloëm-groups, and is completed by division of the pericyclic cells outside the xylem-groups. Thus the cambium (as seen in transverse section) cannot at first form a perfect circle, for it has a depression opposite each phloëm bundle ; but just at these places the cambium starts its work earliest, and is for a time most active, so that the inequalities soon become equalised, and we have a circular zone of actively dividing cells just as in the stem. When once started, there is nothing peculiar about the cambium of the root ; it forms internal xylem and external phloëm just as it does in the stem, but as in this root there is no pith, it is evident there can be no principal medullary rays. One point deserves special attention : the primary xylem of the root differentiates, as we have seen, from without inwards ; the secondary xylem is formed by the cambium from within outwards, so that in roots we have a sharp distinction between *centripetal* primary and *centrifugal* secondary wood, whereas in the stem all the wood arises centrifugally. However old the Wallflower root may be, we can always tell its transverse section from that of the stem, for in the middle, instead of the pith, we find the little diarch xylem-plate unaltered. In the roots of other plants of the same class there are often more than two groups of xylem, and a pith may even be present ; still, if we can once make out the position of the protoxylem elements, we can always be certain whether it is a root or a stem with which we have to do.

Only very young parts of the root retain their root-hairs, and absorb food from the soil. The older parts soon lose their hairs, cease to be absorptive, and now serve only to conduct the food, which the younger part of the root has taken up, towards the stem. These older parts of the root acquire a covering of periderm. The pericycle divides up to form a phellogen, which produces phelloderm internally and corky layers externally, just as in the stem.[1] It is from the outer layer of the pericycle (where it is double) that the phellogen arises. We now see that the pericycle in these roots is an extremely important layer, being the seat of three distinct new formations—

(1) Lateral roots.
(2) Cambium (in part).
(3) Phellogen.

In consequence of the formation of cork by the phellogen, all the external tissues, from the endodermis outwards, wither away, so that the older roots consist entirely of the central cylinder surrounded by periderm. A curious effect of this throwing off of the cortex, which takes place very early in roots, is that they appear thinner, after secondary growth in thickness has begun, than they did before. This is because the root at first loses more by the death of its external tissues than it gains by secondary growth The loss, however, is soon made good.

Now let us consider what purpose all this secondary formation of new tissues serves. So long as a plant like the Wallflower lives, it keeps putting out new branches ; every new branch bears a number of leaves, and so the older the plant grows the greater is the total surface of its foliage, in

[1] In the case of roots the origin of the phellogen from the pericycle is the rule, while in stems it is the exception.

spite of the loss of old leaves. Now the leaves perform two main functions : they give off water vapour through their stomata, and thus keep up the ascending current of sap from the roots ; and also they take up food from the air, namely, the carbon which they obtain by the decomposition of carbon dioxide. Therefore the greater the leaf-surface, the greater will be the amount of water which has to pass up the stem from the roots, and the greater also will be the amount of assimilated food which has to be conveyed away from the leaves to the growing parts. Consequently, as the leaf-surface increases, it is necessary that the amount of conducting tissue should increase also, and this applies both to the xylem, which conducts the sap from the roots to the leaves, and to the phloëm, which conducts the assimilated foods from the leaves to the seats of growth. This increase is provided by the activity of the cambium, and while this is going on the absorptive surface of the root is also increasing. The root branches again and again, and the number of young roots that are able to take up food from the soil is constantly becoming greater.

Thus we see how the growth of all the vegetative organs of the plant works together. As the transpiring and assimilating surface of the leaves increases, the amount of conducting tissue in the stem and older roots increases at an equal rate, while at the same time the branching of the younger roots provides for the greater demands on their absorptive capacity.

Secondary growth in thickness is important in other ways also : as the branching of the shoot goes on, it is obvious that the stem has an ever-increasing load to bear, due not merely to the increased weight, but also to the greater leverage of the spreading branches. Hence the necessity for increased mechanical strength, which is provided for by the formation of additional xylem, especially new fibrous elements, from the cambium.

Lastly, the constant formation of new tissues has the further advantage that old elements which have become past work are replaced by young and active ones. This applies especially to the phloëm, in which it is generally the newer layers which are alone active. This point, however, is not so important in the Wallflower as in some other plants, especially trees, in which the duration of life is longer. Secondary growth, then, has three main functions: to increase the conducting tissues, to increase the mechanical tissues, and to replace old elements by young ones.

III. *INTERNAL STRUCTURE AND DEVELOPMENT OF THE REPRODUCTIVE ORGANS*

We use the term "reproductive organs" here in a wide sense, to include all the parts of the flower, although many of these are only of indirect service in reproducing the plant. We do so because the sole function of these organs is to form a sexually produced embryo or young plant, and to send it into the world as well equipped as possible for its future life. The floral organs are only of service in so far as they contribute to this end.

We have already learnt the general arrangement and outward appearance of the parts of the flower (see p. 16); we will now consider the order and method of their development.

a. Development of the Flower

The very young flower-bud arises from the stem as a rounded outgrowth, which soon acquires a short stalk (see Fig. 38). The calyx is the first of the floral organs to appear. The anterior sepal takes the lead, appearing as a crescent-shaped ridge of tissue (see Fig. 38, A, s^a); next the two lateral sepals (s^l) appear together, and lastly the posterior one, which lies next the main flower-stalk (B, s^p) (see also diagram, Fig. 13). When the calyx has started its growth, the corolla begins to develop. All the four petals show themselves simultaneously, arising as small papillæ immediately within the calyx at the points corresponding to the spaces between the sepals (see Fig. 38, C, p^a and p^p). The petals have no sooner appeared than their growth ceases for a time. Hence we shall find that the next inner whorl, that of the stamens, soon overtakes them.

The stamens appear in the same order as the sepals, opposite which they arise. First an outgrowth is formed opposite the anterior sepal (Fig. 38, D, st^a). It immediately forks into two, and gives rise to the anterior pair of long stamens (E, et^a). Next two papillæ are formed opposite the two lateral sepals. These develop into the two short lateral stamens (D and E, st^l). Finally a posterior papilla is formed, which, like the anterior one, immediately branches and produces the posterior pair of long stamens (D and E, st^p and *et*).

The pistil or gynœcium is the last part of the flower to be formed. The two carpels appear as two slightly elevated ridges opposite the lateral sepals ; they are only distinct at their very first origin. A zone of tissue underneath their insertion begins to grow up immediately, carrying the carpels up with it, and forming a tubular column in the centre of the flower, on the upper edge of which the original carpels can scarcely be distinguished (Fig. 38, E, *cp*). On the inside of the carpellary tube two longitudinal ridges arise ; the position of these ridges is posterior and anterior, *i.e.* they come midway between the centre-lines of the two carpels (F and G). These ridges are the *placentas*, on which the ovules or future seeds are borne. Two elevations soon appear on the edge of the carpellary tube just above the placentas. These elevations become the stigmas (F and H, *sg*), and ultimately close in over the top of the tube, which now forms the cavity of the ovary (see Fig. 38, H). We see,

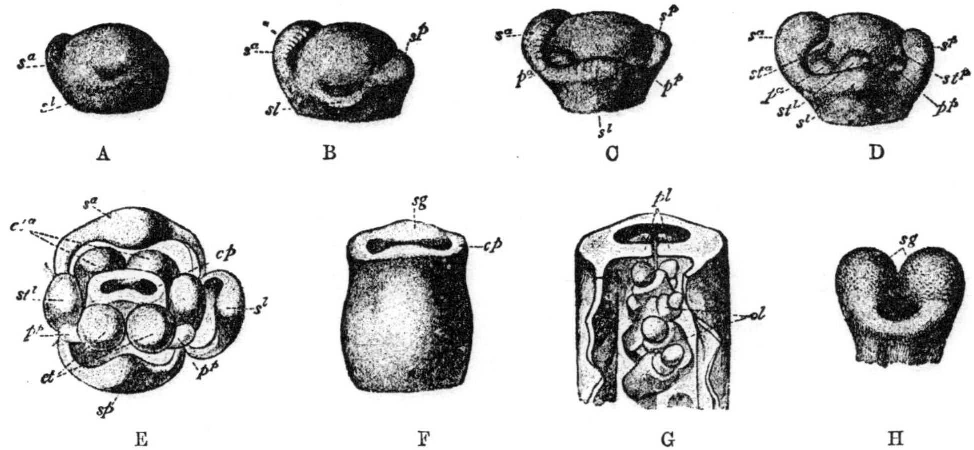

FIG. 38.—Development of the flower in *Cheiranthus*. A, Very young flower-bud, seen from the side: s^a, anterior sepal; s^l, a lateral sepal. B, Slightly older: s^p, posterior sepal. C, Older: p^a, one of the anterior petals; p^p, one of the posterior petals. D, Older: st^l, a lateral stamen; st^a and st^p, stamens of the anterior and posterior pairs. E, Older bud, seen obliquely from above: et^a, anterior pair of stamens; *et*, posterior pair of stamens; st^l, a lateral stamen; *cp*, carpellary tube. F, Pistil somewhat more advanced: *cp*, centre-line of carpels; *sg*, centre-line of stigmatic lobes. G, Older pistil, partly in section: *pl*, placental outgrowth, forming the septum; *ol*, ovules. H, Young stigma: *sg*, stigmatic lobes. Magnified. (After Payer.)

then, how it is that the position of the stigmas is alternate with that of the carpels (see Figs. 10, 11, 12, and 13). The septum appears rather late, after the development of the ovules has begun (Fig. 38, G, *pl*). Each placenta grows out between the rows of ovules ; the two projecting walls thus formed at last meet in the middle to form the septum, dividing the cavity into two. The honey-glands, which lie at the base of the two shorter lateral stamens, arise very late in the development of the flower, when all the more important organs are well advanced.

All floral parts arise like leaves, by the growth and division of the more external cells of the growing-point, namely, the dermatogen and periblem.

b. Structure of the Floral Envelopes

We will now consider more fully the structure of the various organs of the flower, and the respective parts which they play in the process of reproduction. We will first consider the structure of a sepal. The *sepals* form the outer protective covering of a flower while it is still a bud. When the flower opens, the function of the calyx in the Wall-flower is almost at an end.

Of all the floral organs the sepals most resemble foliage-leaves. They are, in fact, merely leaves of simplified structure, having less complex functions to perform than the vegetative leaves.

The sepal has a similar bundle-system to that of an ordinary foliage-leaf, but with simpler ramifications. In each bundle the xylem and phloëm are in the usual position, the xylem being placed on the upper side, *i.e.* on the side turned towards the middle of the flower, and the phloëm on the lower or outer side. The mesophyll is of uniform parenchymatous structure throughout, showing no differentiation between palisade and spongy tissue. The cells contain numerous plastids corresponding to the chlorophyll-granules, but little chlorophyll is present. The dark colour of the sepal is due to purple cell-sap, which is present in the more superficial layers on both the upper and lower sides of the sepal. It is most abundant towards the lower surface, and especially in the hypodermal layer, though it also occurs in the epidermis. Stomata are fairly numerous on both surfaces of the sepal. Hairs of the structure already described and figured (see Figs. 26, 27, and 43) are produced on the lower or outer surface only. There is reason to believe that they serve to protect the young flower-bud from the attacks of slugs.

The *petals*, which form the next floral whorl, are in many respects of very different structure. The petals are at first very slow in their development, and become mature rather suddenly, shortly before the flower is ready to open. They are transitory structures ; they have no protective function, nor do they take any direct part in fertilisation or the production of the embryo, but they are none the less important. The Wallflower is in a great degree dependent on the visits of insects for its fertilisation. The function of the petals is to render the flower conspicuous, and thus to help in attracting insect visitors. Hence, it is generally the corolla which constitutes the chief beauty of the flower in our own eyes. Its business is to be showy. Each of the petals has a vascular system ; its bundles, though slender, are of normal structure, with xylem above and phloëm below. The mesophyll consists entirely of spongy parenchyma without chlorophyll. The epidermal cells of both surfaces contain very numerous yellow chromoplasts—that is, protoplasmic bodies like chlorophyll-granules, but containing a yellow pigment instead of chlorophyll. To these the yellow colour of the petals is due. In cultivated Wallflowers a purple cell-sap, containing an anthocyanin pigment, is also often present, chiefly in the lower epidermis. To the combination of the colours of the yellow chromoplasts and of the purple cell-sap the flower owes those beautiful red-brown shades which we know so well. The wild Wallflower generally has the chromoplasts only, and so its petals are simply yellow.

There are no stomata on either surface of the petal. The epidermis is not smooth, but its cells all grow out into short hairs or papillæ, which are narrower and more prominent on the upper surface of the petal. They give it a peculiar velvety gloss, owing to the varying way in which they reflect the light, and this effect is perhaps enhanced by the fact that the cell-walls of the papillæ are delicately ridged.

The well-known pleasant scent of the Wallflower comes from the corolla. The scent is probably of importance as helping to attract insect guests.

c. Stamens

The six stamens together constitute the *andrœcium*, or fertilising apparatus of the flower. Each stamen, as we already know, consists of two parts—a stalk or filament, and an anther borne upon it. The filament is traversed by two vascular bundles, having two small xylem groups turned towards the inner surface, and two larger united phloëm groups on the outer side. The vascular tissue is surrounded by an endodermis. The mesophyll is of uniform structure all through. The epidermis of the filament has neither hairs nor stomata, but the latter occur on the anther. The filament tapers at the top, and is attached to the anther by a narrow neck. The anther may be compared to the blade of an ordinary leaf, but it shows little trace of the structure of a vegetative lamina, for its functions are entirely reproductive. It is, however, traversed by a midrib (the connective), which forms a direct prolongation of the filament. The connective has a single vascular bundle, which is continuous with the two bundles of the filament (see Fig. 39, *v.b.*).

The anther, when nearly ripe, is divided lengthways into

two halves or lobes, with the connective between them (see Fig. 39, and compare Figs. 9, 10, and 11).

Each lobe of the anther consists of two compartments, called the *pollen-sacs*, so that there are four of these pollen-sacs altogether, running lengthways of the anther, parallel to the connective. Their arrangement, as seen in transverse

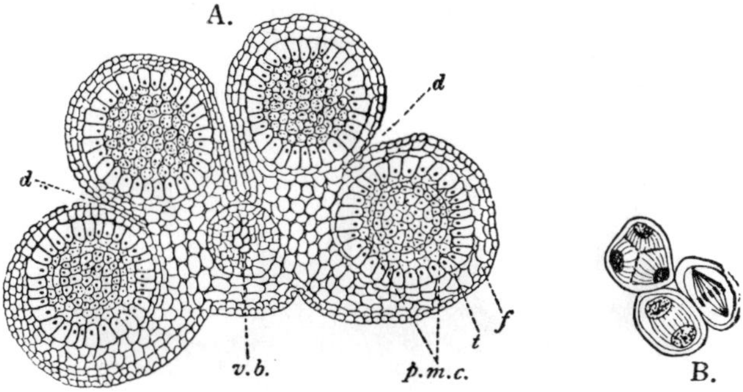

FIG. 39.—A, Transverse section through a young anther of Wallflower, at the stage when the pollen mother-cells are dividing. Observe the four pollen-sacs and the connective. *v.b*, vascular bundle of connective ; *f*, hypodermal layer, which will become the fibrous layer ; *t*, tapetum of radially elongated cells ; *p.m.c*, mass of pollen mother-cells, derived from archesporium of the pollen-sac ; *d*, places where the anther will dehisce. Magnified 110. B, Three pollen mother-cells from a pollen-sac in A, showing three stages of division. Magnified about 450. (R. S.) (Cf. Fig. 10, *st*, p. 17.)

section, is shown in Fig. 39. When ripe, each pollen-sac contains an immense number of isolated cells, the *pollen-grains*, which are set free by the bursting or dehiscence of the anther. The anther bursts by two longitudinal cracks, coinciding with the partition between the two pollen-sacs of each lobe of the anther (see Fig. 40).

The outer wall of the sacs is ruptured along this line, and at the same time the partition separating the sacs is broken down (see Fig. 40, *w*), so that a single opening serves for the escape of the pollen from each pair of pollen-sacs. At the time of dehiscence, therefore, the anther appears to have only two cavities, and we have to go back to unopened anthers to find the four pollen-sacs intact.

We see, then, that the stamen bears four pollen-sacs, which produce in their interior numerous pollen-grains. It will now be well to trace the development of the pollen-sacs, and of the pollen which they contain.

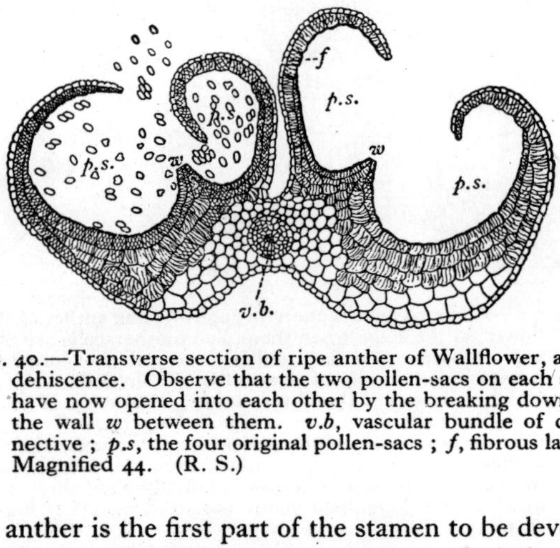

FIG. 40.—Transverse section of ripe anther of Wallflower, after dehiscence. Observe that the two pollen-sacs on each side have now opened into each other by the breaking down of the wall *w* between them. *v.b*, vascular bundle of connective ; *p.s*, the four original pollen-sacs ; *f*, fibrous layer. Magnified 44. (R. S.)

The anther is the first part of the stamen to be developed, so that in a moderately young stamen we find the anther almost full-grown while the filament is still quite short. If we cut a very young anther across, we find that the transverse section is nearly oblong, with rounded corners (Fig. 41, A).

At each of the four corners two or three of the cells lying next below the epidermis divide by walls parallel to the surface (see Fig. 41, A, *p*). The inner cells produced by this division form what is called the *archesporium*, *i.e.* the layer from which the *spores*, here called the pollen-grains,[1] ultimately arise. Each archesporium thus consists at first of a single strand or layer of cells running the whole length of the anther. The archesporial cells grow and divide up

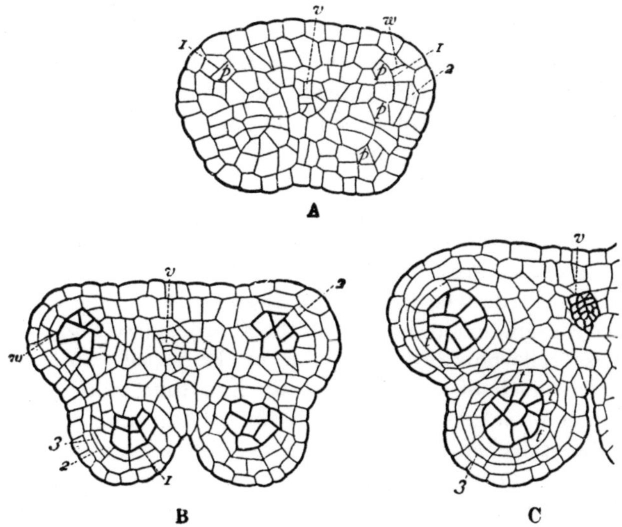

FIG. 41.—A, Transverse section of very young anther. *v*, young vascular bundle ; *p*, *p*, pollen-sac initial cells which divide to give rise to archesporium and cells of wall, *w*. B and C, More advanced stages of the same. The cells with dark outlines belong to the four archesporia. The numbers 1, 2, and 3 refer to the order in which the cell-walls are formed. *t*, tapetum. (After Warming.) Magnified about 200. These figures are reversed in position as compared with Fig. 39.

in all directions, and we soon have, in place of the original layer, a solid cylindrical column of tissue, consisting of large cells (see Fig. 41, B and C), which are densely filled with cytoplasm, and have large nuclei. In the meantime the cells

[1] See Part II., *Flowerless Plants*, Chapter I.

immediately outside the archesporium have undergone several divisions, so that the wall of the pollen-sac becomes about four cells thick (see Fig. 41, A, B, and C), including the epidermis, which does not itself divide. The cells which are actually in contact with the archesporium become elongated radially, and form a well-marked internal lining to each pollen-sac, called the *tapetum* (see Figs. 39 A, and 41 C, *t*). The tapetum extends all round the sac. Its cells are at first rich in protoplasm. The function of this layer is to feed the developing pollen-grains, by which it is entirely used up.

To return to the archesporium : after its cells have divided a great many times, the divisions cease for a while, the pollen-sac continues growing, its cavity becomes more roomy, and so the cells derived from the archesporium become isolated from one another and round off their corners (see Fig. 39, A). These cells now bear the name of the *pollen mother-cells*, because each of them divides up into four pollen-grains. The division takes place as follows: at the time when the mother-cells begin to be isolated, the nucleus in each divides into two, and the two daughter-nuclei at once divide again, so that there are four nuclei in the mother-cell (see Fig. 39, B). New cell-walls are formed between the nuclei, and the mother-cell is thus divided into four distinct cells, each with its own nucleus. These four daughter-cells are arranged like four cannon-balls in a heap, the whole group having roughly the form of a four-sided pyramid or a tetrahedron.[1] Each daughter-cell next forms from its protoplasm a new cell-wall of its own. The wall of the mother-cell is dissolved, and the four pollen-grains are set free. The cavity of each pollen-sac is now filled with a host of free pollen-grains, four times as numerous as the original mother-cells. The grains are at first immersed in a half-fluid substance derived

[1] Hence only three are seen in one plane. See the mother-cell at the top of Fig. 39, B.

from the protoplasm of the tapetal cells, which break down during the formation of the pollen. This substance is soon used up by the pollen-grains to complete their growth.

The ripe pollen-grain is somewhat oval in shape. Its wall is considerably thickened, but there are three thinner bands, indicated externally by shallow grooves, which run lengthways of the grain. The pollen-grain in Fig. 42 is shown as if in cross-section, so that it appears round, with three depressions, and we see the three thin places in its wall at the depressed points. The rest of the wall is thick and cuticularised, except a delicate inner layer of cellulose.

FIG. 42.—A single pollen-grain of the Wallflower, showing the three thin places in the wall, at any one of which the pollen-tube may subsequently grow out. Observe the three nuclei, of which the largest is vegetative and the two smaller are generative. Magnified about 450. (R. S.)

The pollen-grain at first, like most other cells, has a single nucleus. This nucleus divides while the grain is ripening, and one of the two daughter-nuclei divides again. Thus in the mature pollen-grain there are three nuclei, one of which is larger than the other two; the two smaller nuclei have arisen from the second division. From observations on other plants which are more favourable for examination than the Wallflower, it is probable that the cytoplasm around each of the smaller nuclei forms a distinct cell, so that the mature pollen-grain thus consists of three cells. In many other plants the pollen-grain has only two cells at this stage (cf. the Lily, Fig. 81.)

The two layers of cells next outside the tapetum, like it, become absorbed. One of them often disappears even before the tapetum itself, as shown in Fig. 39, A. When the anther is ripe, the wall of each of the four pollen-sacs consists on its free outer side of one or two layers besides the epidermis. These persistent layers are distinguished by a strongly marked spiral or netted thickening on their membranes. These cells form the *fibrous layer* (see Fig. 40, *f*). It completely surrounds each pollen-sac, except at the point just opposite the partition which separates the two pollen-sacs in the same lobe. Here the fibrous layer is interrupted by one or two thin-walled cells. It is thinnest (one cell thick) next the partition, and attains its maximum thickness (about five cells) on the side towards the connective (Fig. 40). This layer is of great importance in the dehiscence of the anther. It is owing to the contraction of the fibrous layer, through loss of water, when the anther is getting ripe, that the wall is ruptured and dehiscence takes place. If the anther is kept from drying, as sometimes happens in very wet weather, dehiscence may fail to take place.

d. The Pistil

The development of the pistil or gynœcium has already been shortly described. The tissue of the carpellary tube is traversed by several vascular bundles, the largest of which are placed opposite the septum. The external epidermis is thickly clothed with hairs of the usual structure ; there are a few stomata. The pistil terminates above in the stigma, the two lobes of which are rough on their inner side, with long papillæ, which, as we shall see presently, serve to catch the pollen-grains.

The ovary contains the *ovules*, or young seeds (see Figs. 10, 11, and 38, G). They are numerous, and grow in four rows, two rows to each compartment, on the placentas.

Each ovule is attached to the placenta by a stalk, the *funicle*.

The body of the ovule consists of a central mass of tissue, the *nucellus*,[1] continuous with the funicle, and of two coats,

[1] In the ovule of Wallflower, when ready for fertilisation, the nucellus is only represented by the embryo-sac (*e*) and a small mass of tissue at its base (*n*), Fig. 43.

the *integuments*, which arise from the lower part of the nucellus and enclose it, leaving only a narrow passage, the *micropyle*, leading down to the free end of the nucellus. The funicle contains a vascular bundle, branching off from the bundle-system of the ovary. The bundle terminates at the base of the nucellus (see Fig. 43, *ch*). The base of the ovule, where the integuments spring from the nucellus, is

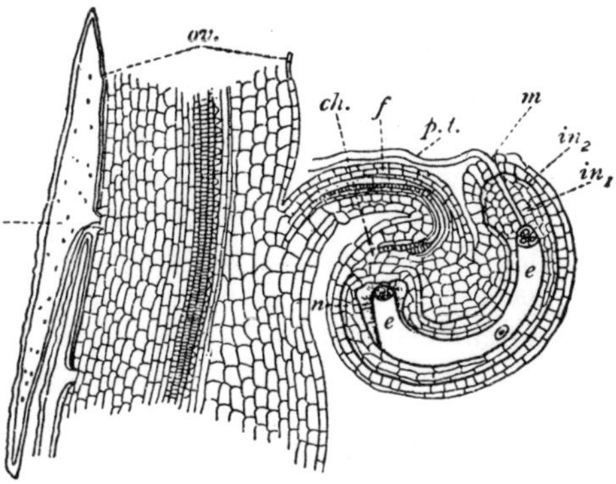

FIG. 43.—Longitudinal section through an ovule and part of the ovary wall of Wallflower. *ov*, ovary wall ; *h*, a hair ; *f*, funicle of ovule ; *ch*, chalaza ; *n*, nucellus (which is only persistent at the base) ; in_1, inner integument ; in_2, outer integument ; *m*, micropyle ; *e, e*, embryo-sac ; *p.t*, pollen-tube, entering micropyle. Magnified about 50. (R. S.)

called the *chalaza*. All these parts, except the vascular bundle, consist of delicate parenchyma.

In the nucellus is a long cell, the *embryo-sac* (see Fig. 43, *e*), which is much larger than its neighbours, and contains in its interior a number of smaller cells. One of these, situated at the end of the sac towards the micropyle, is the *ovum* (Fig. 44, 8 and 9, *o*), the most important part of the whole structure, because when fertilised it develops into the *embryo*, or young plant of the next generation.

The development of the ovule will now be traced with

the aid of observations on the Shepherd's Purse (*Capsella Bursa-pastoris*), a common weed allied to the Wallflower, which will show us all the essential points.

Each ovule arises as an outgrowth from the placenta. Several cells belonging both to the epidermis and to the next lower layer, take part in its formation. This outgrowth is the nucellus of the future ovule. When it has attained the form shown in Fig. 44, 1, that cell of the middle row which lies next below the epidermis (*e.m.* in the Fig.) begins to be distinguished from its neighbours by its larger size and more abundant protoplasm. It is termed the *archesporium* (or embryo-sac mother-cell) of the ovule. It divides by two transverse walls into a row of three cells (see Fig. 44, 2 and 3). The lowest of the three becomes the embryo-sac (Fig. 44, 3, *e*). The two sister-cells above it become completely destroyed (see Fig. 44, 4), and are, in fact, devoured by the growing embryo-sac.

In the meantime the integuments have begun to form ; the inner one arises first, growing out as a ring of tissue at the base of the nucellus, and forming a sheath around it, which soon projects beyond its free end and almost encloses it (see Fig. 43, in_1). The outer integument next appears in like manner, just below the inner one, which it soon overtakes by its more rapid growth (see Fig. 43, in_2). Ultimately the outer integument closes in over the inner, leaving only the narrow micropyle as a passage leading through them both to the nucellus (Fig. 43, *m*). In the meantime the funicle has been formed by the growth and division of the cells at the base of the ovule (Fig. 43, *f*).

Returning to the embryo-sac, we find that it not only absorbs its sister-cells, but, as it continues to grow, soon attacks the epidermis of the nucellus. This also is quickly digested, so that only the basal part of the nucellus, lying below the embryo-sac, remains (Fig. 44, 4 and 5 ; Fig. 43, *n*), while the sides and top of the sac

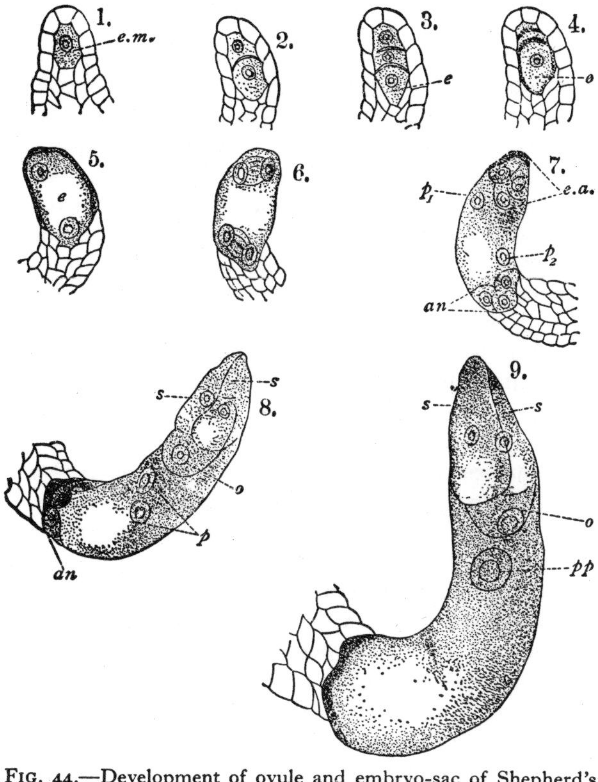

FIG. 44.—Development of ovule and embryo-sac of Shepherd's Purse (*Capsella Bursa-pastoris*). The integuments are not shown. 1, Early stage: *e.m*, archesporium or embryo-sac mother-cell. 2, The mother-cell has divided once. 3, A second division has taken place. 4, The lowest of the three cells, *e* (the embryo-sac), is obliterating the two above it. 5, The nucellus disappears, except at the base; the nucleus of the embryo-sac has divided. 6, Each daughter-nucleus divides again. 7, Four nuclei at each end of sac; three at each end have cells around them: *e.a*, egg-apparatus; p_1, p_2, upper and lower polar nucleus; *an*, antipodal cells. 8, The polar nuclei approach each other; *s*, *s*, synergidæ; *o*, ovum. 9, Fertilisation stage; *pp*, secondary nucleus, arising from fusion of polar nuclei. Magnified about 200. (After Guignard.)

come into direct contact with the inner integument (See Fig. 43).

While these changes are going on, the nucleus of the embryo-sac divides, and the two daughter-nuclei travel, one to each end (see Fig. 44, 5). Each divides again (see Fig. 44, 6), and the four nuclei thus formed divide once more, so that there are now eight nuclei altogether in the embryo-sac, four near each end. As yet the nuclei only have divided, so that all the eight lie free in the cytoplasm of the sac. Now, however, the cytoplasm at each end divides into three distinct cells each surrounding a nucleus. Thus we have six cells in the embryo-sac, three at each end, and also two free nuclei, which lie in the undivided cytoplasm which occupies all the middle part of the sac (see Fig. 44, 7). The three cells at the end towards the micropyle form what is called the *egg-apparatus* (Fig. 44, 7, *e.a*). Of these three, the one farthest from the micropyle is the ovum (Fig. 44, 8, *o*). The two others are termed the *synergidæ* (*s*, *s*), because they co-operate, as we shall see, in bringing about the fertilisation of the ovum. The three cells at the opposite end of the sac are called from their position the *antipodal cells* (see Fig. 44, 7 and 8, *an*). In this particular case the antipodal cells disappear before fertilisation (Fig. 44, 9). Lastly, the two free nuclei are called the *polar* nuclei (see Fig. 44, 7, p_1 and p_2, and Fig. 44, 8, *p*), because one comes from each pole of the sac. Their fate is remarkable. They do not remain in their original positions, but travel through the cytoplasm to meet each other (see Fig. 44, 7 and 8). When they meet they fuse together into a single large nucleus, which may be called the secondary nucleus of the embryo-sac (see Fig. 44, 9, *pp*). Meanwhile the whole embryo-sac has increased considerably in size, and so has the egg-apparatus, the two synergidæ becoming long and pointed (see Fig. 44, 9, *s*, *s*).

The body of the ovule during its development becomes

considerably curved on itself, owing to the more rapid growth of its lower side, so that the micropyle ultimately comes to point almost backwards towards the placenta. The embryo-sac takes part in this curvature (see Fig. 44, 9, and Fig. 43). Ovules in which the nucellus and embryo-sac are bent in this way are called *campylotropous.*

The structure of the ovule, then, when ready for fertilisation, is as follows : it is attached to the placenta by a stalk, the funicle ; the curved body of the ovule is enclosed by the two integuments, which leave a narrow opening at the free end, the micropyle ; the interior of the ovule is almost entirely occupied by the long curved embryo-sac, which is in immediate contact with the inner integument. The rest of the tissue of the nucellus has disappeared except at the base. Within the embryo-sac, at the micropylar end, are three cells, the egg-apparatus. The most deeply-seated of the three is the ovum. In the cytoplasm of the sac is a large nucleus, which has been formed by the fusion of the two polar nuclei. We shall probably find no trace of the antipodal cells,which at an earlier stage occupied the chalazal end of the sac.

The description just given, though taken specially from the Shepherd's Purse, applies to the ovules of the Wallflower also, except that the latter may have several embryo-sacs side by side. Only one of them, however, contains an ovum capable of being fertilised. Fig. 43 is drawn from an ovule in which only one embryo-sac is visible.

e. Fertilisation

We have next to learn how fertilisation is brought about. The ovum by itself is quite incapable of any further development. It can only become an embryo or young plant after union with another cell, namely, with one of the generative cells formed in the pollen-grain. The contents of the pollen-grains, therefore, must in some way be conveyed to the ovules in the ovary, in order that fertilisation may take place.

The first stage in this process is that the pollen-grains have to be deposited on the stigma. This is called *pollination*, and is a necessary preliminary to fertilisation. It is easy to prove experimentally that pollination is necessary for the production of ripe seeds, *i.e.* seeds which contain embryos. In order to prove this, the stamens of some of the flowers are cut off before the anthers have dehisced to shed their pollen. It is, of course, essential to do this in good time, and the flower-buds must be opened artificially in order to perform the operation, for as soon as the flower opens naturally the pollen is shed.

The plant must be kept under a bell-glass or under very fine gauze, so that no insects may reach it which might bring pollen from other flowers. There is no danger of pollen being brought by the wind, as the Wallflower pollen is too sticky to be blown about. It is therefore better to use gauze than a bell-glass, as the plant might become unhealthy if kept from the air. It will be found that in all the flowers which have had their stamens removed, and which are not allowed to receive pollen from other wall-flowers, the pistils wither away without setting their seeds, and if we examine their shrivelled ovules we find that they contain no embryos. We can prove that it is not the mere "shock" of the operation which prevents the seeds from ripening, for if pollen from another flower be placed on the stigmas of the mutilated flowers their seeds will ripen in due course, and when they are sown the embryos will grow into young plants as usual.

We learn, then, that the access of pollen to the stigma is a necessary condition for the formation of fertile seeds. It was by experiments of this kind, performed many years ago, that the existence of distinct sexes in plants was first proved. We thus see that the function of the flower is the sexual reproduction of the plant. We are now in a position to study in detail how the process goes on.

In plants like the Wallflower, which have conspicuous, brightly-coloured, and sweet-smelling flowers, the work of pollination is chiefly performed by insects. The Wallflower is specially attractive to bees, who visit it in search both of honey and of pollen, the "bee-bread," which is collected as food for the young brood. The bees are quite as useful to the Wallflower as the Wallflower is to the bees, for in the course of their visits they do it the essential service of bringing the pollen on to the stigmas of its flowers, and so ensuring fertilisation. We have already seen that there are two large honey-glands or *nectaries* in the flower. The honey formed by these glands collects in the pouches of the inner sepals. As the bee thrusts its proboscis down between the stamens and the pistil in order to reach the honey, its head comes into contact with the inner side of the anthers, on which a mass of loose pollen is hanging, for the anthers are *introrse*, *i.e.* they open on the inner side. When the bee flies to another flower, it is pretty sure that some of the pollen which has stuck to its head will be brushed on to its stigma. If this happens, as it generally will, *cross-pollination* will be effected, *i.e.* the stigma of one flower will receive the pollen of another, and this will often happen between the flowers of different plants. The chief way in which bees and other insects are of use to flowers is in bringing about cross-fertilisation, which, as Darwin's experiments prove, usually produces more and better seed than fertilisation by pollen of the same flower. In the Wallflower, however, self-pollination is also provided for; the anthers of the longer stamens rise up immediately above the stigma, and some of their pollen will be almost certain to fall upon it. It is also, of course, possible that the insect may happen to brush some of the pollen on to the stigma of the same flower. Probably this does not happen very often; for generally, as the bee works round the flower, one side of its head will touch the anthers, while the opposite side brushes against the stigma. In any case, the Wallflower is sure to be cross-fertilised very often ; and if this fails, owing to bad weather and the absence of insects, there is still self-fertilisation to be relied on, which is a great deal better for the plant than no fertilisation at all. Many other flowers have much more elaborate arrangements ensuring cross-pollination, and rendering self-pollination impossible, but the simple mechanism of the Wallflower is effective enough. Some flowers, however, as *e.g.* the Garden Pea (*Pisum sativum*), are habitually self-pollinated.

All that an insect directly does for a flower is to perform the act of *pollination*, *i.e.* to bring pollen on to the stigma. If all goes well, this act has its proper effect, *i.e.* the *fertilisation* of the ovules.

Hence it is usual to speak of bees and other insects as *fertilising* flowers, though they are really only concerned in the preliminary process.

Now we will assume that the pollen has reached the stigma, whether by the agency of some insect or directly from the anther, and will see what happens next.

The surface of the stigma, as we have already learnt, is coated with long unicellular hairs or papillæ, between which the pollen-grains are caught. These cells secrete a slightly sticky fluid, containing sugar in solution, which the pollen-grains absorb. In consequence of this they begin to swell, as can be seen under the microscope, and the thin parts of their walls become bulged out. Under the influence of the food which it obtains from the stigmatic fluid, the pollen-grain begins to germinate. At one of the places where its outer coat is thin the protoplasm grows, and forms a protrusion covered by the thin cellulose membrane ; this protrusion increases in length and becomes the pollen-tube. The greater part of the contents of the pollen-grain, including all the three nuclei, pass into the tube, which continues its growth, makes its way between the stigmatic cells, and penetrates into the tissue of the style below (cf. Fig. 83, from the Lily). It grows on among the cells of the style, and at last reaches the cavity of the ovary. Here the pollen-tube follows the placenta. When it reaches an ovule which is ready for fertilisation, the tube curves away from the placenta, and follows the funicle, until its end reaches the micropyle of the ovule (Fig. 43, *p.t.*). We are at present unable fully to explain the causes which direct the growth of the pollen-tube with unerring accuracy into the micropyle of the ovule. Recent researches make it probable that the apparent attraction is due to some sugary substance secreted by the ovule. The tube grows on through the narrow passage, and eventually its tip reaches the summit of the embryo-sac, and meets the pointed ends of the synergidæ.

And now the growth of the pollen-tube has reached its limit, and its destination is attained. The length of the tube is enormous compared to the size of the pollen-grain which produced it. In the Wallflower the distance from the stigmatic surface to the lowest ovule in the ovary is about 6 mm. ($\frac{1}{4}$ in.) in a straight line. The diameter of the pollen-grain is about ·02 mm.

The tube does not really grow straight, so that we shall be within the mark in estimating its length on the average at about three hundred times the diameter of the pollen-grain before germination. There are other plants in which the growth is ten times as great, or even more. It is evident that this enormous growth is only possible if food is taken up on the way from the cells of the style and ovary. In fact, the germinating pollen-grain lives at the expense of the tissues of the pistil, just as a parasite lives at the expense of its host. We must not, however, suppose that the whole tube is filled with protoplasm. The cytoplasm with the nuclei is at all stages of growth to be found in the growing end of the tube. All the older part of the tube is almost empty.

It is quite easy to cultivate the pollen-grains and make them germinate, if we provide them with an artificial stigmatic fluid. All that is necessary is to put some of the grains into a drop of sugar-solution on a glass slide, so that we can observe them under the microscope. For the Wallflower pollen a 6 per cent. solution of common sugar (cane-sugar) is about right. In this the grains will put out tubes which reach a considerable length, and the passage of the

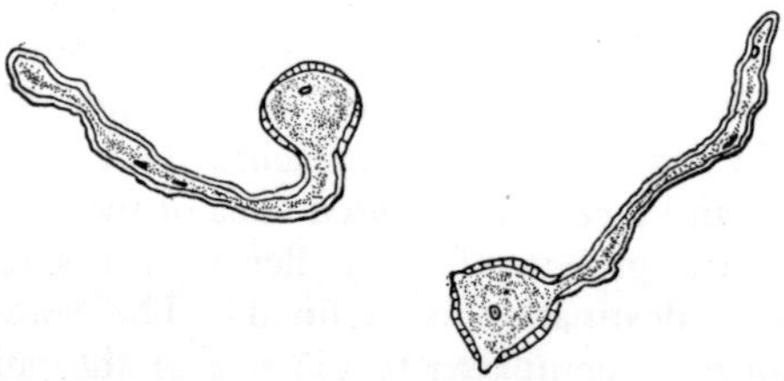

FIG. 45.—Pollen-grains of Wallflower germinating in 6 per cent. sugar solution. Observe the three nuclei, two of which are already in the pollen-tube. Magnified about 200. (R. S.)

nuclei down the tubes can be observed. The pollen-tubes figured (Fig. 45) were obtained in this way.

The details of fertilisation have not been followed in the Wallflower, which is not very favourable for the investigation, as the structures in question are small. What follows is a general description based on observations recorded on other plants. Reference may be made to Figs. 85, 86, and 87 of fertilisation in the Lily, a plant in which all the details have been fully worked out.

All three nuclei travel down the pollen-tube with the cytoplasm. The vegetative nucleus, however, *i.e.* that which was the product of the first division (see p. 99), becomes disorganised and disappears, having nothing further to do with the process of fertilisation. The two generative nuclei alone remain, and when the pollen-tube has penetrated the micropyle of the ovule and reached the embryo-sac, these nuclei are to be found in the cytoplasm at the extreme end of the tube. The cellulose wall of the end of the pollen-tube is very soft, and allows the passage of the generative nuclei out of the tube. One of the two generative nuclei passes on, usually between the synergidæ, and reaches the ovum itself. It must be remembered that these cells have no cell-walls, and so no resistance is offered to the passage of the nucleus. Ultimately this generative nucleus reaches the nucleus of the ovum. *The two nuclei then fuse together into one*, and fertilisation is effected. It is probably that in all cases some cytoplasm passes over from the pollen-tube with the nucleus, and unites with the cytoplasm of the ovum. What, however, we at present know to be constant in fertilisation is the union of the *nuclei* of the sexual cells or *gametes*. This statement holds good for plants generally, and not for flowering plants alone. The other generative nucleus passes farther into the embryo-sac and fuses with the secondary nucleus.

f. Development of the Embryo

The effects of fertilisation are, that the ovum develops into the embryo, the ovule into the seed, and the pistil into the fruit. We will first consider the development of the embryo or young plant, for this is the really important process, to which the changes in the ovule and pistil are subservient.

The immediate result of fertilisation is that the ovum surrounds itself for the first time with a cell-wall. Its nucleus, which, as we have seen, has been formed by the fusion of a generative nucleus from a pollen-grain with the original nucleus of the ovum, next undergoes division. This division of the nucleus is followed by that of the cell, which divides by a transverse wall into two cells. Repeated transverse divisions then go on, until we have a row of several cells (see Fig. 46, A) attached at one end to the wall of the embryo-sac near the micropyle, while the other end of the row hangs freely in the interior of the sac. The last cell at the free end is nearly spherical, and larger than the rest. The whole structure at this stage is called the *pro-embryo.* The round cell at the end is called the *embryonic cell*; the rest of the row is the *suspensor*. From the " embryonic cell " the greater part, but not the whole, of the young plant is formed. This cell first divides by three walls, two longitudinal and one transverse (see Fig. 46, A and B, *e*), cutting each other at right angles, into eight cells, the octants. Next, each octant divides by a wall parallel to the outer surface, into an inner and an outer cell. The eight outer cells thus cut off form the dermatogen of the embryo (see Fig. 46, C, *d*). Their further divisions take place entirely by walls at right angles to the surface, so that the layer remains one cell thick. It gives rise to the entire epidermis of the shoot, and to no other tissue. The eight inner cells then divide in their turn, forming a central group and an intermediate layer (see Fig. 46, D). The central group becomes the plerome, from which the central cylinder of the embryonic stem is developed, while the intermediate layer between plerome and dermatogen becomes the periblem, from which the cortical tissues arise.

The *hypophysis* or last cell of the suspensor, immediately adjoining the embryonic cell, also contributes something to the formation of the embryo. It divides by a transverse wall, and the daughter-cell next the embryonic cell by its further divisions forms the root-cap and periblem of the root (see Fig. 47, *r.c.* and *r*), while the plerome of the root is formed from the same central group as that of the stem.

We see, then, that the apex of the root is turned towards the suspensor and the micropyle, as is always the case in the

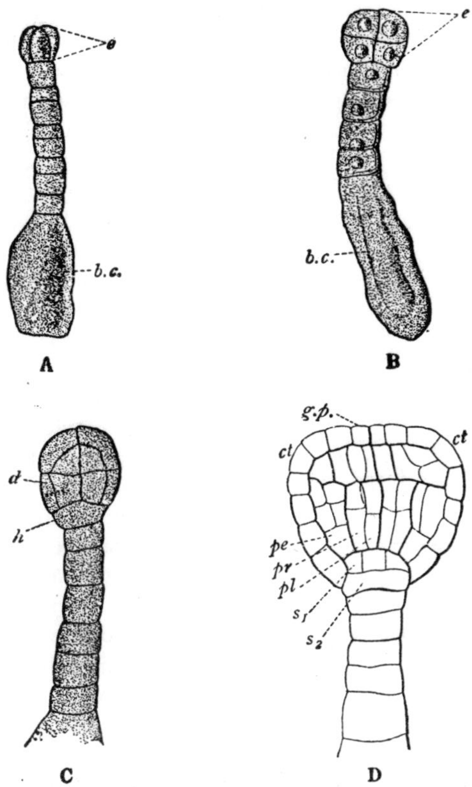

FIG. 46.—Development of the embryo of Shepherd's Purse. *b.c*, basal cell of suspensor ; *e*, embryonic group ; *d*, dermatogen ; *h*, hypophysis ; *g.p.* future growing-point of stem ; *ct*, *ct*, cotyledons ; *pe*, periblem ; *pr*, *pl*, cells of plerome ; s_1, s_2, cells derived from the hypophysis. The micropyle is below the suspensor. (After Hanstein.) Magnified about 200.

embryos of flowering plants. The apex of the stem is placed exactly at the opposite end of the embryo.

As growth proceeds, the embryo ceases to be spherical. It first becomes flattened at the top and then heart-shaped (see Fig. 47), forming two lateral protrusions. These are the *cotyledons* (Fig. 46, D, and 47, *ct*), the two first leaves of the embryo. The growing-point of the stem lies between them (*g.p*), but its development does not begin till later. The Class of the Angiosperms to which the Wallflower belongs derives its name, *Dicotyledons*, from the fact that its embryo normally has two seed-leaves or cotyledons. The growing-point of the embryonic stem in this Class is constantly terminal (*i.e.* in the middle line of the embryo, at the end farthest from the micropyle), and the two cotyledons lateral (*i.e.* one on each side of the growing-point).

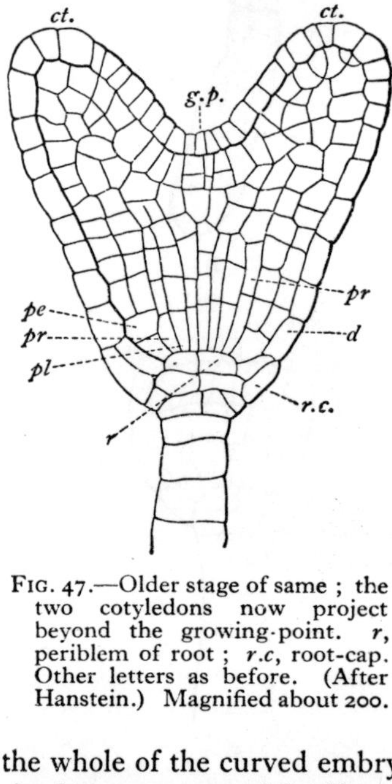

FIG. 47.—Older stage of same ; the two cotyledons now project beyond the growing-point. *r*, periblem of root ; *r.c*, root-cap. Other letters as before. (After Hanstein.) Magnified about 200.

As the embryo attains its full size it gradually occupies the whole of the curved embryo-sac, and therefore of necessity becomes curved itself, so that the cotyledons are bent back and lie parallel to the young root or *radicle*. In the Wallflower the cotyledons are so placed that their edges are turned towards the radicle (see Fig. 49) ; in this case they are said to be *accumbent*. This, however, is a character of no great importance, as it varies in closely-allied plants.

g. The Endosperm

As already pointed out, a *double* act of fertilisation takes place within the embryo-sac ; in addition to the fertilisation of the ovum, the second generative nucleus from the pollen-tube unites with the secondary nucleus of the embryo-sac, itself, as we have seen, the product of a nuclear fusion. To this subject we shall return when we come to our next type, the Lily (Fig. 86). This fertilised nucleus of the embryo-sac undergoes repeated divisions, and around the daughter-nuclei cells are formed. In this way a little tissue is produced, called the *endosperm*, a name given to the tissue thus formed in the interior of the embryo-sac, and outside the embryo, after fertilisation. But in the case of the Wallflower and all its immediate allies, such as the Cabbage, Turnip, Shepherd's Purse, Watercress, and plenty of other well-known plants, the greater part of the endosperm is only a temporary structure. It never attains any considerable size, and is to a great extent used up as food by the growing embryo before the seed is ripe, only the outermost layer being persistent. Seeds which when ripe contain nothing besides the embryo are called *exalbuminous*, because they contain no *albumen*, an old-fashioned name applied to the endosperm and analogous tissues from a fancied resemblance to the white of an egg. The Wallflower seed is commonly described as exalbuminous ; but this is not strictly true, for we know from recent researches that one layer of endosperm at least is persistent. In many plants the embryo-sac becomes

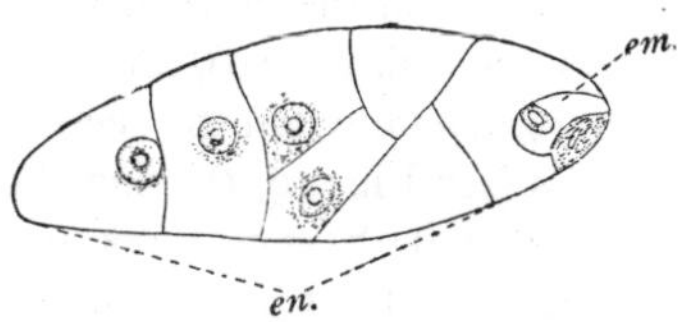

FIG. 48.—Embryo-sac of Elder (*Sambucus*), which has just become filled up with endosperm by cell-division. *en*, endosperm ; *em*, embryo. (After Hegelmaier.) Magnified about 250.

completely filled with endosperm, the embryo itself occupying only a small space. Such seeds are said to be *albuminous*. Fig. 48 shows a young embryo-sac filled with endosperm, from the Elder. (See also Fig. 91.)

h. The Ripe Seed

We will now complete the description of the *seed*, which is the fertilised ovule, with its contents. The nucellus, as we have seen, disappears in great part even before fertilisation. By the time the seed is ripe scarcely a trace of it remains except a little group of cells at the chalaza, in contact with the tips of the cotyledons of the embryo. The integuments, on the other hand, remain (though some of their layers are obliterated), and develop into the hard seed-coat or *testa*, which when the seed is ripe consists of several layers of cells (see Fig. 49, *t*). On the extreme outside is a layer of clear cells with very thick walls, so thick that there is no cell-cavity left at all. These walls are mucilaginous, and swell up in water, rendering the seed-coat sticky, and causing it to adhere to the soil or to anything which it touches. Beneath this are two or three layers of brown cells, more or less flattened, and on the inside of the seed-coat next the embryo is a layer of thin-walled cells, with very abundant contents, consisting chiefly of nitrogenous food-materials in the form of protein granules (see p. 33). This innermost layer belongs to the endosperm, of which it is the only remnant. The outermost mucilaginous layer is the epidermis of the outer integument ; the brown cells in between belong partly to one integument and partly to the other. The funicle remains as before, and its vascular bundle can still be traced in the ripe seed. The micropyle is closed while the seed is ripening, but it still has a part to play, as we shall see presently. The ripe seed, then, consists essentially of the funicle, the testa or modified integuments, and the embryo. The whole interior of the seed is occupied by the large curved embryo. At the end next the micropyle is the *radicle* of the embryo, and above this is the *hypocotyl*. At the upper end of the latter the two broad *cotyledons* are borne, and between them is the *plumule*, or growing-point

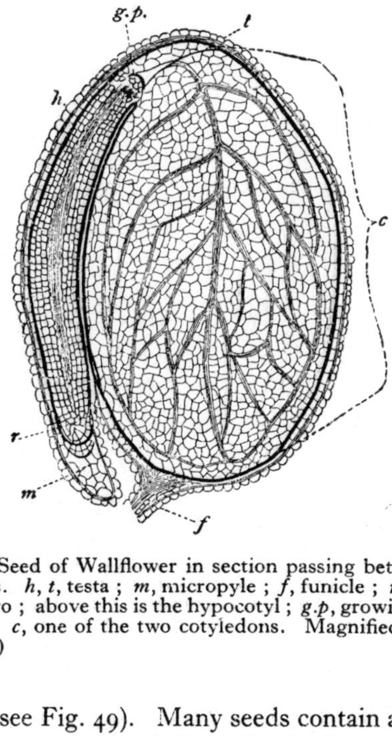

FIG. 49.—Seed of Wallflower in section passing between coty-ledons. *h, t*, testa ; *m*, micropyle ; *f*, funicle ; *r*, radicle of embryo ; above this is the hypocotyl ; *g.p*, growing-point of stem ; *c*, one of the two cotyledons. Magnified about 20. (R. S.)

of the stem (see Fig. 49). Many seeds contain a much larger amount of permanent food tissue, in the form of endosperm within the embryo-sac, to be used during germination, *e.g.* Buttercup, Foxglove, Castor-oil seed ; while in a few the nucellus is persistent, forming the *perisperm*, *e.g.* Pepper, Water Lily.

i. The Fruit

The pistil ripens into the fruit or seed-vessel. In doing so it increases greatly in bulk, growing from a quarter of an inch to two inches or more in length (Fig. 50, 1). Its tissues show changes of some interest. Perhaps the most important change is the development of a layer of very thick-walled, elongated cells (*sclerenchyma*) from the inner parenchyma, adjoining the layer which lines the cavity of the fruit. The parenchyma lying between the sclerenchymatous zone and the epidermis contains chlorophyll. The outer epidermis is densely clothed with overlapping hairs of the usual Wallflower type, and has numerous stomata.

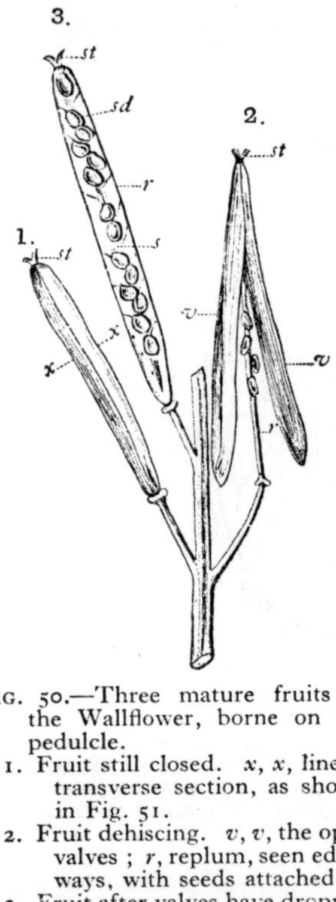

FIG. 50.—Three mature fruits of the Wallflower, borne on the pedulcle.
1. Fruit still closed. *x*, *x*, line of transverse section, as shown in Fig. 51.
2. Fruit dehiscing. *v*, *v*, the open valves ; *r*, replum, seen edgeways, with seeds attached.
3. Fruit after valves have dropped off, seen in plane of replum. *r*, *s*, septum stretched across replum ; *sd*, seeds, showing parietal placentation ; *st*, stigma, in all three fruits.
⅔ of natural size. (R. S.)

As the fruit matures, its tissues give up their water and become dry and rigid. When quite ripe, it splits open, or *dehisces*, owing to the unequal contraction of the harder and softer layers of the carpellary walls. The sides of the fruit split away from the *replum* or persistent placenta, the severance beginning below and extending upwards. The valves thus formed remain for a time attached at the top (Fig. 50, 2). When they fall the replum is left intact, with the seeds attached to it, and the papery septum stretched across it (Fig. 50, 3). A fruit of this kind is called a *siliqua*.

The lines along which dehiscence is to take place are marked out beforehand as thinner bands in the wall (see transverse section, Fig. 51, *d*, *d*). A special layer of separa-

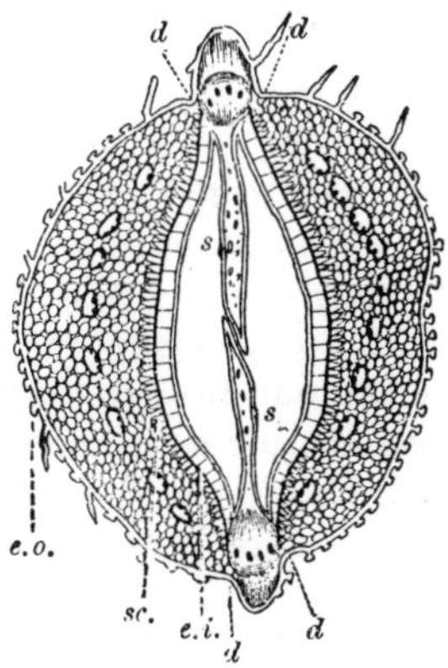

FIG. 51.—Transverse section of young fruit of Wallflower (seeds omitted). *d*, *d*, thin portions of ovary wall, where dehiscence takes place ; *s*, *s*, septum, which is in two parts ; *e.o*, epidermis ; *e.i*, layer lining cavity ; *sc*, sclerenchyma. The thick masses of tissue at the two ends of the septum belong to the replum. Magnified about 8 (R. S.) (Cf. Fig. 50, and Figs. 9, 10, 11, pp. 16–18.)

tion is formed at these places, the cells of which split apart during dehiscence, and so release the valves.

The thin, flat seeds hang on to the replum for a time and then fall, or are blown off by the wind, on to the ground. The seeds are light and are easily blown about by the wind. In this manner they may become widely dispersed from the parent plant. This illustrates an important biological principle, viz. that nearly all organisms have some effective means of dispersal. This results in less intense competition between individuals of the same species. If a number of wallflower seeds germinated together, the seedlings would compete for light and nutrients, and only a few would survive, whereas if the seeds were scattered widely on to suitable soil, all the seedlings would have a good chance of survival.

k. Germination and Life-History

When the temperature is sufficiently high, and the weather is moist enough for the seeds to take up a supply of water, the process of *germination* begins.[1] By germination is meant the development of the seedling from the embryo contained in the seed.

The mucilaginous layer of the testa takes up water in large quantities, swells into a slimy coating, and thus helps to fix the seed to the soil. Part of the water which is thus taken up is passed on to the embryo inside, and stimulates it after its long rest to renewed growth. The hypocotyl and radicle begin to elongate rapidly, and the apex of the latter is forced out through the micropyle. In whatever position the germinating seed may lie, the radicle as soon as it becomes free curves, so that its tip is directed straight downwards, and at once penetrates the soil (see Fig. 52). As soon as this has happened, the young root immediately begins to develop root-hairs, which take a firm hold of the particles of soil, and serve both for nutrition and for the fixing of the plant during its subsequent growth. Meantime the hypocotyl grows very rapidly, and curves so that the apex of the stem comes to point directly upwards. The cotyledons for a time remain imprisoned in the seed-coats, which are lifted

[1] When some seeds are shed they are not *immediately* capable of germination. In many seeds germination is dependent also upon access of oxygen and ready removal of carbon dioxide from the seed. In other seeds, *e.g.* legumes, the seed coat remains impermeable to water for a considerable time despite soaking. If, however, the seed coat is broken, water can enter and germination proceeds. Most seeds begin to germinate as well in darkness as in light, but a few will only germinate when exposed to light.

up with them above the level of the ground, but they soon spread themselves out flat, and in doing so strip off the seed-coats. Cotyledons which are raised above the ground are said to be *epigeal.*

We see, then, that in germination the root grows vertically downwards and the stem vertically upwards, while the cotyledons take an approximately horizontal position. The apex of the stem (*the plumule*) now begins to develop further and form leaves. The root at the same time puts out the first lateral rootlets, and we may now regard germination as completed; the embryo has become a seedling (see Fig. 30).

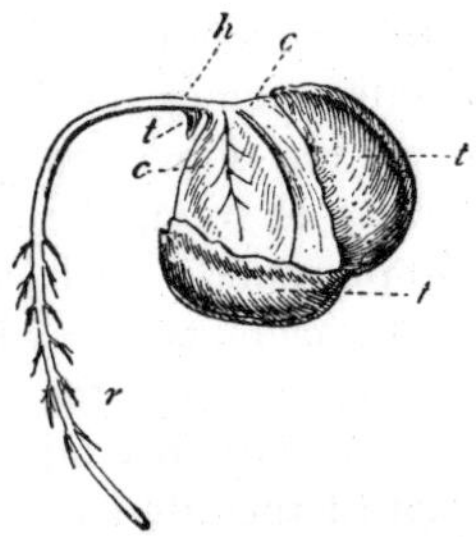

FIG. 52. — Germinating seed of Wallflower. *t, t,* burst testa; *c, c,* cotyledons; *h,* hypocotyl; *r.* root. Magnified about 5. (R. S.)

The food-supply necessary for germination is obtained at first from the organic substances (starch, etc.) stored up in the cotyledons themselves, and in part also from the remaining layer of endosperm, which, as we saw above, is rich in protein granules. The food in most seeds is stored as solids, many of which are insoluble in water. Before this reserve food can be used for growth during the process of germination, it must be rendered soluble by the action of enzymes.[1] Insoluble starch, for instance, is converted into soluble sugar by the action of *diastase.* As soon as the root-hairs have taken hold of the soil, the seedling can begin to supply itself with mineral food, but it is only when the cotyledons have expanded and become green that carbon-assimilation can

[1] Enzymes are substances of complex composition, in the presence of which certain chemical changes are rapidly effected, changes which under other conditions proceed slowly and perhaps so slowly as to be negligible. Individual cells of animals and plants may contain many kinds of enzymes.

begin and new organic substances be formed. We see, then, that in this plant the cotyledons have a double function ; they serve at first as stores of ready-made food for the seedling, and afterwards they act as assimilating leaves, and can manufacture new food on their own account.

In most of the points described the germination of the Wallflower may be regarded as typical of that of the Dicotyledons generally. For example, the exit of the radicle through the micropyle is very constantly the first stage in germination. The behaviour of the cotyledons, however, differs much in different members of the Class. In some plants, such as the Broad Bean (*Vicia Faba*), the cotyledons never rise above the ground at all, but serve simply as storehouses of food, the first assimilating leaves being those of the plumule. Such cotyledons are said to be *hypogeal.* In seeds, again, which when ripe contain abundant endosperm, the first duty of the cotyledons is to absorb the food which the endosperm contains, and they do not emerge from the seed (if they do so at all) until this has been accomplished.

One important feature concerning the biology of seeds remains to be mentioned. Many seeds, including those of the Wallflower, can retain their vitality for several years in the dry state, or even if they are buried at some depth in damp soil. Such seeds are said to be dormant. In them the life processes, including respiration, are almost completely in abeyance. When conditions become favourable, even after the lapse of years, germination proceeds normally. There is a limit, however, to the period during which seeds can retain their vitality in a dormant condition, and there is no truth in the claim sometimes made that grain recovered from the tombs of Egyptian mummies is now capable of germination.

Seeds are extremely effective reproductive units, and it is probably largely on account of their efficiency that the Flowering Plants are now the dominant features of the vegetation of the world. Seeds can survive under conditions of drought and temperature which would be fatal to the plants which produce them.

The Wallflower has been described as an herbaceous perennial in its natural habitat. It flowers in its second and each succeeding year until it dies. In British gardens, however, the Wallflower is usually treated as a *biennial* plant. The seed is sown in late spring or early summer, and after being transplanted once or twice the young plants are finally set in the autumn in the beds where they flower in the following spring. After flowering they are usually discarded. This is because older plants grow somewhat lanky and look ungainly, and also because old plants in Britain seem to be much more damaged by frost than are young ones. A *biennial* plant is one which grows only vegetatively and stores up food in its first year, flowers in its second year, like the Carrot, and then dies. Other kinds of perennials, *e.g.* shrubs and trees, have already been mentioned, but there are many plants whose life-cycle is completed within a single year or even within a few months : these are called *annuals* or *ephemerals*. For example, the common weed Shepherd's Purse (*Capsella Bursa-pastoris*) completes its life-cycle in the course of a few weeks under favourable conditions.

CHAPTER II

TYPE II

THE WHITE LILY (*Lilium candidum*, L.)

It matters little which of the common kinds of Lily we take for our type. The following description refers in the main to the White Lily, but the general Figs. 53 and 54 are of *Lilium auratum*, and some of the other figures refer to different kinds.

The White Lily grows wild in Southern Europe, Persia, and the Caucasus, and is one of the commonest Lilies cultivated in gardens. Unfortunately it is very liable to disease in Britain.

I. *EXTERNAL CHARACTERS*

A. Vegetative Organs

In cultivation, Lilies are usually raised from *bulbs*. A bulb is really a large underground bud. It consists of a short conical stem, covered with densely-crowded, spirally-arranged, thick scale-leaves, which completely coat the surface, leaving no internodes between them. The base of the bulb produces fibrous roots. The bulb is that part of the plant which persists at all times, in winter as well as in summer. During the winter the scale-leaves contain stores of food-material, which were laid by during the previous summer, while the green foliage-leaves were assimilating. When growth begins in spring, the short conical axis of the bulb grows straight upwards into a tall stem, producing the foliage-leaves, and ultimately the flowers. After blossoming, the whole of the flowering shoot of the White Lily dies, and for a month or two there is no sign of life above ground.

The White Lily is peculiar, inasmuch as it produces two sets of green foliage-leaves. One set appears in autumn, arising at the base of the rudimentary flowering stem of the following year. The other set is borne in summer on the tall flowering stem itself. If we examine a plant of the White Lily in autumn, after the flowering stem of the past summer has died down, we see a group of broad lanceolate green leaves, the bases of which are underground, the stem on which they grow being hidden in the earth. These leaves we will call the *ground-leaves*.

FIG. 53.—General view of flowering stems of *Lilium auratum*. From a photograph. Much reduced; the plant was about 6 ft. high.

If we dig up the whole plant and look at it closely, we find the following structure: on the outside of the bulb there are a number of broad, thick scales, each of which has a blunt withered tip. These scales (*b*, Fig. 55) are the persistent bases of the ground-leaves of the year before. They are spirally arranged, and if we remove them one by one, we find opposite the innermost of them the decayed base of last summer's flowering stem (*a*, Fig. 55). After these scales are removed, we come upon bud-scales of another kind; these are thick and fleshy like the former, but differ from them in having pointed and uninjured tips (*c*, Fig. 55). The pointed scales are complete leaves which have never developed any further; they also are spirally arranged. If we remove these, we next come to the green ground-leaves of the present season (*d*, Fig. 55). Their bases are already somewhat swollen, but not nearly so much so as they become

FIG. 54.—Flowers of *Lilium auratum*. From a photograph. Reduced.

later. Removing these again, we find that they surround a large bud (Fig. 56), which is the young flowering stem ready to develop next summer. The leaves of this bud (*b*, Fig. 56) are simply the future stem-leaves of the flowering stem. In the axil of the innermost of the *ground-leaves*, and therefore at the base of the young flowering stem, we find a tiny bud (*c*, Fig. 56), with only two or three very minute scales. This is destined to develop into the next year's bulb and stem. We find, then, that the White Lily bulb consists of (1) the base of the old stem, bearing the remains of its ground-leaves, (2) the new stem which arose in the axil of the innermost ground-leaf. On this new stem the first leaves are the pointed bulb-scales which never develop any further ; then the ground-leaves, which come up in autumn ; and lastly,

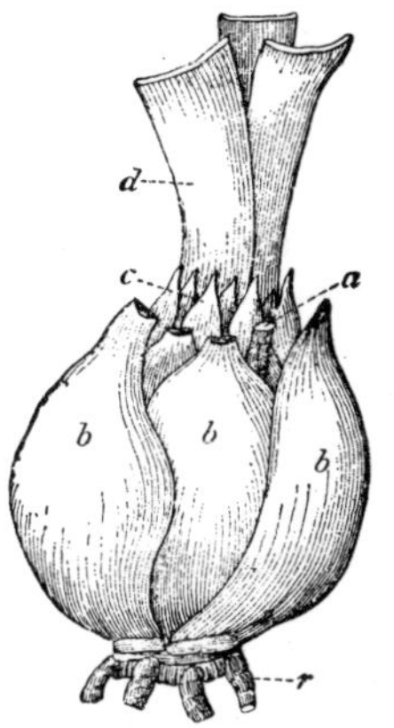

Fig. 55.—Bulb of White Lily in autumn. *r*, roots ; *b*, blunt bulb-scales formed from bases of old ground-leaves ; *a*, remains of old flowering stem ; *c*, pointed bulb-scales of new stem ; *d*, ground-leaves of new stem.

Fig. 56.—Young flowering stem removed from the middle of bulb in Fig. 55. *a*, surface of stem ; *b*, undeveloped stem-leaves ; *c*, bud from which next year's bulb will arise. (Both after Irmisch.)

the stem-leaves, which can only expand when the flowering stem develops next summer. The little bud (*c*, Fig. 56) will repeat this whole process a year hence. In vigorous plants more than one of these buds may be formed, each giving rise eventually to a new bulb.

All Lily bulbs are not so complicated as this. In the

Turk's Head (*L. Martagon*) and the Royal Lily (*L. regale*), for instance, there are no ground-leaves, and consequently no scales formed from their bases ; all the bulb-scales here are of the pointed kind, *i.e.* they are all complete but simplified leaves.

The leaves of the flowering stem are arranged spirally. The phyllotaxis is not very regular, but a divergence of $\frac{3}{8}$ is common, *i.e.* if we trace the spiral upwards, each leaf is in this case separated from the one next above it by $\frac{3}{8}$ of the circumference of the stem.

The leaves are lanceolate in form, without petioles, and have a base which embraces about $\frac{1}{3}$ of the circumference of the stem. The autumnal ground-leaves chiefly differ from the stem-leaves in being stalked, and in having a swollen base, and also in their broader ovate-lanceolate form.

The arrangement of the veins in the leaf is different from what we found in our dicotyledonous type. In the Lily all the principal veins run from end to end of the leaf, and are approximately parallel to each other. The veins near the middle of the leaf run almost straight, while those towards the edges are more and more curved, so as to follow the outline of the leaf. There are generally three veins more strongly marked than the rest, the midrib being most prominent of all. These larger veins form projecting ridges on the lower surface. The longitudinal veins are connected by oblique branches. The whole system, therefore, forms a network, but of a simpler form than that in the Wallflower. The longitudinal veins are, on the whole, stouter than the oblique connections. There are, as a rule, no blindly ending branches within the meshes, another point of difference from the dicotyledonous type. The "parallel" type of leaf venation is widespread in the Class of the *Monocotyledons* to which the Lily belongs. The position of the blade of the leaf is roughly horizontal in the stem-leaves, and more oblique in the ground-leaves.

As a rule the flowering stem does not branch ; there is a bud in the axil of each leaf, but it generally remains undeveloped. If, however, the main stem be injured, it may be replaced by a branch developed from one of these buds. In some other species, such as *L. bulbiferum*, these axillary buds grow into little bulbs, which become detached and give rise to new plants, thus affording a good example of *vegetative propagation.* We speak of vegetative propagation when a plant possesses some means of producing descendants without the help of the sexual organs, the stamens and pistil.

In most Lilies such propagation takes place by means of the bulbs. We have already seen that a vigorous old bulb in the White Lily may give rise to several buds, which develop into new bulbs. While the leafy stem is assimilating, the young bulbs receive the food which it forms ; part of this they use for their own growth, the rest they store up, especially in the thickened scale-leaves, for future use. Every new bulb can form roots and a leafy stem of its own, and thus ultimately produce flowers, giving rise, in fact, to a complete new plant. In our own country, Lilies are usually propagated from bulbs, seedlings being more rarely met with.

A striking peculiarity of the Lily as distinguished from the Wallflower is the absence of any main root. All the roots of a mature Lily plant are borne on the stem, and are called *adventitious*, as they do not arise from the radicle of the embryo. Usually it is the base of the stem, at the bottom of the bulb, which bears them ; sometimes, however, roots also arise from the lower part of the flowering stem above the bulb. When a Lily plant is very young, it has a main root formed from the radicle of the embryo, but this root soon dies away, and the root-system is henceforth entirely produced from the stem. Most members of the Class to which the Lily belongs (*Monocotyledons*) lose their main root early, and replace it by adventitious roots.

B. The Flowers

The flowers are borne in a *terminal raceme*, *i.e.* they spring from the elongated main axis—the older flowers below, the younger above ; each flower is stalked, and grows in the axil of a small leaf called a *bract*.

The flowers are large, and very easy to examine. The first great difference which we notice, from the Wallflower, is that the whole *perianth*, including both calyx and corolla, is coloured alike, and its leaves are all of about the same size. In the bud, when just about to open, the three outer perianth leaves or sepals enclose all the other organs, only the midribs of the petals showing between them. A good idea of the arrangement of the parts of the flower is obtained by examining a transverse section across a bud (see Fig. 57 ; also diagram, Fig. 79).

The *sepals* are three in number, and are all inserted at one level, forming a single whorl. They are lanceolate in shape, and their venation is much like that of the green leaves, only less conspicuous. In the bud each sepal is somewhat hooded, so as to arch over the other organs. At the base of each sepal, on its inner side, is a shallow groove with raised sides, into which honey is secreted.

The three *petals* also form a whorl, and stand opposite the spaces between the sepals, or, in other words, are alternate with them. The petals are narrower at the base than the sepals, but rather broader above. A petal differs from a sepal in having a well-marked midrib, which projects as a ridge on its outer or lower surface. On either side of this ridge is a deep furrow, into which in the bud the edge of the adjoining sepal fits (see Fig. 57). The side veins of the petal diverge from the midrib, running obliquely upwards and outwards to the edge, these veins being connected by oblique branches to form a network. At the base of each petal is a honey-bearing groove like that on the sepal.

The *andrœcium* consists of six *stamens*, in two whorls of three each. The outer whorl is alternate with the petals, and therefore opposite the sepals. The inner whorl is alternate with the outer, and therefore opposite the petals. In fact, all through the flower the successive whorls of three alternate regularly with each other (see Figs. 57 and 79).

The stamens have long filaments and large *versatile* anthers, *i.e.* the filament in the mature state is attached to

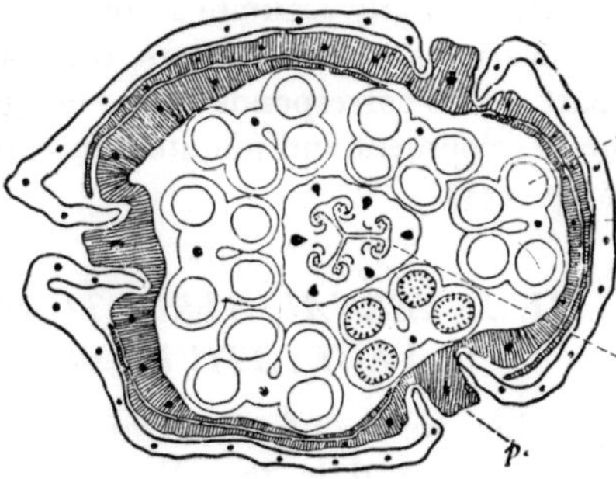

FIG. 57.—Transverse section across a flower-bud of the White Lily. *s*, one of the three sepals (unshaded); *p*, one of the three petals (shaded); *st*, one of the six stamens in two whorls of three; *O*, pistil consisting of three carpels. Magnified about 8. (R. S.) (Cf. Fig. 79, p. 153.)

the back of the anther near its middle, so that the anther swings freely upon the filament. In all essentials the structure of the stamen is like that of the Wallflower, but its large size in the Lily makes it much more favourable for study. The anther consists of a connective and four pollen-sacs. Here, as in our former type, the two pollen-sacs on the same side of the anther become fused when it is ripe, so that the mature anther has only two cavities. Here also the anther is *introrse*, *i.e.* it opens on its inner side.

In the middle of the flower is the *pistil* or *gynœcium*. The ovary is cylindrical and ribbed, the style about twice as long as the ovary and surmounted by a large three-lobed

10

stigma (see Fig. 82), the velvety surface of which is visible even without a lens. The pistil is syncarpous ; the carpels are three [1] in number, and alternate with the inner whorl of stamens. The ovary is three-chambered ; the ovules are numerous, and are inserted in two rows in each chamber, arising at its inner angle ; in other words, the placentation is *axile* (see Figs. 57 and 79).

If we suppose each of the carpels folded inwards until its edges meet, and then all three carpels closely joined laterally, the united edges of all three meeting at the middle, we shall have a good idea of the structure of the ovary, though not one which is strictly accurate according to the development. In this ovary, as in many others, the ovules are inserted on that part which corresponds to the margins of the carpels.

The development and minute structure will be dealt with below (p. 151).

II. *INTERNAL STRUCTURE OF THE VEGETATIVE ORGANS*

In the Lily we shall find the same great systems of tissue which we observed in the Wallflower, and their general arrangement is also similar. There are, however, some broad differences between the two types, chiefly connected with the course and development of the vascular bundles.

a. Stem

We shall find it best to begin with the leafy flowering stem (see Fig. 58). If we examine a transverse section through an internode, the first thing that strikes us is the very well defined central cylinder, limited on the outside

[1] On the same view of the morphology of the ovary as that indicated in a footnote on p. 22, E. R. Saunders considers that the ovary of the white lily consists of six carpels, an outer whorl of three which are sterile alternating with an inner whorl of three which are fertile.

by a zone of thick-walled cells (*pc* in Fig. 58). In the central cylinder we see the transverse sections of a large number of vascular bundles (*b*). They do not form a single ring, as in our dicotyledonous type, but appear at first sight to be irregularly scattered. We can, however, make out an arrangement in concentric rings, though they are not very regular. Towards the middle of the stem we find the larger

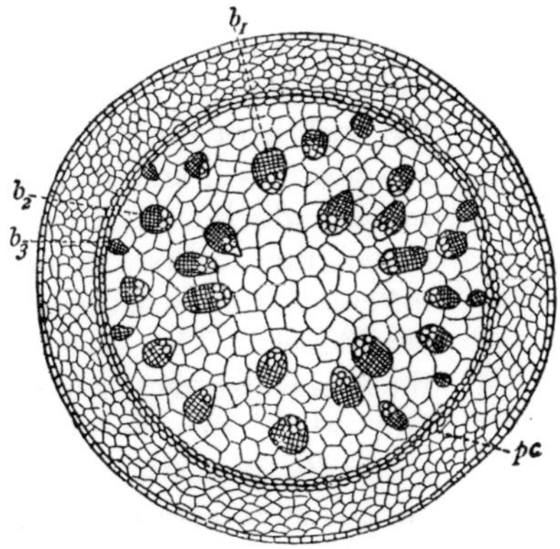

FIG. 58.—Transverse section of the flowering stem of White Lily. Somewhat diagrammatic. *pc*, thick-walled pericycle, forming outer limit of central cylinder ; b_3, one of the small vascular bundles on its way from a leaf ; b_2, bundle cut through on its inward course ; b_1, one of the principal bundles which form the inner ring. Magnified about 20 times. (R. S.)

vascular bundles forming an irregular ring (b_1). Surrounding these is a zone of smaller bundles (b_2) ; while quite at the outside of the cylinder, next the thick-walled tissue, we find the smallest bundles (b_3), which do not, however, form a complete ring. Now, this arrangement in irregular concentric rings is characteristic of the vascular bundles of most Monocotyledons as distinguished from the Dicotyledons, in which we usually find a single and regular ring.

This difference in the transverse section depends on the course of the vascular bundles through the stem. Here, just as in Dicotyledons, all the bundles are *common* to leaf and stem—that is to say, each bundle in the stem is the direct downward continuation of one which comes from a leaf. The diagram (Fig. 59) will help us to understand this, though it is much simpler than the reality. We see that each bundle on entering the stem from a leaf (l) passes very gradually inwards, until it is only a short distance from the middle of the stem. It then turns straight down, and at last joins on to a bundle which starts from a lower leaf. Now, this course is not essentially different from what we found in the Wallflower. The great difference in the transverse section depends on the fact that the bundles pass very gradually inwards, so that when we cut the stem across, our section passes through a number of them at various points of their inward course. Thus the small outer bundles (Fig. 58, b_3) in the transverse section are those which are nearest the leaf and have only just entered the cylinder. (The outline of the cylinder is indicated by dotted lines in the diagram, Fig. 59.) The intermediate bundles (b_2) are cut through at a point nearer the interior; while the largest (b_1) are those which have approached nearest to the centre of the stem and have turned straight down.

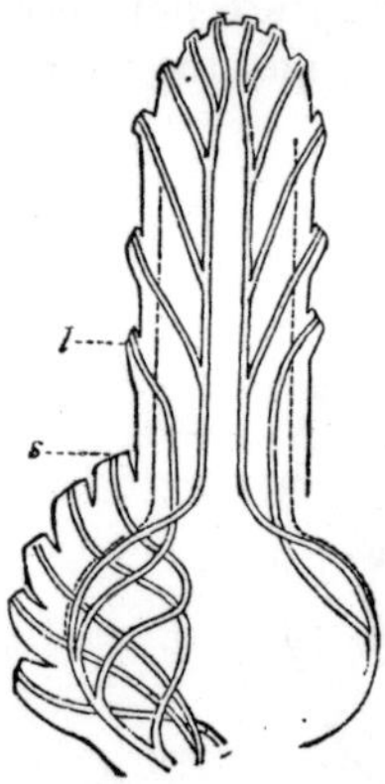

FIG. 59.—Diagram to show course of the bundles in the flowering stem and bulb of the Lilies, as seen in longitudinal section. l, insertion of a foliage leaf; s, insertion of a bulb-scale. (After Falkenberg.)

In the bulb the course is still more complicated, for here each bundle first bends in towards the middle and then out again, and only joins on to other bundles near the outside of the stem (see lower part of Fig. 59). In the White Lily three bundles enter the stem from each leaf.

If we now consider the structure of the central cylinder or stele more in detail, we find that the vascular bundles are

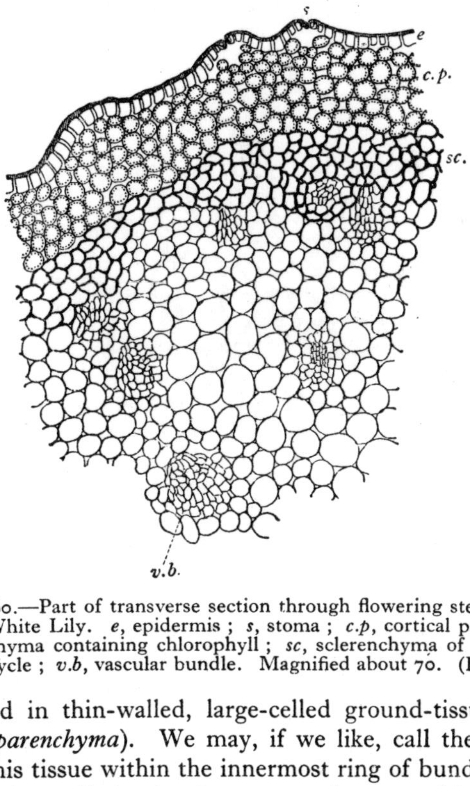

FIG. 60.—Part of transverse section through flowering stem of White Lily. *e*, epidermis ; *s*, stoma ; *c.p*, cortical parenchyma containing chlorophyll ; *sc*, sclerenchyma of pericycle ; *v.b*, vascular bundle. Magnified about 70. (R. S.)

embedded in thin-walled, large-celled ground-tissue (*conjunctive parenchyma*). We may, if we like, call the central part of this tissue within the innermost ring of bundles *pith*, but there is no distinction in structure between this and the rest of the conjunctive tissue, all the cells of which have the form of vertical cylinders with horizontal ends. They are always living cells, and besides the cytoplasm and nucleus generally contain starch granules.

The outer layers (pericycle) of the cylinder, however, are different from the rest. They have thick, lignified walls, and are very long, with pointed ends (*pc* in Fig. 58, *sc* in Fig. 60). As the pericycle forms a rigid case round the cylinder, it is evident that no further growth in thickness is

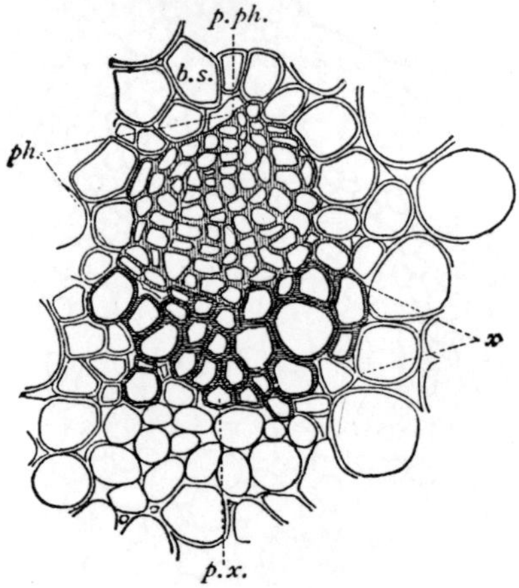

FIG. 61.—Single vascular bundle from the last section. *x*, xylem ; *p.x*, protoxylem ; *ph*, phloëm ; *p.ph*, protophloëm; *b.s*, sclerenchymatous bundle-sheath. Magnified 210. (R. S.)

possible. The pericycle here represents the chief *mechanical tissue* of the stem, serving to give it the necessary stiffness. The vascular bundles themselves have on the whole a similar structure to those of the dicotyledonous type. They are collateral, the inner half being the xylem and the outer half the phloëm (see Figs. 61 and 62). Each bundle is surrounded by a special sheath of long thick-walled lignified cells (*b.s.* in Fig. 61, *sh* in Fig. 62). There is often a group of thin-walled cells between the sheath and the inner edge of the xylem, which itself consists chiefly of vessels with a few living cells among them. The vessels are of the same kinds as those of the Wallflower. Starting from the side nearest the middle of the stem, we first find annular or loosely-coiled spiral vessels (often with a double spiral band), then more closely-coiled spiral vessels of larger size, and

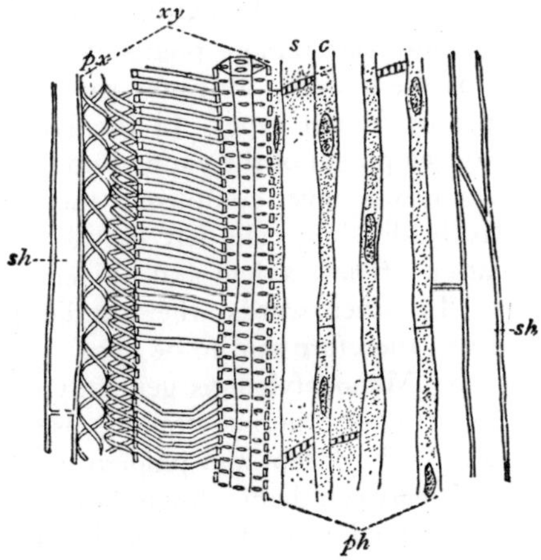

FIG. 62.—Longitudinal radial section through a vascular bundle of the stem of the White Lily. *sh*, bundle-sheath; *xy*, xylem, or wood; *px*, protoxylem; *ph*, phloëm; *s*, sieve-tube; *c*, companion-cell. Magnified 220. (R. S.)

finally, next the phloëm, vessels with simple pits. We do not usually meet with *bordered* pits in the Lily.

The phloëm is of simple structure, and is entirely made up of sieve-tubes and companion-cells, without any fibres or phloëm-parenchyma. (Fig. 62, *ph.*) Thus both xylem and phloëm are composed of the same elements as those of the Wallflower, but with rather less variety. There are no mechanical tissues in the vascular bundle.

The *development* of the vascular bundle is also such as we are already familiar with. In the xylem the first elements to be differentiated are those on the extreme inner edge of the bundle (*px* in Figs. 61 and 62). As usual, this proto-xylem may be known by its annular or loosely spiral vessels, adapted to undergo much stretching after their walls are thickened. Conversely, in the phloëm, the first cells to become mature are those on the extreme *outer* edge of the bundle. From these two remote points the differentiation advances towards the middle. But now we come to a great difference from our former type. In the Dicotyledon we found that the development of the young bundle into mature xylem and phloëm was never complete ; there was always a band of actively dividing cells left between the two, so that the formation of new xylem and phloëm could go on without limit. In other words, the bundle possessed a cambium, and was therefore said to be *open*.

In the Lily and Monocotyledons generally, on the other hand, the *whole* of the procambial strand is used up to form permanent tissue. The xylem and phloëm meet ; no layer of cambium is left between them ; the bundle is a *closed* one, and when once formed is quite incapable of any further growth.

This is one of the most constant distinctions between the two great Classes of which we have taken the Wallflower and the Lily as types. The one has *open* bundles, capable of indefinite growth by means of cambium, while the other (with the rarest exceptions) has *closed* bundles with no cambium, and therefore with limited growth.

Almost all Monocotyledons have closed vascular bundles, and the great majority have no cambium at all, so that the stem or root, when once developed from the growing-point, increases no more in thickness. There are, however, a few members of the Class, such as the Dragon-tree and *Yucca*, which have a kind of cambium, and in which secondary growth in thickness consequently takes place. In these plants the cambium arises altogether outside the vascular bundles, so that the growth is quite different from that in typical Dicotyledons. We need not, however, concern ourselves here with these exceptional cases.

The *cortex* requires little description. Its cells have thin cellulose walls, and in their cytoplasm chlorophyll-granules are embedded, by means of which the stem is able to do some assimilation. The leaves, however, are the chief organs for this function.

The *epidermis* has elongated cells, and rather numerous large stomata, placed lengthways, so that a transverse section cuts across both guard-cells.

The base of the stem, which bears the scale-leaves of the bulb, has a rather different structure. The distinction between central cylinder and cortex is less sharp ; there are no chlorophyll-granules in the cortical cells, because this part of the stem is never exposed to light. The vascular bundles have a different course, as explained above. The ground-tissue contains quantities of starch and also oil. The reserve substances of the bulb stem are, however, used up in summer, when the flowering stem develops, and then we find its cells almost empty. The bulb-scales, which we shall describe later on, are more important than the stem as storehouses of food.

b. Leaf

It matters little whether we take an autumnal ground-leaf or a summer leaf from a flowering stem. They are different, as we have seen, in shape, but in internal structure they are very much alike.

We have already learnt the general course of the veins. In the midrib and principal veins of the larger leaves several vascular bundles run side by side ; the smaller veins correspond each to a single bundle. If we trace the bundles outward from the stem into the leaf, we find that they immediately begin to branch, the main branches taking a parallel course, while they are connected by minor oblique strands (see Fig. 63, *v.b.*).

Thus the whole surface of the lamina is provided with an elaborate system of conducting channels, chiefly differing from that of our dicotyledonous type in the absence of

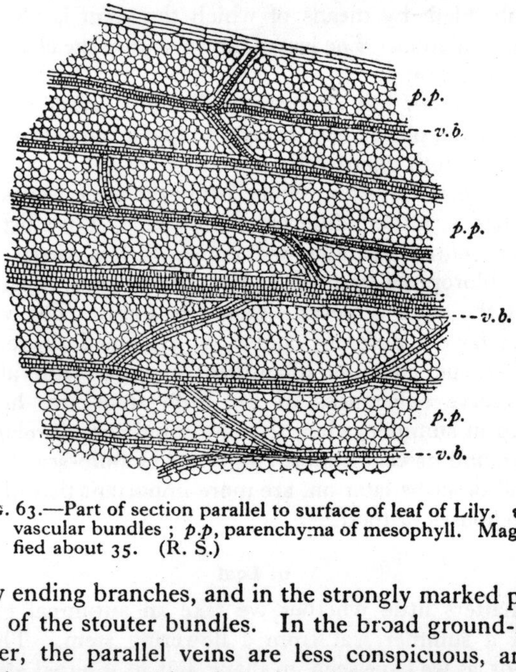

FIG. 63.—Part of section parallel to surface of leaf of Lily. *v.b*, vascular bundles ; *p.p*, parenchyma of mesophyll. Magnified about 35. (R. S.)

blindly ending branches, and in the strongly marked parallel course of the stouter bundles. In the broad ground-leaves, however, the parallel veins are less conspicuous, and the network more resembles that of a Dicotyledon.

In these leaves, as in those of most plants, the rule holds good that the xylem of the bundle is directed towards the upper and the phloëm towards the lower surface.

The minute structure of the bundle in the leaf is like that

in the stem. The finer bundles, however, are much reduced, and here we usually find tracheids instead of vessels (cf. p. 57). The bundles are accompanied by thin-walled parenchyma, and the whole strand, bundle and parenchyma together, is enclosed by a well-marked sheath of cells containing abundant starch (see Fig. 64). The parenchyma accompanying the bundles is continuous with that of the central cylinder of the stem.

In this leaf, as in most horizontal leaves, we find that there is a marked difference between the upper and lower

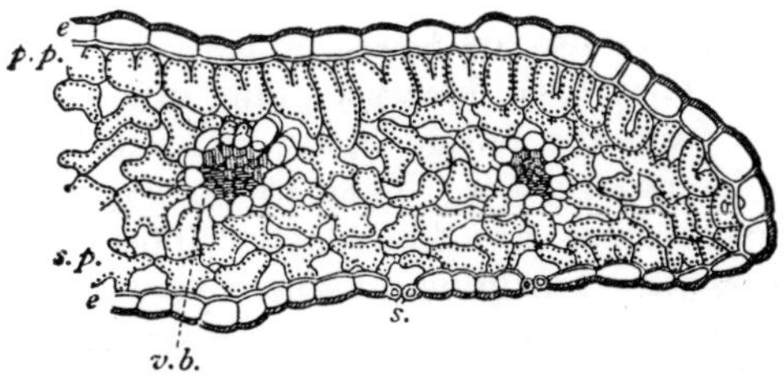

FIG. 64.—Part of transverse section of foliage-leaf of White Lily. *e*, epidermis ; *s*, stoma ; *p.p*, palisade parenchyma ; *s.p*, spongy parenchyma ; *v.b*, vascular bundle, with xylem above and phloëm below. Magnified about 70. (R. S.)

parts of the ground-tissue ; in fact, we have here, as in the Wallflower, to distinguish between palisade and spongy tissue. The palisade cells, however, are of a peculiar form (see Fig. 64, *p.p.*) ; they are branched, and the branches (two or three to each cell) are placed at right angles to the surface of the leaf, forming what is called *arm-palisade* tissue. The effect of the branching is that extra vertical wall-surface is gained for the disposal of chlorophyll-granules. We must not suppose that this kind of tissue is peculiar to Monocotyledons.

The lower part of the tissue of the leaf is spongy ; its cells are branched irregularly in all directions (see Fig. 64, *s.p.*), so as to leave large intercellular spaces between them.

The *epidermis* consists, on both surfaces of the leaf, of large elongated cells, with a thick outer wall, limited by a well-marked cuticle. The stomata are mostly on the lower surface, though we may find a few scattered about in the upper epidermis as well. Each stoma consists of two guard-cells, with a pore between them, leading to an air-chamber below. In fact, their structure is precisely the same as in the Wallflower, but in the Lily they are much larger and easier to observe.

The *bulb-scales* also are leaves or bases of leaves, as we have seen. They have, however, quite different functions to perform from the foliage-leaves, and accordingly have quite a different structure. They are not exposed to light, and therefore have no chlorophyll. The epidermis is without stomata. The ground-tissue is very abundant, and is all alike, there being no need for any special palisade or spongy tissue. The cells are crammed with reserved food-material, especially starch, which is gradually consumed as the flowering stem develops. The vascular bundles are much fewer than in the foliage-leaf, and their xylem is small, for there is no active water-current to be conducted.

To sum up : while the foliage-leaves are adapted to the functions of carbon-assimilation and transpiration, the scale-leaves of the bulb have only the work of storing up the food which has been formed elsewhere.

c. Root

The main root, as we have already seen, only lasts for a short time, though it grows to a considerable length. After the seedling stage is past, the plant has adventitious roots only. Indeed, as seedling Lilies are not common in our gardens, the only roots which we generally see on these plants are adventitious ones. They usually arise from near the base of the stem, below the bulb-scales (see Fig. 55, *r*). They are somewhat sparingly branched. The roots are of large size, and very favourable for investigation. The general structure follows the same lines as that of dicotyledonous roots, but there is this difference : a root of the Lily, like that of nearly all Monocotyledons, has no secondary growth in thickness, but has to depend entirely on its

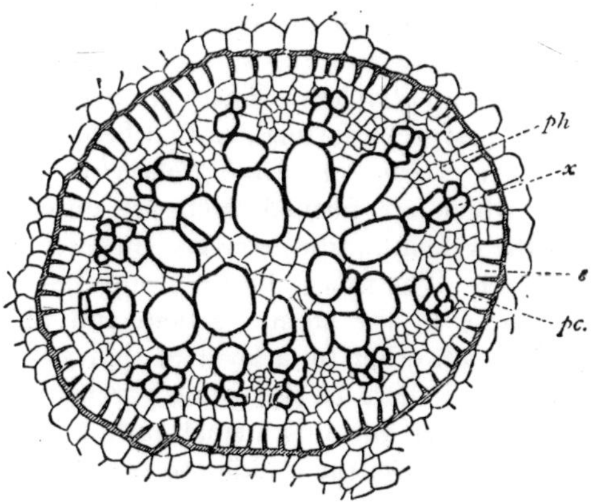

FIG. 65.—Transverse section across the central cylinder, and inner layers of cortex of an adventitious root of the White Lily. *e*, endodermis ; *pc*, pericycle ; *ph*, a phloëm group ; *x*, a xylem group ; *protoxylem* at the outer end of each xylem group. Magnified about 150. (R. S.)

primary tissues ; the latter are accordingly much more highly developed, in the case of the large adventitious roots, than is necessary in a dicotyledonous root, where there is a cambium ready to make good the deficiencies of the original structure.

An adventitious root of a Lily is traversed by a large central cylinder, which includes a great many distinct bundles of xylem and phloëm, often as many as a dozen of each (see Fig. 65). They are arranged alternately all round the cylinder. We thus see at once that the structure is essentially the same as in the young root of a Wallflower, but in the Lily there are many bundles of xylem and phloëm, while in the Wallflower there were only two of each. In other words, the root of the Lily is *polyarch*, that of the Wallflower *diarch* (cf. Fig. 28). The main root of the Lily seedling, however, which is little more than an embryonic organ, is diarch, like that of the Wallflower.

We notice that here also the smallest vessels lie to the outside of each group of xylem. These are the spiral vessels, which were the first to be formed, so that here, as indeed in all roots, we have *external* or *exarch protoxylem*. The inner large vessels, which are formed later, are pitted. Sometimes two of the xylem bundles converge inwards, and meet in a single vessel. The small phloëm groups have the same structure as those in the stem. Xylem and phloëm are separated from each other by a layer of thin-walled tissue. In the root figured there is also some parenchyma in the middle of the cylinder, forming a small pith, but sometimes the xylem groups meet in the middle, so as to leave no room for a pith.

The outermost layer of the cylinder, surrounding the whole ring of xylem and phloëm, is a very definite thin-walled pericycle (Fig. 65, *pc*). Outside this, again, we come to the innermost layer of the cortex, the very conspicuous endodermis (*e* in Fig. 65), the cells of which fit closely together, and have those walls cuticularised by which they are in contact with each other. In the older roots the endodermal cell-walls become thickened, especially on the radial and external surfaces, as shown in Fig. 65. The endodermis serves especially to cut off all communication between the intercellular spaces of the cortex and the interior of the central cylinder. These spaces contain air, and if this air could make its way into the vessels of the wood, the water-current through the latter would soon be stopped. Hence the importance of the isolating endodermis.

The *cortex* is many layers of cells in thickness, and is persistent instead of being thrown off, as in Dicotyledons. Outside this, again, is the absorptive or *piliferous* layer, which forms the external surface of the root. Many of its cells grow out into unicellular root-hairs (see Fig. 66, *r.h*), which are the special organs for absorption, taking up water and dissolved salts from the soil.

It is only, however, the young parts of the roots which absorb food ; sooner or later the piliferous layer dies away,

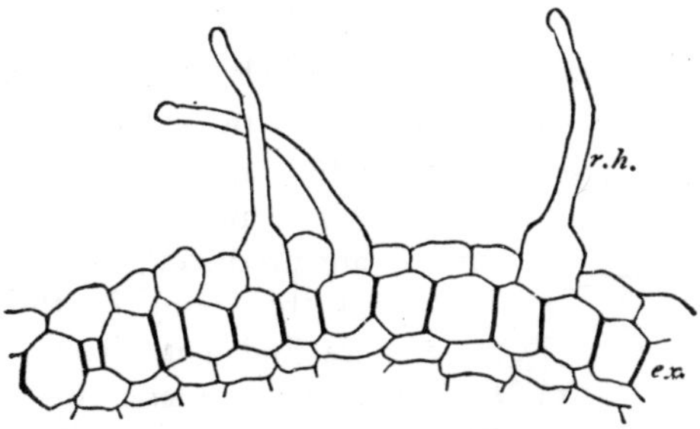

FIG. 66.—Transverse section of tissues at the *extreme outside* of the same root. *r.h*, root-hair ; *ex*, exodermis, forming outermost layer of cortex. Magnified 330. (R. S.)

absorption ceases, and the root henceforth becomes merely an organ of conduction, which serves to pass on the water towards the stem. This conducting part of the root requires some kind of protective layer on the outside to guard it against abrasion, or the more serious danger of attack by parasites, and also to prevent possible loss of water if the soil should become dry. The protective function is performed by the outermost layer of the cortex, which soon becomes distinguished from all the inner layers by its cuticularised walls. For a time it has the structure of an endodermis ; subsequently the corky change extends to all its walls alike. Thus the older part of the root is clothed by a complete protective covering, similar to the epidermis of the stem, while the piliferous layer, which formed the original outer surface, disappears. This protective covering, which is called the *exodermis* (*ex* in Fig. 66), is found in many dicotyledonous roots also, but in them its functions are very temporary, and it is usually soon replaced by secondary periderm. In most monocotyledonous roots there is no periderm, and here the exodermis continues to be the only protective layer throughout the life of the root.

The tip of the root is enclosed in a root-cap, the development of which will be described below.

d. Growing-Points and Mode of Branching

The growth of the stem follows the same general course as that of a Dicotyledon. Here also the growing-point, where new tissues and organs are produced, lies at the apex of the stem. In this position we find a meristem, which at first is made up of quite uniform cells. The description given of the growing-point of a Dicotyledon holds good here with only slight differences (see Fig. 67). The external layer of the apex is the dermatogen, from which the epidermis is formed. The periblem and plerome probably have a common initial group. From the plerome, which is the inner part of the meristem, the vascular cylinder, including all its bundles, is developed. The development of the vascular bundles begins at the interior of the cylinder and

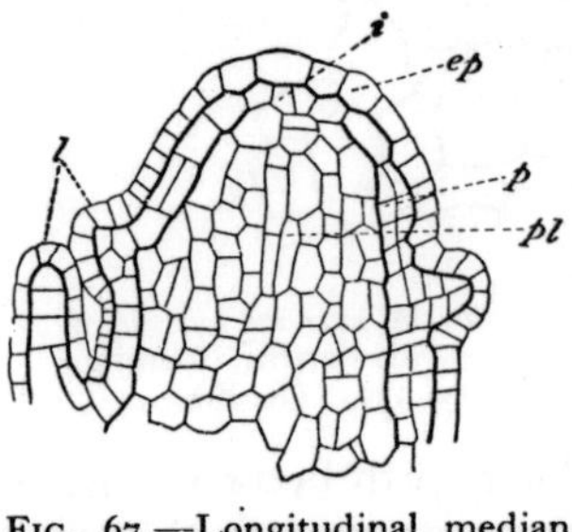

FIG. 67.—Longitudinal median sections of the growing-point of the stem in a Monocotyledon. *ep*, epidermis; *p*, periblem; *pl*, plerome; *i*, initial cells common to periblem and plerome; *l*, young leaves. Magnified about 200. (After Douliot.)

advances outwards. When the outermost bundles have been formed, the pericycle thickens its cell-walls, and all further development ceases.

The leaves here, as usual, develop from below upwards, so that the youngest are always nearest the apex. It is mainly the outer layers of the growing-point which take part in the formation of the leaves and branches of the stem (see Fig. 67, *l*). The growth of each leaf after its first origin goes on chiefly at its base. This is characteristic of leaves and accounts for the continued growth of grass after mowing.

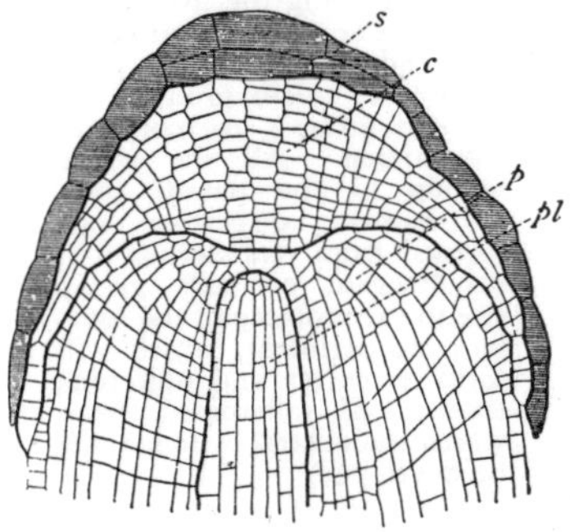

FIG. 68.—Longitudinal median section of apex of young root of a Monocotyledon (*Amaryllis*). *s* (shaded), "digestive sac," formed from endodermis of parent-root; *c*, calyptrogen, giving rise to root-cap only; *p*, periblem (the outermost layer formed from this bears the root-hairs); *pl*, plerome. Magnified about 150. (After Van Tieghem and Douliot.)

The apex of the root is on the whole similar to that of a Dicotyledon, and we find the same three layers of meristem, giving rise to the root-cap, cortex, and vascular cylinder respectively. The superficial layer of the root, which bears the root-hairs, arises here, however, from the same group as the cortex, instead of having a common origin with the root-cap, as in Dicotyledons (see Fig. 68, and compare it with Fig. 33).

The *branching* of the root can also be dismissed in a few words, as in all important respects it takes place in the same way as in the Wallflower.

Here, just as in Dicotyledons, the rootlet arises from a deeply-seated layer of the parent-root, namely, from its

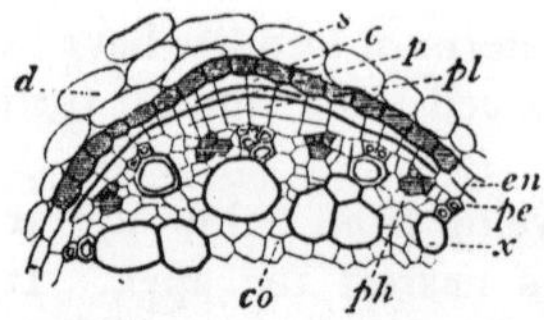

FIG. 69.—Part of transverse section of root of a Liliaceous plant, showing the first origin of a rootlet. *d*, cortical cells; *en*, endodermis; *pe*, pericycle; *x*, xylem group; *ph*, phloëm group; *co*, conjunctive tissue. All the above belong to the parent-root. *s* (shaded), disgestive sac, formed from endodermis; *c*, calyptrogen, or cells giving rise to the root-cap of the rootlet; *p*, periblem, which will form its cortex and piliferous layer; *pl*, plerome, from which the central cylinder will be developed.

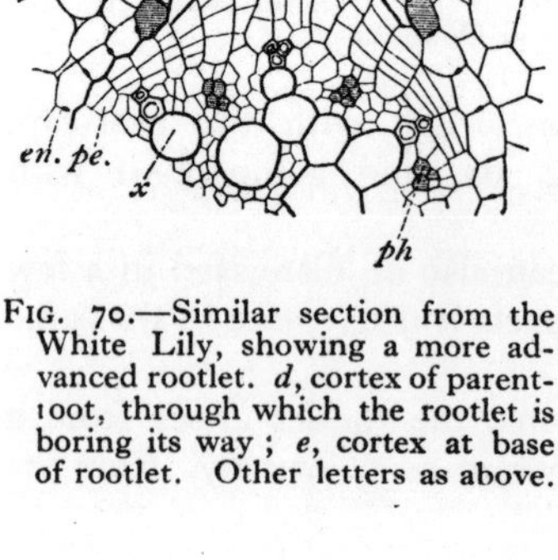

FIG. 70.—Similar section from the White Lily, showing a more advanced rootlet. *d*, cortex of parent-root, through which the rootlet is boring its way; *e*, cortex at base of rootlet. Other letters as above.

pericycle. The details of the development are sufficiently shown by Figs. 69, 70, and 71. The endodermis of the parent-root here takes some part in the process, as it forms a sheath around the developing rootlet

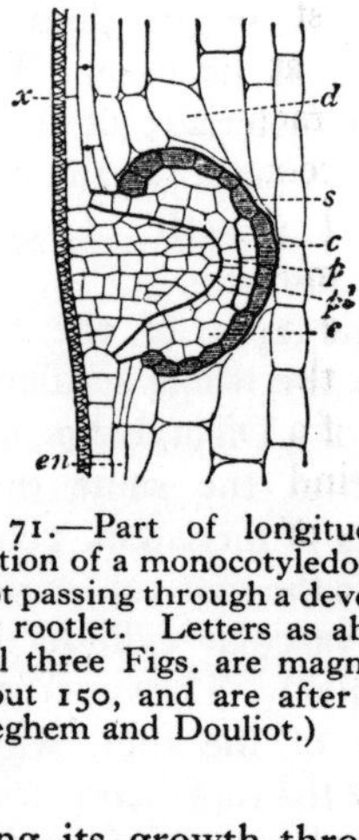

FIG. 71.—Part of longitudinal section of a monocotyledonous root passing through a developing rootlet. Letters as above. (All three Figs. are magnified about 150, and are after Van Tieghem and Douliot.)

during its growth through the cortex. As this endodermal sheath serves to absorb the cortical tissues, through which the rootlet as it were eats its way, it bears the name of the *digestive sac*. When this function has been performed, the sac is cast off, the whole permanent part of the rootlet being thus formed from the pericycle.

The origin of an *adventitious* root is similar to that of a rootlet. Just as a rootlet arises endogenously from the parent-root, so does an adventitious root arise endogenously from the stem, and in both cases it is the pericycle from which the new organ is formed.

Summary of Monocotyledonous Anatomy

The general working of the vegetative organs in the Monocotyledons goes on in the same way as in our first type. In both, the green leaves, by means of their palisade and spongy tissue, are the great organs of assimilation and of transpiration. The leaves are in communication with the stem through the vascular bundles and the tissues immediately surrounding them. The conducting tissues divide up their work as follows: the xylem serves chiefly to convey the water and mineral salts taken up by the roots; the phloëm helps to conduct to other parts of the plant, especially growing and storage organs, the food which has been assimilated in the leaves, including that which contains nitrogen. Both in Monocotyledons and in Dicotyledons all these tissues are continuous throughout the central cylinder of the stem and its branches, and ultimately with the corresponding tissues of the root. Neither as regards structure nor function can we point to any profound difference between the two Classes, *so far as the primary tissues are concerned.* The real distinction lies in the fact that in typical Monocotyledons there are *no secondary tissues*; the entire structure is developed from the growing-points once for all, and henceforth no further additions are ever made, however long the plant may live. Consequently, we generally find that the primary structure of stem and root is more complex in Monocotyledons than in the other Class, because in the former the primary tissues have to do all the work, while in the Dicotyledons they are supplemented, or even entirely superseded, by the secondary tissues formed from the cambium. The few Monocotyledons which have a kind of cambium are here left out of consideration.

This difficulty, however, arises: many Monocotyledons, when mature, have a stem which is of great thickness compared with that of the same plant when a seedling. Take the Maize, for instance: a well-grown plant may have a stem two inches in diameter, while the stem of the seedling is perhaps an eighth of an inch thick; yet there is no cambium and no formation of secondary tissues. How is this possible? The explanation is that the thin stem of the seedling never does become any thicker, but for a time the vigour of the growing-point constantly increases, so that each node and internode which is added to the stem is thicker than the last. Thus for a time the stem has the form of a cone standing on its apex. But sooner or later the growing-point reaches its full strength. From this time onwards the portions of the stem which it successively produces are of equal diameter; thus all the upper part of the stem is cylindrical. The first-formed part of the stem, which alone has the shape of an inverted cone, generally becomes buried in the ground, or like the tap-root it may die away altogether. While the whole plant is still very young, the stem and main root have a simple structure, comparable to that of a dicotyledonous seedling at a corresponding stage. In typical Monocotyledons, then, the increasing diameter of the stem is due to the addition of successively thicker nodes and internodes, and not to the growth in thickness of those already formed.

The Palms, which are found in the tropics and the subtropics, are also Monocotyledons. Some of them become large trees with obvious growth in thickness of the trunk with age. There is the usual increase in size of the apical meristem from the seedling stage onwards for some time, but in addition to this there are various processes of growth in the stem which are too complicated to discuss in an elementary book.

III. *INTERNAL STRUCTURE AND DEVELOPMENT OF THE REPRODUCTIVE ORGANS*

The flower of a Lily is not only very favourable for study, from the large size and simple arrangement of all its parts, but it has the more important advantage that it is an excellent type of the flowers of Monocotyledons generally, most of which are built on the same ground-plan. We already know the external characters of the flower ; we will now shortly trace its development.

a. Development of the Flower

The flower-bud arises, like any other branch of the stem, from the external layers of tissue of the growing-point, with which it is directly continuous. It is formed in the axil of a bract, and in some species, as in that figured, the axis of the flower itself produces a small lateral vegetative leaf, termed a bracteole (see Fig. 72, B, *b*). In the White Lily, however, the latter is absent. The future flower has at first a broadly rounded shape. The earliest stage shown in the figures is that just after the sepals have appeared. All the three sepals do not arise at the same time, but one after the other (see Figs. 72 and 73, *s*). Next, the three petals appear, all at the same time, and exactly between the sepals (see Fig. 74, *p*). After this, three fresh outgrowths show themselves just inside the petals, and alternating with them, so that they lie exactly opposite the sepals ; these are the three outer stamens (st^a, Fig. 75). Very soon afterwards the three inner stamens appear (st^o, Fig. 76) exactly between the first three, and therefore opposite the petals. Lastly,

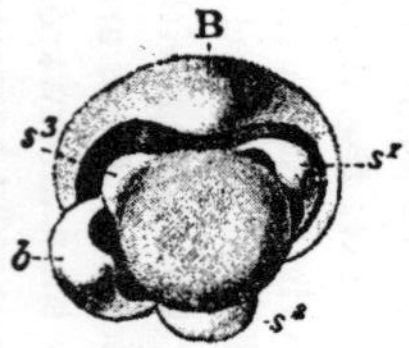

Fig. 72.

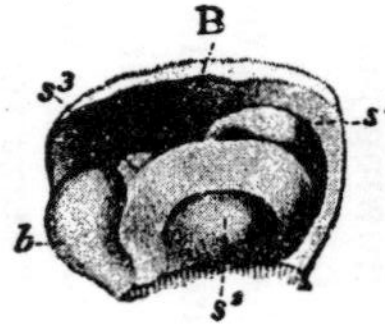

Fig. 73.

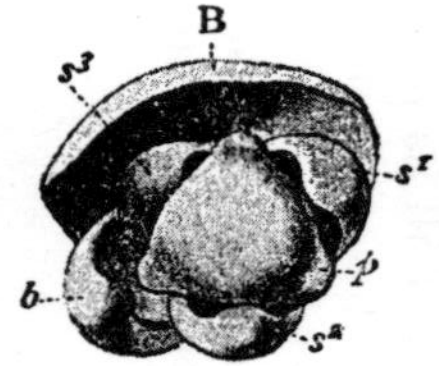

Fig. 74.

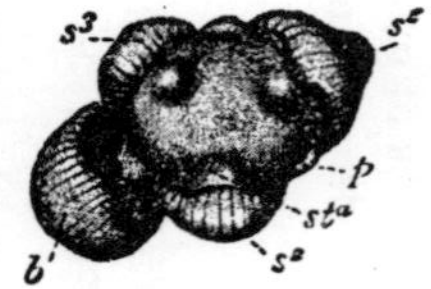

Fig. 75.

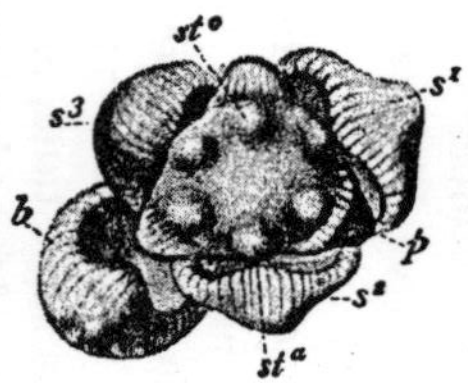

Fig. 76.

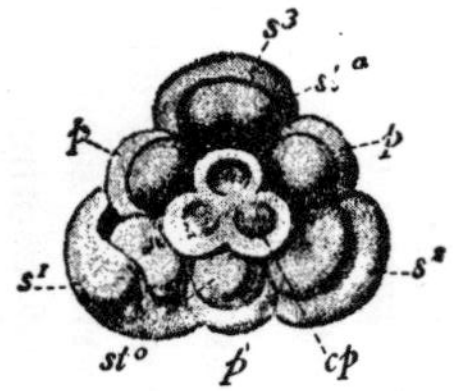

Fig. 77.

Fig. 78.

FIGS. 72 to 78.—Development of flower of a Lily. B, Bract ; *b*, bracteole ; s^1, s^2, s^3, sepals in order of origin ; *p*, petals ; st^a, outer whorl of stamens ; st^o, inner whorl of stamens ; *cp*, carpels. FIG. 72.—Very young flower-bud seen from above ; only the sepals have been formed. FIG. 73.—Similar bud seen from the side. FIG. 74.—Older bud, showing the beginning of formation of petals, seen from above. FIG. 75.—Bud in which the three outer stamens have just appeared. FIG. 76.—Bud in which the three inner stamens have also appeared. FIG. 77.—Bud in which all the carpels have appeared. FIG. 78.—Young pistil seen from above while still open : *cp*, carpel ; *pl*, placenta. (All the above after Payer.) Magnified.

around the centre of the growing-point, yet three other structures arise, each shaped like a horseshoe, open towards the inside of the flower. These are the *young carpels*, forming the pistil or gynœcium, which is the innermost whorl of the flower (Figs. 77 and 78, *cp*). The carpels are alternate with the inner whorl of stamens, and are therefore opposite the outer stamens. During the development of the floral parts, the stalk, on which the whole flower-bud is borne, goes on growing.

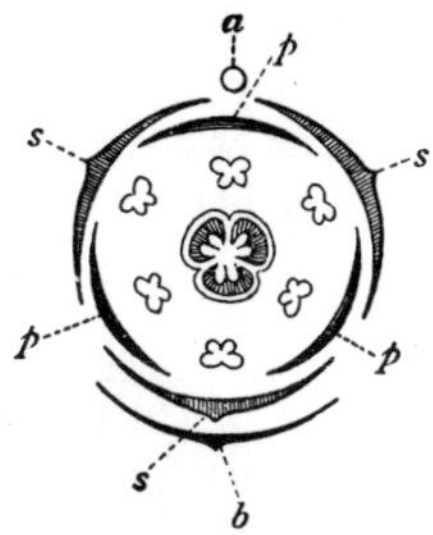

FIG. 79.—Floral diagram of a Lily. *a*, axis of inflorescence; *b*, bract; *s*, sepals; *p*, petals. Within this are seen the two whorls of stamens, and in the middle the three carpels united to form the ovary with axile placentation. (After Eichler.)

We see, then, the extreme regularity and simplicity of the Lily flower. All the whorls consist of three members each, they arise in regular succession, and the members of each whorl always alternate with those of the whorl next outside them. The diagram (Fig. 79), representing a ground-plan of the flower, illustrates this regularity (cf. Fig. 57).

b. Structure of the Perianth and Stamens

The calyx and corolla are much alike in their structure, in which they resemble foliage-leaves, but are somewhat simpler. The same systems of tissue, vascular bundles, mesophyll, and epidermis, are present, and the epidermis possesses stomata. The mesophyll, however, is destitute of chlorophyll-granules, and in the White Lily has no colouring matter at all. In the yellow Lilies, such as the Tiger Lily, chromoplasts are present, and where we get a distinctly red tint, as in the spots on *Lilium auratum*, this is due to the presence of red cell-sap.

The groove at the base of each sepal or petal is lined by a tissue of thin-walled cells, rich in protoplasm, by which the nectar or honey is secreted.

The stamens require no detailed description, as they have essentially the same structure as those of the Wallflower, though they are much larger. The development, both of the stamen as a whole, and of the anther with its four pollen-sacs, takes place in the same way as in Dicotyledons.

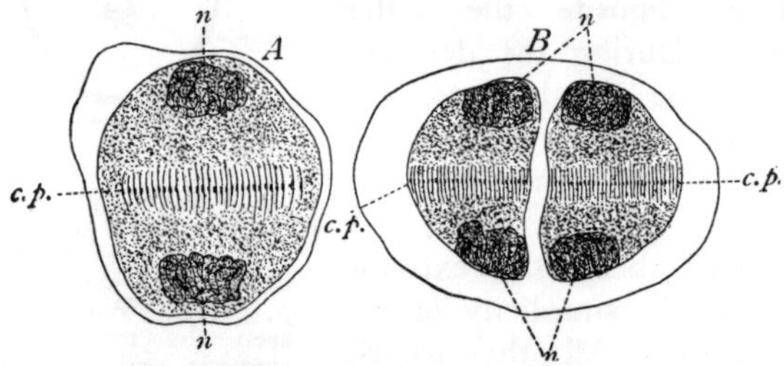

FIG. 80.—Division of the pollen mother-cell in a Lily. *A*, First division : *n*, *n*, daughter-nuclei ; *c.p*, granular cell-plate, from which the new cell-wall will be formed. *B*, Second division : *n*, *n*, the four granddaughter-nuclei ; *c.p*, *c.p*, the cell-plates from which the second cell-walls will be formed. The four cells in *B* become the pollen-grains. (After Strasburger.) Magnified 800.

In each of the pollen-sacs there is a layer of tissue (*archesporium*), which gives rise to a great mass of pollen mother-cells. The division of a pollen mother-cell to form its four pollen-grains, which takes place in a slightly different way from that usual in Dicotyledons, is shown in Fig. 80.[1]

In the pollen-grain itself further changes take place. The nucleus divides, and around one of the two nuclei thus produced a small cell is formed (see Fig. 81, *g*).

This, as in the Wallflower, is the *generative* cell, so called

[1] For the whole development of the stamen careful reference must be made to the description given in the case of the Wallflower (pp. 94–100).

because it is directly concerned in fertilisation. It is much more easily seen in the large pollen-grain of a Lily than in the small one of a Wallflower. The membrane of the ripe pollen-grain is cuticularised, and its surface is rough with raised ridges, which form a delicate network. Along one side of the grain there is a fold, the membrane of which is thicker than the rest of the wall; it is smooth, and consists of pure cellulose. In the contents of the grain there are many starch granules.

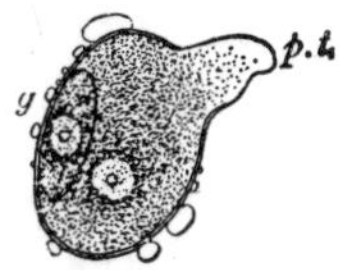

FIG. 81.—Pollen-grain of Lily at the beginning of germination. *g*, small generative cell with its nucleus; *p.t*, first outgrowth of pollen-tube. Oil-drops are shown outside the grain. Magnified several hundred times. (After Dodel-Port.)

The dehiscence of the anther takes place as described in the Wallflower. Here also the wall separating the two pollen-sacs on the same side of the anther breaks down at the moment of dehiscence (see Fig. 40, of Wallflower), so that the perfectly ripe anther appears to have only two compartments instead of four. The abundant pollen is set free, but most of it remains hanging on the burst anthers, for the grains are made sticky by the presence among them of innumerable drops of oil, of an orange colour.

c. Structure of the Pistil

The pistil or gynœcium has a very different structure from that of the Wallflower, though it resembles that of many other Dicotyledons, so that the differences in question are not characteristic of the two Classes. We have already seen that each of the three carpels arises as a ridge, shaped like a horse-shoe, on the growing-point of the flower; very soon the ridge is completed on the inner side of each carpel, which thus becomes a closed ring. Where the carpels join on to each other they are completely united from the first, the wall separating their cavities being a single, and not a double one ; the pistil is, in fact, completely *syncarpous*. Henceforward growth goes on at the base of the pistil, and so the whole structure grows up into a triple tube. At last the upper edges of each carpel join together at the top, so as to close the tubes. The upper closed part of the pistil develops into the style. At the extreme tip the three carpels remain partly distinct, forming the three lobes of the stigma (see Fig. 82, *stig*), which here correspond in position to the carpels, unlike the arrangement which we found in the Wallflower.

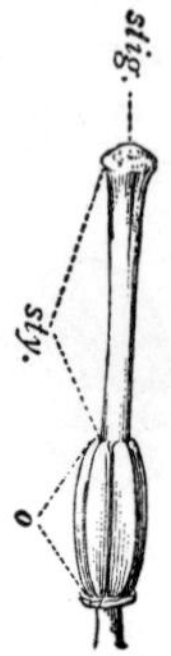

FIG. 82.—Young pistil of a Lily. *o*, ovary ; *sty*, style ; *stig*, stigma. Natural size (After Dodel-Port.)

The ovules develop from the inner angle of each of the united carpels, forming two rows in each carpel (see diagram, Fig. 79). This *placentation* is known as *axile*.

d. Ovules

Passing on now to the ovules themselves, we find that the first stages of their growth are like those of the Wallflower, and here also the ovule soon becomes curved. The curvature, however, takes place in a different manner. In the Lily the nucellus and integuments remain straight ; the bending takes place at the chalaza, so that the whole body of the ovule turns through an angle of 180°, and so comes to lie parallel to the funicle, with the micropyle directed towards the placenta, which it closely approaches (see Fig. 84). Such an ovule is called *anatropous*. The difference between this form and the *campylotropous* ovule of the Wallflower, in which the nucellus and integuments are themselves bent, will be evident on comparing Fig. 84 with

Fig. 43. The anatropous ovule is the commonest form both among Dicotyledons and Monocotyledons.

In the Lily the second or outer integument does not become closed, but remains widely open, the narrow micro-

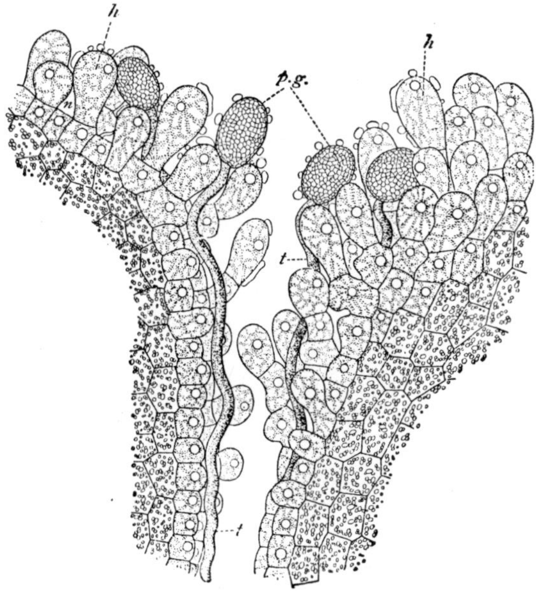

FIG. 83.—Longitudinal section through part of the stigma of a Lily, showing germinating pollen-grains. *h*, papillæ of stigma; *p.g*, pollen-grains; *t*, pollen-tubes. Highly magnified. (After Dodel-Port.)

pyle being thus formed by the inner integument alone (see Fig. 84). This is common among Monocotyledons. In other respects there is a general agreement between our two types; in fact, the main points in the development of the ovule are very uniform throughout the great majority of the members of both Classes.

When we come to the origin of the embryo-sac itself, however, there are some differences to note, though they are not characteristic of the two Classes, as similar variations are found within the limits of both Monocotyledons and Dicotyledons. In the Wallflower we found that the original

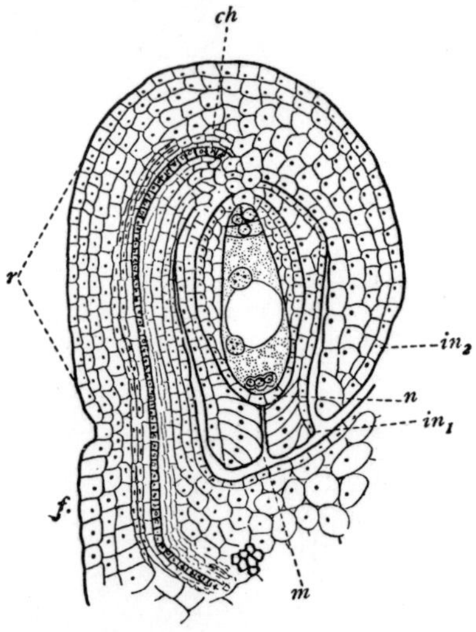

FIG. 84.—Longitudinal section of ovule of Lily. *f*, funicle; *r*, raphe; *ch*, chalaza; in_1, inner integument; in_2, outer integument; *m*, micropyle; *n*, nucellus. Within the nucellus (*n*) is seen the long pear-shaped embryo-sac containing egg-apparatus, polar nuclei with vacuole between them, and antipodal cells. Magnified 82 diameters. (R. S.)

archesporial cell underwent two divisions before the embryo-sac was finally marked off. In the Lily the matter is simpler: a cell of the middle row of the nucellus, next under its epidermis, grows larger than the rest, and itself becomes the embryo-sac, usually without any further divisions. It grows to a comparatively large size, so much so that there are no plants more favourable for the study of the embryo-sac than the Lilies. The divisions of the nucleus follow the same general course as that described above (see p. 102). We find as the final result three nuclei at the end of the sac towards the micropyle, three at the opposite end, and two near the middle. The three nuclei next the micropyle are in cells which, however, are devoid of walls ; these three cells are the *egg-apparatus*, the *ovum* being the deepest seated of the three, while the two others are the helpers or synergidæ (see Fig. 85, *s*). In like manner the three nuclei at the other end become the centres of the antipodal cells, which, unlike their opposite neighbours, have cell-walls. The two polar nuclei are of exceptionally large size, especially the one from the antipodal end (see Fig. 85, p_2). A more interesting fact is, that these two nuclei do not fuse till very late—in fact, not till after fertilisation. On the whole, however, the processes inside the embryo-sac are closely similar to those in the dicotyledonous type, the description of which should be carefully compared (pp. 102–105).

The nucellus of the Lily does not disappear, but encloses the embryo-sac until after fertilisation (see Fig. 84).

e. Pollination and Fertilisation

The pollination of Lilies is carried on by moths and butterflies, which come in search of the honey secreted by the nectaries on the perianth. The flowers are made very conspicuous to the visitors by their great size and bright colours, and in some species the sweet smell is an additional attraction. Generally the style projects so far that an insect comes into contact with the stigma before touching the stamens. Hence it will usually bring pollen from the flower which it has last visited, and will thus effect cross-pollination. But it may sometimes happen that the anthers and stigmas of the same flower brush against each other, and that self-pollination may result. As the visits of the insects are rather uncertain and very dependent on weather, this possibility of fertilisation by its own pollen may be of use to the Lily as an extra chance of setting its seeds.

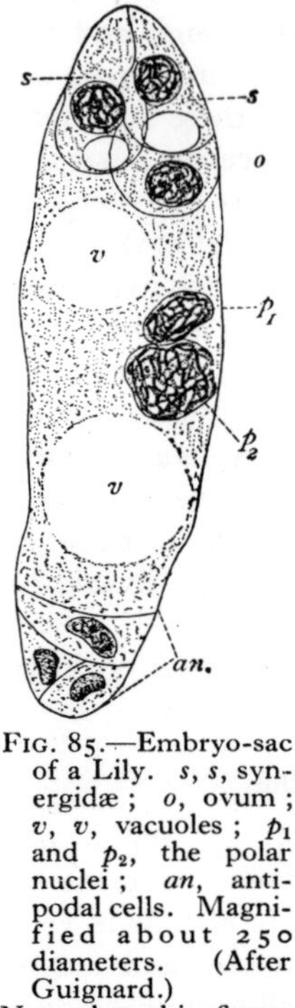

FIG. 85.—Embryo-sac of a Lily. *s*, *s*, synergidæ; *o*, ovum; *v*, *v*, vacuoles; p_1 and p_2, the polar nuclei; *an*, antipodal cells. Magnified about 250 diameters. (After Guignard.)
Note that this figure is reversed as compared with Fig. 84.

When the pollen-grains, by one means or the other, are brought into contact with the stigma, they stick to its viscid surface and soon begin to send out their tubes. The grain takes up water from the stigma, and begins to swell. The fold of pure cellulose (see p. 155) bulges out, and, following the growth of the cytoplasm within, forms the membrane of the pollen-tube (see Fig. 81).

The tube at first creeps along among the papillæ of the stigma, and then grows down between the lobes and penetrates the middle of the style, until its growing end reaches the ovary (see Fig. 83). Here it grows along the surface of the placenta, and at last turns aside into the micropyle of an ovule. Its growth still continues, until it makes its way between the cells of the nucellus to the embryo-sac, which it enters, usually penetrating for a short distance between the synergidæ to the ovum (see Fig. 87). Here the growth of the tube ceases. The length which it has attained is enormous compared to the size of the pollen-grain, for in a large Lily the nearest ovule may be as much as six inches from the surface of the stigma.

The large size of the pollen grains and tubes has rendered it possible to trace exactly what is going on inside, and these changes have perhaps been followed more minutely in the

Lilies than in any other plants. When the pollen begins to germinate, the vegetative nucleus belonging to the large cell of the pollen-grain is the first to enter the tube. It takes the lead all through so long as it exists, but it becomes gradually disorganised, and before the ovule is reached is altogether lost in the cytoplasm. The small generative cell follows the vegetative nucleus into the tube and immediately divides, so that there are now two such cells, one behind the other, each with a large elongated nucleus (see Fig. 86). When the embryo-sac is reached, the nuclei of both generative cells, accompanied by part of their cytoplasm, pass out through the yielding membrane at the end of the tube, assuming a curved, vermiform shape. One of the male nuclei, probably by its own active movements, enters the ovum, and reaches its nucleus. The two nuclei—male and female—then apply themselves closely together, and probably some of the cytoplasm from the generative cell also unites with that of the ovum. Fertilisation is now accomplished (Fig. 87, *g.n.*[1], *o.n.*), and the ovum next acquires a cell-wall of its own.

FIG. 86.—Pollen-tube of a Lily on its way down the style. *g*, *g*, the two generative cells, each nearly filled by its large nucleus; *nv*, vegetative nucleus, already becoming disorganised. Magnified about 450 diameters. (After Guignard.)

In the meantime, a second act of fertilisation has taken place within the embryo-sac. The other generative nucleus passes down through the cytoplasm of the sac; it increases in size, and its long, worm-like form is very characteristic, suggesting a power of wriggling motion. It then attaches itself to the two polar nuclei, which have approached each other, but not yet fused (Fig. 87, *g.n.*, p_1, p_2). Ultimately all three nuclei unite into one, so that the secondary nucleus of the embryo-sac is the product of a triple fusion, one of the component nuclei belonging to the male, and two to the female parent. It is to this triple union that the endosperm owes its origin.

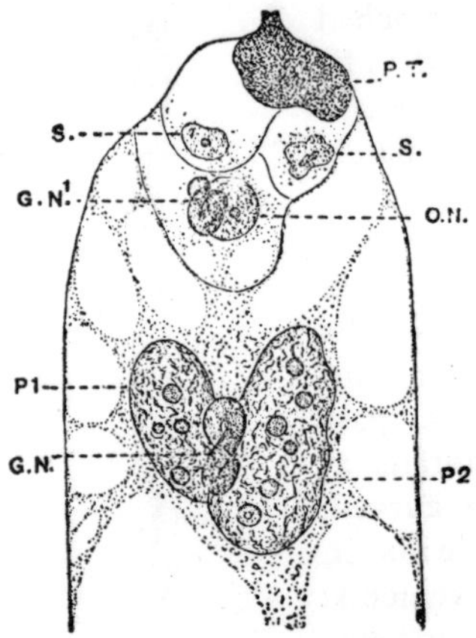

FIG. 87.—Upper part of embryo-sac of a Lily, showing double fertilisation. *p.t*, end of pollen-tube ; *s, s*, the two synergidæ ; *g.n.*[1], one of the vermiform generative nuclei from the pollen-tube in contact with the nucleus of the ovum, *o.n* ; *g.n*, the other generative nucleus from the pollen-tube, in contact with the two polar nuclei, p^1 and p^2. Magnified about 400. (After Guignard.)

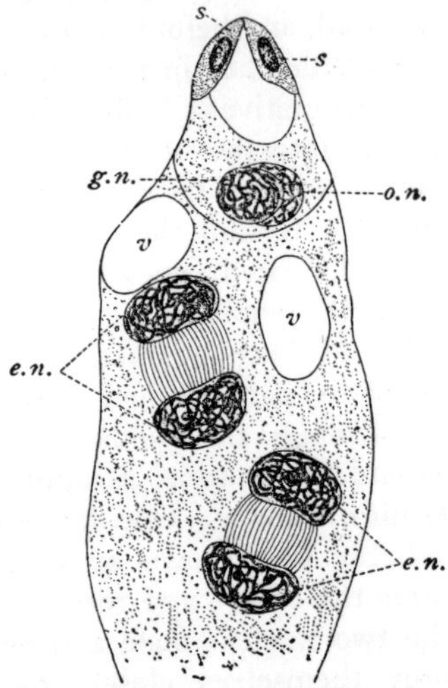

FIG. 88.—Part of the embryo-sac of a Lily, just after fertilisation. *s, s*, remains of the synergidæ ; *g.n*, generative nucleus, *o.n*, nucleus of ovum (these are now closely united) ; *v.v*, vacuoles ; *e.n*, *e.n*, endosperm nuclei in course of division. Magnified about 250 diameters. (After Guignard.)

Thus fertilisation in Angiosperms is double, the one act giving rise to the embryo, the other to the endosperm.

The synergidæ, having done their work, which probably consists in guiding the end of the pollen-tube to the ovum, now disappear. The antipodal cells also undergo no further development. The embryo-sac is in future shared between the embryo and the endosperm, which latter structure is much more important here than in the Wallflower.

f. Development of the Embryo

We need only describe fully those points in the formation of the embryo in which the monocotyledonous and dicotyledonous Classes differ. These differences are important, and among them we shall find one of the most constant points of distinction between the two divisions.

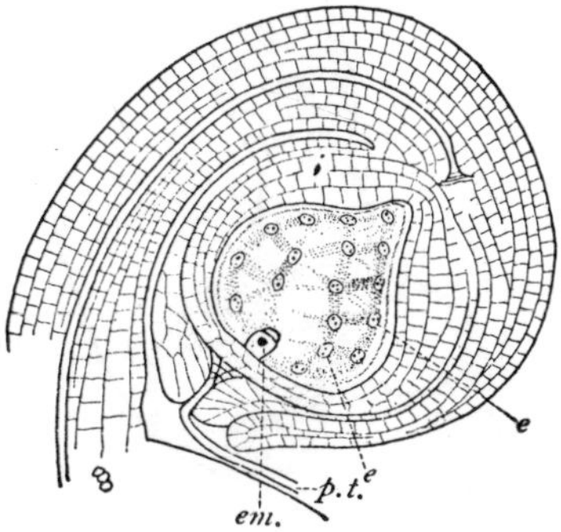

FIG. 89.—Longitudinal section of ovule of *Scilla* (a plant allied to the Lily) after fertilisation. *p.t*, pollen-tube; *em*, embryo; *e*, *e*, endosperm nuclei still free in the cytoplasm of the embryo-sac. Other parts as in Fig. 84. Magnified 55 diameters. Somewhat diagrammatic. (R. S.)

The nucleus of the fertilised ovum, consisting of the united male and female nuclei, divides, and this is followed by the division of the ovum as a whole. The first two walls formed are transverse. Of the three cells thus produced, the one next the micropyle (which we will speak of as *the lowest*) undergoes no further divisions, but often grows to a great size, as in the embryo of *Alisma Plantago* (Water Plantain) shown in Figs. 90 and 91, and then forms a temporary store of food for the embryo.

The *uppermost* cell, instead of forming nearly the whole of the embryo, as in our dicotyledonous type, here only gives rise to one part, namely, the single seed-leaf or cotyledon, which in this case forms the free end of the embryo.

The middle cell of the three divides by further transverse walls, forming, after some longitudinal divisions have taken place, the groups marked *st* and *r* in Fig. 91.

The group of cells marked *st* is a very important one, for from the cells on one side of this group arises the growing-point of the stem, which in Monocotyledons is a lateral structure. The group marked *r* forms the hypocotyl and root, and also contributes something to the suspensor.

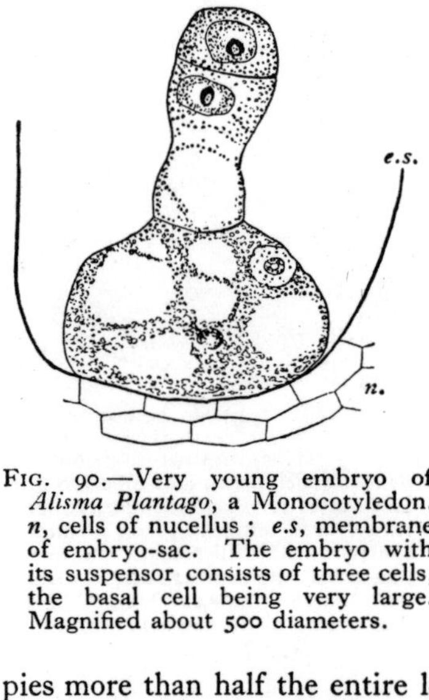

FIG. 90.—Very young embryo of *Alisma Plantago*, a Monocotyledon. *n*, cells of nucellus ; *e.s*, membrane of embryo-sac. The embryo with its suspensor consists of three cells, the basal cell being very large. Magnified about 500 diameters.

The form of the more advanced embryo, after a great many cell-divisions have taken place, is shown in Fig. 92. Towards the top of this figure we see the radicle (*r*), with its growing-point directed towards the micropyle (*m*). On one side of the embryo is a depression, in which the lateral growing-point of the stem (*g.p.*) is situated. All the part beyond this is the cotyledon (*c*), which at this stage occupies more than half the entire length of the embryo.

The essential differences, then, between the embryo of a Monocotyledon and that of a Dicotyledon are these. In the Monocotyledon there is a single terminal cotyledon, and the apex of the stem is lateral, while in the Dicotyledon there are two lateral cotyledons, with the terminal apex of the stem between them. These distinctions are on the whole constant, in spite of great variations in the embryology of both Classes.

The dimensions of the embryo in the Lily, as compared with the whole seed when ripe, are shown in Fig. 92. At the stage figured the chief systems of tissue in the young plant are already marked out. The embryo now enters on a period of rest. Its further development is postponed until the germination of the seed, and will be described below. In the meantime we must shortly consider the changes which take place in other parts of the ovule during its conversion into a seed.

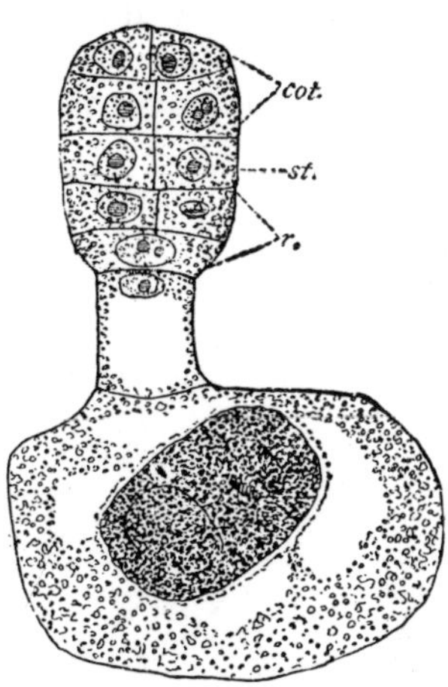

FIG. 91.—Older embryo of same, seen in section. The basal cell is still larger than before. Its enormous nucleus has two nucleoli. *cot*, cells which will form the cotyledon ; *st*, disc of cells from one side of which the stem arises later ; *r*, radicle. Magnified about 500 diameters.

g. Endosperm

In the Lilies the union of the two polar nuclei of the embryo-sac takes place comparatively late — in fact, not till after they have been joined by the male, vermiform, nucleus from the pollen-tube (Fig. 87). The further changes about to be described, leading to the formation of the endosperm, only take place if this curious process of fertilisation has been accomplished. The development of the endosperm usually starts before that of the embryo.

The triple nucleus of the embryo-sac divides, but without any formation of a cell-wall between the daughter-nuclei. The division is repeated again and again in the same manner (see Fig. 88). Soon the nuclei thus formed arrange themselves in the layer of cytoplasm lining the wall of the sac, and there they still go on dividing. All this time the embryo-sac is increasing enormously in size, and the increase in the number of the nuclei keeps pace with its growth (see Fig. 89). For a long time no cell-walls are formed between them. They lie scattered in the undivided cytoplasm. Only after hundreds of free nuclei have been produced does tissue-formation begin. Then for the first time cell-walls make their appearance between the nuclei,

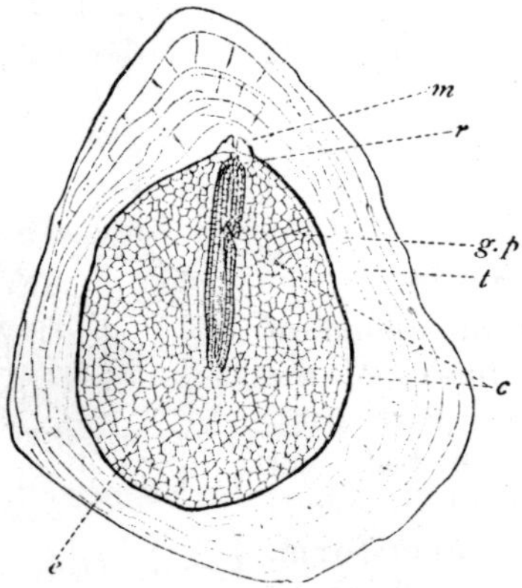

FIG. 92.—Longitudinal section of the ripe seed of a Lily. *t*, testa ; *m*, micropyle ; *e*, endosperm, in which the embryo is embedded ; *r*, radicle of embryo ; *g.p*, growing-point of stem ; *c*, cotyledon. Magnified 6½ diameters. (R. S.)

so that the protoplasm is partitioned up into as many distinct cells as there were free nuclei before. The endosperm tissue, as we may now call it, at first forms a single layer lining the sac. Growth and cell-division, however, go on continuously, until by the time the seed is ripe the whole cavity of the sac, except the small part occupied by the embryo, is filled by a mass of endosperm (see Fig. 92). Its cells contain reserve starch and proteins, and have very thick cell-walls. The whole tissue forms a great storehouse of food, from which the young plant can draw supplies during germination, before it is ready to shift for itself. This is the most striking difference between the seeds of the Lily and of the Wallflower : in the latter the greater part of the endosperm is a transitory and unimportant structure, which is used up long before the seed is ripe ; in the Lily it forms a massive tissue constituting the bulk of the ripe seed, and is only consumed during germination. The Lily thus has a typically *albuminous* seed, while the Wallflower is nearly, though not quite, *exalbuminous.* As regards the presence or absence of endosperm in the ripe seed, there is no constant difference between the two Classes. While, as we have seen, a large number of Dicotyledons have albuminous seeds, many Monocotyledons are exalbuminous, as, for example, the Orchids and the Water Plantains. From one of the latter (*Alisma Plantago*) our Figs. 90 and 91 were taken. In both Classes a majority of the members have albuminous seeds, and in almost all there is at least a temporary formation of endosperm.

The albuminous seeds of some of the Monocotyledons have great practical interest. In Wheat, Rye, and the other grains, it is the endosperm of the seed from which flour is prepared. In fact, when we eat bread, we are converting to our own use the stores of food which the Wheat or Rye plant had provided for the benefit of its seedlings.

h. The Ripe Seed and Fruit

The seeds of the Lilies are arranged in six rows in the fruit, two rows in each compartment ; they are packed very close together, and are flattened, so that each row has somewhat the appearance of a pile of coins (Fig. 93, *s″*). Fig. 92 shows the section parallel to the broad side of the seed. Each seed has a firm, thick testa, formed from the two integuments of the ovule, the outer of which constitutes the thicker and harder layer.

The ripe seed, therefore, consists of three essentially different structures: (1) the embryo, developed from the fertilised ovum; (2) the endosperm, or tissue formed inside the embryo-sac, also in consequence of an act of fertilisation; (3) the testa, formed from the integuments of the ovule (Fig. 92). The nucellus has by this time been absorbed. The point where the seed is attached to the funicle is called the *hilum*, and as the ovule from which the seed was developed was anatropous, it follows that the hilum and the micropyle are close together.

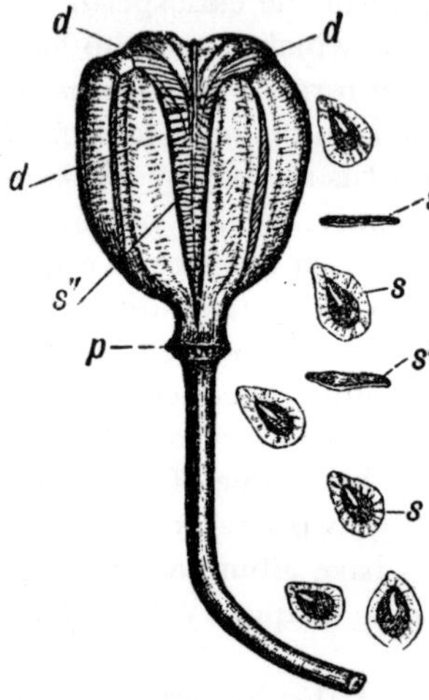

FIG. 93.—Ripe capsule of a Lily (*L. pyrenaicum*) seated on the pedicel. The capsule is dehiscing by three longitudinal splits (*d*), each corresponding to the middle of a carpel. On the right are seen the seeds being shed; *s*, seeds in surface view; *s′*, from the edge; *s″*, seeds still contained within the capsule. For the seeds, cf. Fig. 92. *p*, ridge of receptacle, from which the perianth has fallen. ⅔ natural size. (R. S.)

The fruit is called a *capsule*, *i.e.* it is *syncarpous*, *dry*, and *dehiscent*. The structure shows very little difference from that of the ovary, but there is an enormous increase in bulk, and some change in shape and texture, the fruit being thicker in proportion to its length than the ovary, while its tissues have become harder (Fig. 93). The style withers away during the ripening of the fruit. The increase in size of the ovary in becoming the fruit is a consequence of the fertilisation of the ovules, but how exactly this influence is brought to bear on the ovary is unknown.

The opening or *dehiscence* of the capsule to set free the

seeds takes place by three longitudinal splits, one running along the middle of the outer wall of each compartment throughout its whole length (Fig. 93, *d*). The fruit is thus cleft into three divisions, each representing two half-carpels. The flat, light seeds are well adapted to being scattered by the wind, as they are shed from the dehiscent capsule (Fig. 93).

The White Lily (*L. candidum*) rarely produces ripe fruits and seeds in Britain. Other lilies, however, *e.g.* the Royal Lily (*L. regale*), which comes from China, produce abundant seed.

i. Germination and Life-History

Lily seeds germinate readily under suitable conditions of moisture, temperature, and oxygen supply. *L. regale* is a good species in which to observe the stages in germination. Seed of this species is readily available.

The organ which grows most actively in the early stages of germination is the cotyledon, which elongates to an enormous extent, attaining a length of several inches. The radicle is first pushed out through the micropyle, and then grows down into the earth. The tip of the cotyledon remains immersed in the endosperm, and acts at first as an absorbing organ, by means of which the food substances in the seed are taken up by the young plant. The cellulose, starch, and proteins of the endosperm have to be rendered soluble before they can be absorbed. This is brought about by means of various *enzymes*, which are probably secreted by the tip of the cotyledon (p. 121).

In its lower, sheathing part the cotyledon encloses the bud of the stem, and below this again are the rudimentary hypocotyl and the main root, which is developed from the radicle and grows straight downwards. Henceforth the seedling is able to provide its own mineral food. The seed-coat is lifted high above the ground on the tip of the cotyledon,

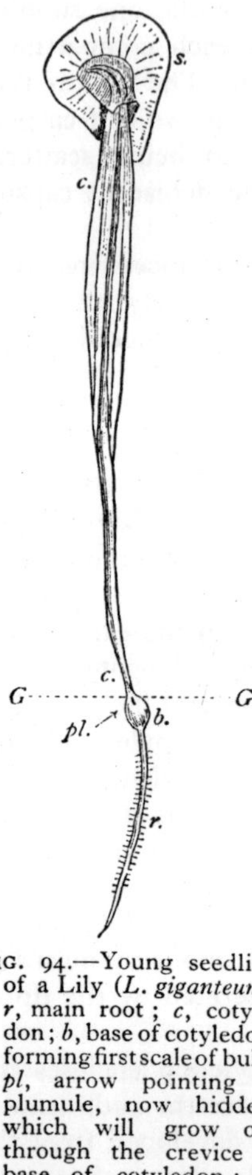

FIG. 94.—Young seedling of a Lily (*L. giganteum*). *r*, main root; *c*, cotyledon; *b*, base of cotyledon, forming first scale of bulb; *pl*, arrow pointing to plumule, now hidden, which will grow out through the crevice in base of cotyledon; *s*, seed-coat, still attached to tip of cotyledon; *G...G*, level of ground. Natural size. (R. S.)

which becomes a long narrow green leaf, and thus constitutes the first assimilating organ of the seedling (see Fig. 94).

The young stem develops into the bulb, the first bulb-scale being formed by the sheath of the cotyledon. Other scale-leaves are soon formed from the growing-point. The bulb appears at first as a lateral growth at one side of the base of the cotyledon. The main root may attain a length of a couple of inches, but it sends out few, if any, rootlets. Its function is only temporary, for it is soon replaced by adventitious roots arising from the base of the bulb. After a time the bulb sends up green foliage-leaves (Fig. 95); the cotyledon decays, leaving only its base; the seed, the endosperm of which is now exhausted, is reduced to an empty shell, and germination is now ended.

In some Lilies, and in many other plants, both Monocotyledons and Dicotyledons, the base of the stem is brought to its proper depth in the ground by the contraction of the roots. In *L. Martagon* the young bulb of the seedling is gradually drawn down year by year, owing to the shortening of the adventitious roots. As the end of the root attaches itself firmly to the soil, the effect of the contraction is to exert a downward pull on the bulb. The upper part of the root is alone capable of contraction, and is much thicker than the rest. The inner cortex is the actively contractile tissue; as it contracts the external layers are thrown into transverse wrinkles, as shown in Fig. 95. New roots of this kind are formed each year, until the bulb has reached its normal depth (three inches or more in the case of *L. Martagon*).

In *L. regale* and many other lilies it is not until the third or fourth year after germination of the seed that the bulb produces a flowering stem.

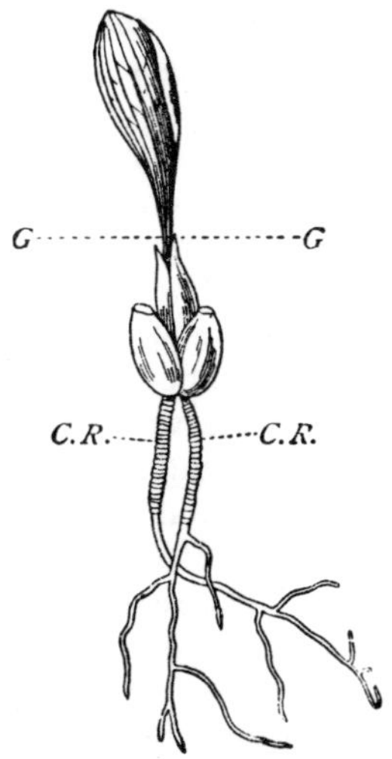

FIG. 95.—Seedling of a Lily (*L. pardalinum*) in its second year, bearing a foliage-leaf. *C.R*, contractile adventitious roots, showing transverse wrinkling; *G...G*, level of ground. Natural size. (R. S.)

Lilies are herbaceous perennial plants which persist for many years by means of their subterranean bulbs. The bulbs of some plants, *e.g.* Tulip and Bluebell (*Scilla nutans*) are of somewhat simpler structure than those of lilies. Closely similar to bulbs are *corms* (*e.g.* of the Crocus) in which the storage tissue belongs to the stem instead of to highly modified leaves. *Rhizomes*, which are horizontal subterranean stems, also frequently contain large food reserves, as *e.g* in many Irises. Plants which perennate by means of subterranean storage organs are said to be of *geophilous* habit. In Nature they often occur in regions which are arid during one part of the year or in woods in which the light conditions are only favourable for photosynthesis during spring. Bulbs and corms can withstand desiccation in the dormant condition.

Summary

If we now shortly sum up those essential points in the life-history of the Lily in which it serves as a type of the Monocotyledons generally, we find that the following are the chief :

(1) In the external morphology the most striking point is the predominance of adventitious roots, the main root serving a merely temporary purpose, namely, the nutrition of the seedling. The explanation of this peculiarity is to be found in the mode of growth of the plant as a whole. The main root is a slender structure, corresponding to the small dimensions of the thin basal portion of the stem, with which it develops simultaneously. Hence, as the stem becomes for a time larger and larger in each successive node and internode, the root becomes more and more inadequate to supply its needs. The root has no secondary growth in thickness, and therefore cannot make good its deficiencies. Thus the requirements of the plant can only be satisfied if the later-formed bulkier portions of the stem produce new and larger roots proportioned to the organs which they supply.

(2) The primary bundle-system in the mature stem and its adventitious roots is on the whole more complex than in Dicotyledons, the individual bundles being more numerous, and their course, in the stem at any rate, less simple. In the leaf we have, as a rule, what is called " parallel venation," which means that there are a number of large longitudinal bundles with more delicate transverse connections. The presence of numerous main bundles in the leaf goes together with the complex conducting system of the stem. In many Monocotyledons the leaf-bases are broad, with

numerous bundles entering them from the stem, though this does not apply to the White Lily.

(3) The closed structure of the vascular bundles, and the general, though not universal, absence of secondary thickening in Monocotyledons, have been sufficiently dwelt upon above. These characters render necessary the greater complexity of the primary tissues, and also explain the absence of a persistent tap-root, which without cambium cannot satisfy the needs of the plant as it grows in bulk.

(4) Perhaps the most constant character of all is found in the structure of the embryo, with its single terminal cotyledon and lateral growing-point. It is this peculiarity, more than any other single character, which separates the Monocotyledons from the Dicotyledons as a distinct Class. We have now considered the distinctions between the two Classes ; their points of agreement scarcely need to be emphasised.

The general morphology is alike in both : stem, leaf, and root show the same distinctive characters, both internal and external.

The structure of the tissues, as distinguished from their arrangement, is similar in the two Classes.

The most striking agreement, however, is in the floral organs. Speaking broadly, we may say that there are *no* characteristic differences between the two Classes here. Their similarity is equally marked, whether we consider the general plan of the flower, the structure of its several organs, or the minute details of their development. Apart from one or two trivial and not very constant differences, we may say that every detail in the development of the perianth, of the stamens, of their pollen-sacs and pollen, of the pistil, the ovules, the embryo-sac and the endosperm, as well as the process of fertilisation, shows the most minute correspondence in both Classes. In fact, so close is the agreement in all important characters, that if we want to find a ready means of distinguishing these Classes by their flowers, we have to fall back on arithmetic, and make what use we can of the rule that in Monocotyledons the parts of the flower are usually in threes, and in Dicotyledons in fours or fives !

On the whole, we may say that the points of agreement between our first two types much outweigh the points of difference.

The classes Dicotyledons and Monocotyledons together constitute the sub-kingdom *Angiosperms*, characterised by the fact that the ovules are contained within a closed ovary, and that consequently fertilisation has to take place through the intervention of a stigma.

CHAPTER III

OUTLINES OF PHYSIOLOGY

In studying our first two types we have obtained a general idea of the form and structure of the root and shoot in green plants. We have also learnt something as to their functions. It will be well for us, however, before proceeding further, to endeavour to gain a more connected view of the mode of working of the vegetative organs. We have to learn what food the plant needs ; from what sources its various kinds of food are obtained ; what changes they have to undergo in order to be made use of by the plant ; we have also to inquire how plants respire ; and generally, what are the actions which a plant carries on as a living organism. All these are physiological questions, for we have already learnt that physiology attempts to answer what is the work done by plants and their organs. We shall only try here to give a very short and simple answer to these inquiries ; anything like a full answer would require an advanced knowledge of physics and chemistry, which we must not take for granted. A little elementary knowledge of these two subjects is indeed absolutely necessary, if we wish to understand even the simplest facts about the life of plants.

I. *The Food of Plants.*—The first step necessary in order to understand anything about the nutrition of plants is to learn what chemical elements a plant contains. Probably most of the elements which exist have been detected at some time or other in plants ; those, however, which are *constantly* present in appreciable amount, are only thirteen in number, namely, carbon, hydrogen, oxygen, nitrogen, sulphur, phosphorus, potassium, calcium, magnesium, iron, sodium, chlorine, and silicon. But although these elements are constantly found, on chemically analysing green plants, it does not follow that they are all necessary to the life of the plant. Many of them we know to be so, because they enter into the chemical composition of the substances of which the tissues of the plant consist. For example, the three elements, carbon, hydrogen, and oxygen, constitute cellulose and starch ; the proteins, of which the cytoplasm is composed, contain, in addition, nitrogen and sometimes sulphur ; while phosphorus is regularly present in the substance of the nucleus. As regards the other seven elements, analysis does not tell us whether they are necessary to the life of the plant, or whether they may not be taken up simply because they are present in the soil.

This question can only be answered by experimental cultures, so arranged that we know exactly what chemical bodies are supplied to the plant.

The simplest method is by means of cultivation in water. We must take care that the water used is pure to begin with, and therefore carefully distilled water must always be chosen. To this water we add small quantities of the chemical compounds with which we wish to experiment. These compounds when in solution are broken up into *ions*, *e.g.* potassium nitrate (KNO_3) in water becomes dissociated into potassium (K) and nitrate (NO_3) ions. Hence the same four types of ions will be present in a solution of potassium nitrate (KNO_3) and sodium chloride (NaCl) as in a solution of potassium chloride (KCl) and sodium nitrate ($NaNO_3$). In the dilute solution used in water cultures the salts are practically completely dissociated. It is in the form of ions that the substances dissolved in the water pass into the plant. It will be well to choose such compounds as are known to be present in a natural soil.

We may begin with a food-solution composed thus :

Water (H_2O)	1000 cubic centimetres
Potassium nitrate (KNO_3)	1·0 gram
Sodium chloride (NaCl=common salt)	·5 gram
Calcium sulphate ($CaSO_4$)	·5 gram
Magnesium sulphate ($MgSO_4$)	·5 gram
Calcium phosphate ($Ca_3(PO_4)_2$)	·5 gram
Iron chloride (Fe_2Cl_6)	a trace.

It will be seen that the amount of the various salts is very small in comparison with the bulk of water. Plants in nature take up their food, in so far as it is supplied by their roots, from very weak solutions.

A seed (say, of the Maize, or Bean, or Pea, or Buckwheat) is allowed to germinate in damp sawdust. The seedling is washed in distilled water, and is then fixed in a split cork fitted into a wide-mouthed bottle, with its root dipping into the food-solution with which the bottle is filled; the young plant is exposed to light. We have spoken of a food *solution*, but one of the salts, calcium phosphate, is only very slightly soluble in water, and most of it will remain as a powder at the bottom of the bottle.

Now, in a water culture of this kind, most kinds of young plants will remain healthy and continue to grow; they will even in time flower and ripen seed. In fact, under these artificial conditions, the plant may be brought successfully through its entire life-history, while the amount of its organic substance is increased, it may be several hundred-fold, as compared with the quantity contained in the seed to start with. The success of this experiment proves that the compounds presented to the plant contain all that is necessary for its nutrition. Yet, among the elements contained in the water and salts of the food-solution, one is conspicuous by its absence; we have given the plant no carbon whatever. We know, however, that the *dry-weight* of a plant—that is to say, its weight after evaporating the water which it contains—is half made up of carbon. We know also that every organic substance—cellulose, starch, sugar, protein, or anything else which goes to build up an organism—is a compound of carbon ; we know, then, that as the organic substance of a plant has increased some hundredfold, so also has its carbon increased in like proportion. Our specimen, then, has obtained abundant carbon, but not from the food-solution, and therefore not by means of its roots. The only remaining source is the atmosphere, and the carbon can only have been obtained by means of those organs which are exposed to the air, namely, the leaves and stem. If a seedling plant is kept in air devoid of carbon dioxide it will soon cease to increase in weight.

We will return later to the question how the plant obtains its carbon ; our immediate object is to determine which of the other elements are essential. It will be noticed that no compound of silicon was added to the water, and this element certainly cannot be obtained from the air. Some trifling quantity might possibly be derived from the glass of the bottle, but the amount to be thus obtained is too small to be of any importance. We infer, then, that silicon, in spite of its general presence in plants, is *not* essential to their nutrition. This does not prove that it is of no use. To many plants, such as grasses, silicon is of great importance in hardening the outer surface of the epidermis, and thus rendering it a better protection against parasites, but *as food* silicon is not necessary to the plant.

Of the salts offered to the plant, sodium chloride can easily be proved not to be essential. If it is left out altogether, the plant will still flourish and come to maturity. The sodium, then, can certainly be dispensed with. It is said that for some plants chlorine in some form or other is needed, but certainly this is not the case with most ; yet these two elements are constantly present in plants, simply because they are constantly present in the soil.

If, however, we omit any of the other elements from our food-solution, the plant suffers. Suppose, for example, that no compound of iron be added to the water ; for a time the seedling grows normally ; the new leaves which are formed, however, are not green, but yellowish—in fact, they contain no true chlorophyll.[1] Hence no assimilation of carbon can take place, and no further increase in the organic substance of the plant can be produced. The reason why the first few leaves are green is that a certain amount of iron is contained in the seed. The disease caused by the want of iron is called *chlorosis*, which means pallor, and is very easily cured. If a few drops of a solution of an iron salt be added to the water, the pale leaves soon begin to turn green, and the same result is produced if the surface of the leaf be painted with the dilute iron compound. It is very important to add no more iron than is necessary. In quantities sufficient to colour the water it is poisonous to the plant. The way in which iron is necessary is thus made quite clear. Without iron no true chlorophyll can be formed.

Carbon, hydrogen, and oxygen are the constituent elements of carbohydrates (*e.g.* sugar, starch, and cellulose) which enter largely into the living processes and structure of plants. The necessity for nitrogen, sulphur, and phosphorus is easily proved experimentally. Suppose, for example, that we wish to determine whether phosphorus is essential. All we have to do is to leave out the calcium phosphate from the food-solution and substitute for it the same quantity of calcium sulphate. The growth of the seedling soon ceases—as soon, in fact, as the phosphorus in the seed has been used up. In like manner we can prove the necessity for nitrogen and sulphur. The presence of potassium, calcium, and magnesium is also essential. If we substitute sodium nitrate for potassium nitrate, the seedling stops growing ; and the same is the case if we replace the calcium and magnesium compounds by those of any other metals. These elements are in some way or other necessary

[1] See p. 29.

for growth ; in their absence the formation of new organic substance is checked. Their exact mode of action, however, is uncertain, although magnesium is known to be an essential constituent of true chlorophyll.

We learn, then, by this method of water culture and by other considerations that the *essential* elements for the nutrition of most green plants are only ten in number, namely, carbon, hydrogen, oxygen, nitrogen, sulphur, phosphorus, potassium, calcium, magnesium, and iron.[1]

Something must be said about the source of nitrogen. Nitrates, such as potassium nitrate, are, as we have seen, quite successful as a source of nitrogen to the plant. If we use ammonium compounds, such as ammonium phosphate ($(NH_4)_3PO_4$), instead of nitrates, the plant will also flourish, but it is possible that in this case the ammonium compound is converted into a nitrate before being made use of. There is some evidence that more complex nitrogenous compounds can be absorbed by certain plants. If, however, we leave out nitrogen compounds from the food-solution altogether, most plants will starve. Yet we know that the atmosphere consists of $\frac{4}{5}$ of nitrogen to $\frac{1}{5}$ of oxygen, approximately. In most cases, then, the free nitrogen of the air cannot be assimilated by plants. It is known, however, that there are important exceptions to this rule ; plants of the Pea family (Leguminosæ), for example, are able, under certain conditions, to supply their demand for nitrogen entirely from the air, while no nitrogenous compounds are offered to their roots. The conditions under which this takes place, however, can only be understood after we have made ourselves acquainted with some of the simplest vegetable organisms, for it is only by the help of certain Bacteria living in their roots that leguminous plants can make use of

[1] It has been shown recently that minute amounts of some other elements, *e.g.* boron, zinc, and manganese, are necessary for the normal growth of certain plants. These are called *trace-* or *micro-elements*.

atmospheric nitrogen, and so enrich the soil for future crops (see *Flowerless Plants*, Chapter V).

All crop plants absorb considerable quantities of nitrogen, potassium, and phosphorus from the soil, and of the elements obtained by plants therefrom these are the most likely to be deficient in adequate amount, especially when several successive crops have been taken from the land. Different crops absorb N, K, and P from the soil in differing degrees, so the well-known principle of the *rotation of crops* is generally applied in arable farming. One of the best-known systems in eastern England is the four-course rotation of mangolds or sugar-beet, barley, clover, and wheat, grown in successive years. Clover is a leguminous plant and therefore tends to enrich rather than to impoverish the soil in nitrogen when the roots are ploughed in. This is followed by wheat which requires much nitrogen. The land is usually heavily dressed with farmyard manure, which is rich in nitrogen, potassium, and phosphorus, before mangolds are grown. Farmyard manure also contains much partly decayed plant material which has a beneficial effect on the soil apart from its manurial value, increasing the water-holding capacity of the soil and improving its texture. Barley of good quality for brewing is obtained from a relatively poor soil so it is not usually manured. Other nitrogenous manures are sulphate of ammonia and sodium nitrate (Chili saltpetre). Potassium is applied manurially to such crops as potatoes in the form of potassium salts, and phosphorus is added to the land for certain crops, *e.g.* turnips, in the form of phosphates, especially superphosphate. Boron in very small amount is necessary for the healthy growth of sugar-beet (see Footnote, p. 180), so on land known to be deficient in this element small quantities of borax are distributed over the soil when this crop is to be grown.[1]

[1] Lime is also often applied to land in considerable quantity ; it is an important factor in maintaining fertility as it corrects soil acidity, which is detrimental to many plants. Lime also improves soil texture.

II. *Assimilation of Carbon.*—We will now consider the question how a plant containing chlorophyll obtains its supply of carbon, which is essential for nutrition. We have already learnt that the plant certainly does not obtain this element by its roots, and that the air is the only source of carbon available.

Now, the atmosphere contains carbon in the form of carbon dioxide. The proportion of this gas contained in the air is very small, amounting only to about three volumes in 10,000 volumes of air. Yet from this relatively small concentration the entire supply of carbon for most green plants is obtained.

Carbon dioxide is formed whenever carbon is completely burnt, or, in other words, completely oxidised. It contains the maximum amount of oxygen with which carbon can combine. Now we know that the substance of plants is combustible ; in fact, all our fuels are directly or indirectly of vegetable origin. It follows, then, that the tissues of plants must be comparatively poor in oxygen, or they would not burn. Hence we see that the carbon dioxide of the air must lose the whole or part of its oxygen when it is assimilated by a plant.

That green plants in sunlight give off oxygen is easily proved. Some pieces of any green water-plant (such as the *Elodea*, mentioned on p. 36) are put into a glass of ordinary water, which always contains some carbon dioxide in solution. The glass with the Waterweed in it is allowed to stand in a sunny place, and we soon see that bubbles are given off from the plant in a constant stream. If we catch these bubbles, as they rise to the surface, in an inverted test-tube filled with water, we find that they consist chiefly of oxygen, as can be easily shown by plunging a glowing match into the gas thus collected ; the match will at once catch light again. If, however, the plant be placed in water which has been previously deprived of its carbon dioxide by boiling, no oxygen will be given off.

If a plant, or simply some green leaves, be placed in a bell-glass, dipping into mercury below, and containing a measured quantity of carbon dioxide, and left for some time exposed to light, the volume of the gas in the jar remains unaltered. If the gas be analysed after some hours, the carbon dioxide will be found to have diminished, and the oxygen to have increased, and in fact the volume of carbon dioxide lost will be exactly equal to the volume of oxygen gained. In other words, the plant has given out exactly as much oxygen as it has absorbed carbon dioxide.

The decomposition of so stable a compound as carbon dioxide is a very difficult process to carry out in the laboratory. Thus, in order to decompose carbon dioxide by heat, a temperature of no less than 1300° C. is required. Yet decomposition of carbon dioxide is carried on by the green plant at the ordinary temperature of the air—say 10° to 30° C. The energy which enables a green leaf to do its work is derived from the rays of the sun ; plants will only assimilate carbon in the light.[1] This is easily proved ; if we repeat the experiment just described in a dark place, no carbon dioxide is absorbed, and no oxygen is set free. Other sources of light may be substituted for the sun. Thus, plants will assimilate actively and continuously when exposed to electric light. As in Nature sunlight alone is in question, we may say that all the organic material of the world, all the animal and vegetable substance which it contains, has been directly or indirectly produced through the assimilation of carbon from carbon dioxide by green plants in sunlight. There are not many rules in science without an exception : it has lately been shown that certain very simple organisms (Bacteria) are able to obtain their carbon from inorganic sources without the help of light ; the amount of organic substance thus formed, however, is too small to affect our statement appreciably. It is only

[1] Hence carbon-assimilation is now often termed *photosynthesis*.

green plants, and green parts of plants, which can decompose carbon dioxide in sunlight. If we had used for our experiments coloured or white flowers, or colourless roots, or any part of a plant containing no chlorophyll, we should have found that no absorption of carbon dioxide or evolution of oxygen took place. It is not necessary that the plants should appear green externally, it is enough that they contain chlorophyll. For instance, the leaves of the Copper Beech are of a brownish-red colour, yet they assimilate like ordinary green leaves, because they contain chlorophyll, which, however, is disguised by the presence of a red colouring matter (anthocyanin) in the sap.

We already know that chlorophyll is contained in the protoplasmic chlorophyll-granules or plastids. The whole apparatus is necessary for the work to be done ; the chlorophyll is useless for the purpose of assimilation without the protoplasmic plastid, and the plastid cannot assimilate without the chlorophyll. We are not at all clear at present as to the special duties performed by the two partners. Most probably the chlorophyll serves to catch certain rays of light, by the energy of which the plastid is enabled to do the work of decomposing carbon dioxide and water. It has been proved that the red rays of light are the most effective in promoting carbon-assimilation. The absorption spectrum of chlorophyll shows a conspicuous dark band in the red. Hence we know that these red rays are absorbed by the chlorophyll. It is very probable that it is the energy of these absorbed rays which is employed in the work of assimilation.

We have now seen that carbon-assimilation is dependent on light ; it is also dependent on temperature. Although all other conditions may be favourable, yet if the temperature be too low no assimilation takes place. Thus, in our experiment with the *Elodea*, we should find little or no oxygen given off unless the temperature of the water was at least

6° C. The minimum temperature varies very much for different plants, but for every plant there is a certain temperature below which no carbon dioxide can be decomposed.

Light is not only necessary in order that the chlorophyll-containing organs may perform the work of assimilation, but also, with few exceptions, for the formation of chlorophyll. Although the plastids are always present, the pigment chlorophyll is not generally formed in them in the absence of light. If a Potato plant be grown in the dark, it remains a yellowish colour in the absence of the two green pigments from the mixture of pigments which we have called chlorophyll (see p. 29). We must be careful to distinguish between these two actions of light, which are quite distinct : a plant will become green in the light even though it be kept in air containing no carbon dioxide, so we see that assimilation is not necessary for the production of chlorophyll, though the presence of chlorophyll is one of the necessary conditions of assimilation.

A green plant kept in the dark cannot decompose carbon dioxide, and therefore under these circumstances it forms no new organic substance. A potato kept in a warm and damp but dark cellar will often grow. It produces a pale, unhealthy, *etiolated* plant, with long stems and small leaves. If such a plant be dried and weighed, we find that its solid substance is not greater, but rather less, than that of the tuber from which it has developed. It has grown entirely at the expense of the food-substances ready stored in the tuber ; a great deal of water has been absorbed, but no new organic substance has been formed ; on the contrary, some has been lost, for a reason which we shall explain presently. The plant, in fact, has earned nothing, but has been living on its capital, and has naturally become poorer by doing so.

Again, no gain would accrue if a plant were allowed to grow in the light, but in an atmosphere destitute of carbon dioxide. For the purpose of this experiment the specimen is grown under a bell-glass, into which air is admitted through a tube, containing pieces of pumice-stone soaked in a solution of caustic potash (KOH). The potash absorbs the carbon dioxide, so that none reaches the plant. Here also the result is that no fresh organic material is formed ; if the plant grows at all, it does so at the expense of the food-substances which it had previously stored up.

We may sum up our conclusions up to this point as follows :

Green terrestrial plants, as a rule, obtain all their carbon by the decomposition of the carbon dioxide of the atmosphere. Similarly, green submerged aquatic plants obtain their carbon from the carbon dioxide dissolved in the water.

Consequently, in the absence of carbon dioxide, no fresh organic substance is formed. The decomposition of carbon dioxide can only take place under the action of light (the red rays being the most effective) and at a sufficient temperature.

It can only be performed by those parts of plants which contain chlorophyll in their plastids.

This process of the assimilation of carbon from carbon dioxide may fairly be considered the most fundamental of all biological phenomena, for on it the whole existence of plants and animals is dependent. Herbivorous animals eat plants, and are in their turn eaten by carnivorous animals and by man. Our own life, and that of the whole animal kingdom, would thus be absolutely impossible without the work done by plant chloroplasts.

We will now inquire what becomes of the carbon after it is taken up by the plant. We know in a general way that it is employed with other elements to build up the protoplasm, starch, cellulose, and other bodies of which a plant consists. Can we trace the stages of the process ? Our knowledge here is still very imperfect, but yet something has already been made out.

The facts are that carbon dioxide and water are converted eventually by the chloroplasts under the influence of light into carbohydrates, with the evolution of oxygen. A simple suggestion from the chemical standpoint has been made that in the first place carbon dioxide and water combine to make formaldehyde (CH_2O), as indicated in the following equation—

$$CO_2 + H_2O = CH_2O + O_2.$$

It is further suggested that the formaldehyde is converted into a sugar, *e.g.* one having the formula $C_6H_{12}O_6$, which is the same formula as that for glucose, thus—

$$6\ CH_2O = C_6H_{12}O_6.$$

On exposure to light the formation of sugar commences, the concentration of sugar in the mesophyll cell rises, and the rate of diffusion of sugar from the cells also increases. In many plants starch appears in the plastids when a certain concentration of sugar is attained.

That starch is formed in certain plants as the result of carbon assimilation can be shown by simple experiments. Take a Pelargonium (" Geranium ") plant growing in a pot and keep it in darkness for two days to ensure that all the starch previously in the leaves has disappeared. (One of the leaves can be detached and tested for starch as outlined below.) Then, after covering the halves of several leaves with silver-paper to ensure that these parts remain in darkness, expose the plant to bright light for a period of four to six hours. At the end of this time cut off the leaves that were partly covered, remove the silver-paper, mark the halves distinctively that were exposed, and plunge the leaves in boiling water. Then place them in alcohol until they have been decolorised. Place the leaves on a white plate and pour over them a solution of iodine in potassium iodide, which stains starch grains a deep blue. Then transfer the leaves to water. It will be seen that the halves of the leaves exposed to light are coloured bluish-black because of the concentration of starch-grains in the plastids, whereas the halves which were kept in darkness are yellowish (iodine stains proteins yellow). If a similar plant had been exposed to bright light in an atmosphere devoid of CO_2 no starch would be formed in the leaves.

To demonstrate this, it is best to use a shoot cut from a Pelargonium plant which has been kept in darkness for two days. Place the shoot immediately in a small beaker of water and stand it under a bell-glass on a plate covered with lumps of soda-lime, which should also be packed round the edge of the bell-glass. The soda-lime will absorb the CO_2 in the air under and diffusing into the bell-glass, rendering it unavailable to the plant. The reason for using a cut shoot in this experiment instead of a plant growing in soil is to avoid complications from the CO_2 which may be produced in appreciable quantities by certain soils. A control experiment should be carried out at the same time, in which another shoot, similarly exposed to bright light, is placed in water under a bell-glass standing on stone chips, to allow of the diffusion of CO_2 into the atmosphere surrounding the shoot. When, subsequently, the leaves are tested with the iodine solution as indicated above, it will be found that only those kept in the atmosphere containing CO_2 will give the reaction for starch.

In green submerged aquatic plants, carbon is obtained by photosynthesis from the carbon-dioxide dissolved in the water. Filaments of the Green Alga, *Spirogyra*, commonly found in ponds (see Part II., *Flowerless Plants*, Chapter III.), begin to form starch within an hour in their large chloroplasts when they are exposed to bright sunlight or an electric lamp. This alga is particularly convenient for microscopic examination as the plant consists of a long filament of cells arranged end to end, and after staining with a solution of iodine in potassium iodide the starch can be readily seen in the chloroplasts. Before placing the filaments in bright light they should be kept in darkness for forty-eight hours to ensure that no starch remains in the chloroplasts.[1] To demonstrate that carbon-dioxide is necessary for photosynthesis some filaments, exposed to bright light, are kept in water which has been freed of carbon-dioxide by boiling. In filaments kept in water devoid of carbon-dioxide no starch will be formed.

In some plants, *e.g.* certain Monocotyledons, starch is never found in the leaves under natural conditions. That this is due, not to any inherent inability of the protoplasm to form starch, but to the sugar-concentration never reaching a sufficiently high level, is shown by the fact that starch formation can be induced in such leaves by placing them in a strong sugar solution.

The change from sugar to starch can be reversed. On the coming of darkness, when sugar formation by photosynthesis ceases, the diffusion of sugar away from the mesophyll cells continues for some time and more sugar is produced at the expense of the starch. When the concentration of sugar has fallen to a certain level all the starch will have disappeared from the plastids. This concentration is practically the same as that at which starch formation began to take place during photosynthesis. Although under some conditions the amount of starch increases and under other conditions decreases in the mesophyll cells, there are reasons for supposing that the formation of starch from sugar and the formation of sugar from starch take place side by side, the net result depending upon which of the processes is the faster.

[1] It is best to carry out this experiment in spring or summer, when the filaments of *Spirogyra* are usually vigorous. In winter the filaments are often moribund, and it is then practically impossible to de-starch them by keeping them in darkness.

It must be clearly understood that there is yet no proof that formaidehyde is first formed in the process of photosynthesis. If it is so formed, it must be *immediately* converted into some other compound, *e.g.* a sugar, as formaldehyde itself is poisonous to plants. Many of the simpler kinds of sugar have been made in recent years by chemists from inorganic compounds in the laboratory, and it is an interesting fact that bodies of the nature of aldehydes are important links in the chain of reactions by which sugars have been artificially produced.

Starch is an insoluble form of carbohydrate, which is convenient for storing purposes. Both in the chloroplast and in the leucoplast starch is produced at the expense of sugar. The difference between them consists in the origin of the sugar, which in the former case is the direct result of photosynthesis, while in the latter the sugar has been transported away, sometimes after conversion from starch, to other parts of the plant where the leucoplasts occur.

If starch is formed, it has to be rendered soluble in order to be transportable to the other parts of the plant. It is well known that starch can be converted into sugar by means of a complex of enzymes called *diastase*, which is derived from the protoplasm, and is found, for example, in large quantities in germinating seeds, such as those of barley during the process of malting. It has been proved that diastase is present in the leaf, and this probably serves to convert into sugar the starch which is formed in the chloroplasts.

The sugars pass from the mesophyll cells to the phloëm cells of the small veins and are then transported along the sieve-tubes to other parts of the plant. This process is known as *translocation*. The sugars may go to the seats of growth, such as the apical growing-points, or the cambium, to be employed chiefly in the formation of new cell-walls, or they may be conveyed to organs which serve as storehouses of reserve food, such as the endosperm or the cotyledons in a seed, the tuber of a potato, or the root of a turnip. In these regions the sugar is very often reconverted into starch by the leucoplasts ; it may, however, be stored up in a soluble form, as inulin in the dandelion, or cane-sugar in the beet. In some cases it is employed to form enormously thick cellulose walls, as in the endosperm of many Palms ; these cell-walls are dissolved again on germination, and so rendered available as food for the seedling. In other cases, again, as in oily seeds, the non-nitrogenous food-materials are stored, not as carbohydrates, but as fatty oils.

The most complex of the nitrogenous food materials are the *proteins*, which are essential for increasing the protoplasm itself. Proteins are formed by combining nitrogen with organic compounds, derived from carbohydrates, which contain carbon, hydrogen, and oxygen only. Some proteins contain sulphur and phosphorus in addition. The elements N, S, P are absorbed by the roots in the form of the ions of nitrates (or of ammonium salts), sulphates, and phosphates. Combination to form protein may occur in any cell well supplied with carbohydrate, but is particularly active in green leaves in light. Green leaves export much of the protein so formed to other parts of the plant, *e.g.* growing-points and seeds. For this transport, which takes place along the sieve-tubes, the proteins are partly converted into simpler soluble nitrogenous compounds. In seeds the proteins may be stored as granules such as the aleurone grains (see p. 33).

III. *Absorption.*—We have now obtained some idea of the chief nutritive processes of a chlorophyll-containing plant, but we have still to inquire how the various food-substances are absorbed.

It is by way of the stomata that the bulk of the gaseous interchange between the green parts of plants and the atmosphere takes place. During photosynthesis the carbon dioxide dissolved in the water contained in the mesophyll cells is used up ; more carbon dioxide dissolved in the wet walls surrounding the cells passes into the cells to take its place ; in turn more carbon dioxide in the intercellular spaces of the mesophyll becomes dissolved in the wet cell walls, and this is replenished from the external atmosphere. Although illustrated by a step-to-step process, in reality there is a constant diffusion of carbon dioxide from the outside atmosphere through the stomata into the chloroplasts within the mesophyll cells.

The extremely small size of the stomatal pores, through which the gases have to pass, may seem a difficulty, but it has recently been proved that the rate at which a gas will diffuse through an opening varies, not with the *area* of the opening, but with its *diameter*. Hence a large number of minute pores is far more effective as a means of diffusion than a smaller number of larger pores having the same total area, so the epidermis with the stomata widely open does not offer a great resistance to the passage of gases.

The mineral substances taken up by the root are absorbed through the root-hairs, where these are present, and must be taken up from solution. Yet we know that many plants flourish well in a comparatively dry soil. In such a soil, however, each particle of earth is surrounded by a film of water, which adheres to it firmly. The root-hairs attach themselves to such particles, and are thus enabled to absorb a part of the watery film which surrounds them (see Fig. 29). The root-hairs, however, are not dependent entirely on salts which are ready dissolved in the water of the soil. They are able to render other compounds soluble. Calcium phosphate, for example, is only soluble to a very small extent in water. The root-hairs, however, give off carbon dioxide in solution, by which minute quantities of salts are dissolved, and can then be absorbed as ions. This solvent action of the root-hairs can be proved by growing plants in a thin layer of earth covering a slab of polished marble. After some time, an exact print of the course of the roots will be found on the smooth surface of the marble, which has become corroded wherever the hairs of the roots have been in contact with it.

The passage of water into a root-hair has been explained on purely physical grounds. If water containing substances in solution is separated from pure water by a membrane, *e.g.* parchment, which lets water pass through, then water will tend to pass into the solution, the *attraction* of the solution for the water being greater the stronger the solution The hydrostatic pressure on the solution which will just prevent the movement of water through the membrane is called the *osmotic-pressure*, hence we speak of the osmotic-pressure of the solution. In a root-hair the concentration of substances dissolved in the cell-sap is usually greater than that in the surrounding soil water. Hence the cell-sap tends to draw water from the soil through the cell-wall and protoplasmic lining of the cell. As water passes into such a cell, the cell-sap increases in volume and exerts an outward pressure on the cell-wall. This is stretched and presses inwards on the cell-sap. As more water passes in and the stretching increases, a point is reached at which the inward pressure of the wall (*wall-pressure*) just prevents the further entrance of water, although the concentration of substances dissolved in the cell-sap is still greater than that in the soil water. The actual pulling force of the root-hair on the water in the soil is called the *suction-pressure*. This is equal to the osmotic-pressure of the cell-sap in the root-hair minus the wall-pressure. A living cell, *e.g.* a root-hair, which is extended and is made relatively rigid in this way, is said to be *turgid*. When the root-hair is fully turgid the suction-pressure is zero. The turgidity of the living cells in young stems and leaves is the most important factor in maintaining the rigidity of these tissues. To return to the root-hairs : in the cell adjoining the root-hair there are also substances in solution in the cell-sap. This cell tends to draw water from the root-hair in the same way that the root-hair does from the soil. As the adjoining cell draws water from the root-hair, the wall pressure of the latter decreases, and so the root-hair can now draw more water from the soil to make up for that lost to the adjoining cell. *In this way the root-hair acts as a channel for the water from the soil to the cell adjoining the root-hair.* This adjoining cell continues to take up water *via* the root-hair until its wall is stretched to the point at which further passage of water is prevented. In this way we can proceed all over the plant. Since the chain of cells between any one cell and the soil merely acts as a way for the passage of water, the chain could be replaced by a column of water in a tracheid or even a wet tracheid wall. It should be realised that there are many channels from the soil water to any one cell, so that other paths remain open even if one is rendered unavailable, *e.g.* if a tracheid becomes empty or its walls dry.

When the plant is losing no water all the living cells have their walls stretched to the point at which no more water can be drawn from the soil. When, however, water is lost in transpiration by a mesophyll cell of a leaf, the wall pressure decreases and so the cell can now draw water from its neighbouring cells. The wall pressure in these is reduced and they can draw water from more remote cells, and so on, until the soil is reached. This goes on continuously and not step by step. It is rather like the start of a train in which the trucks are connected by means of iron bars instead of slack chains. In such a train the last truck would start almost simultaneously with the engine. The water passes up to a transpiring mesophyll cell along many paths. Most of it passes along the path of least resistance, which is probably the tracheids and vessels of the vascular system.

We now turn to the passage of salts into the plant. The

cell-wall and protoplasm of the root-hairs are permeable to some of the salts in solution, *i.e.* to their ions, in the soil water. Different kinds of protoplasm are permeable to ions in varying degree. The concentration of these salts, by diffusion, tends to become the same inside the root-hairs as outside in the soil water. As these substances are used up in the growth of the plant, the meristems draw on the surrounding cells, which eventually draw on the root-hairs, into which more of the substances pass from the soil. Substances in solution may also pass into and be distributed over the plant without going through the cytoplasmic membrane into the vacuoles of the living cells, the path traversed being the wet walls and perhaps the cytoplasmic lining of the living cells and the walls and lumina of dead vessels and tracheids. In those parts of the plant such as tracheids, where the solution can move bodily, the salts move along with the water, and hence transpiration, which causes the water movement, will favour the movement of the salts. When there is practically no water movement (*e.g.* into protected buds) the salts move by diffusion only.[1]

IV. *The Ascent of the Sap.*—The ascent of the sap through the plant is a subject which is by no means thoroughly understood at present. There are two distinct processes which have to do with it: one is *root-pressure*, the other is *transpiration.* The water taken up by the root through its hairs is forced upwards, as can be easily proved by cutting off the stem or branches of a plant, especially in spring, when the ascent of the sap has begun, and before there is any appreciable loss of water from the surface of the leaves. From the cut stumps of the stem or branches great quantities of water, amounting in some cases to several litres, are exuded. This has long been observed in the case of the Vine in spring, when it is said to *bleed.* The same thing can be seen at any time during active vegetation, but in the

[1] Some translocation of salts may take place through the phloëm.

summer the plant is losing water rapidly by its leaves, with the result that the water in the vessels is often under tension. Consequently, when a stem is cut off, air is drawn into the open vessels and some time elapses before exudation of liquid from the stump begins.

This exudation of liquid is due directly or indirectly to the activity of the living cells of the plant. Various suggestions have been made as to the mechanism which is involved. It is thought by some that the living cells of the endodermis in some manner take in water from the cortex of the root and force it out on the side towards the vessels. The cortical cells from which the water is drawn will absorb water from the cells nearer the root surface, and this will continue until the root-hairs draw water from the soil. The water forced into the vessels will result in exudation. Another suggestion is that substances such as sugars are produced by living cells in the wood, and that this sugar passes into the vessels and draws water from the soil as would a sugar solution in an osmometer. In this case a sugary solution would be forced out of the cut surface of a stem.

If root-pressure acted alone, the pressure of liquid in the wood might result in water being forced out of the plant. This actually happens in certain cases ; after a damp, cool night, when little or no transpiration can take place, we find that the leaf margins of many plants are studded with drops of water, which we might very easily mistake for dew. If wiped off, however, the drops soon reappear in the same spots, and, in fact, they come from within, not from without, being excreted by the plant. When no transpiration is possible a plant may lose water in the liquid form. That such loss is not always due to root-pressure is suggested by the fact that in some plants the point where the water is exuded is a gland ; if this gland is poisoned, then excretion ceases.

This state of things, however, is exceptional. In the daytime, under the influence of light and heat, the stomata are active, and through them water is given off in the form of vapour. That this is so is easily proved. If a plant be placed under a bell-glass, the soil in which it is growing being completely covered with tinfoil, so that no evaporation from the earth is possible, we find that dew is soon deposited on the inside of the glass. This dew can only have been produced by the condensation of the aqueous vapour given off by the plant. Under ordinary conditions the air in the bell-glass soon becomes saturated with vapour, and no further transpiration is possible. If, however, we place under the bell-glass some calcium chloride ($CaCl_2$), or other substance which absorbs water vapour, we can obtain a measure of the transpiration of the plant, for the increase in weight of the calcium chloride in a given time is due to the water which it has absorbed and which was previously transpired by the plant. The quantity of water given off by transpiration is very large, amounting, as has been calculated, to from 50 to 100 litres a day in the case of a large tree.

Now, this emission of water vapour goes on mostly through the stomata. In very young leaves, before the cuticle has been fully formed, the surface of the epidermis itself can give off vapour ; but in mature functional leaves the cuticle is nearly impermeable to water, and the stomata are the chief means for its exit. The stomata, therefore, by their opening and closing can regulate the transpiration of the plant. Generally speaking, they open in light and close in darkness, *i.e.* they open when the conditions are favourable for assimilation. Thus transpiration and assimilation go on together, an arrangement which evidently has advantages, for the current set up by transpiration raises within the plant the water and mineral matter without which growth cannot take place.

A few words must be said as to the mechanism of the stomata, though the subject is too difficult to be discussed fully here. As a general rule, the stomata open when the guard-cells are turgid, and close when they are flaccid. This can be proved simply enough in the case of large stomata, such as we find in many Monocotyledons. A piece of epidermis is removed from a living leaf and mounted under the microscope, first of all in distilled water. The guard-cells curve more and more, and the pore between them opens to its full extent (see Fig. 96, *B*) ; next we replace the water by a solution of common salt—a strength of 2 per cent. is sufficient. The curvature of the guard-cells now diminishes, and as they straighten themselves the pore between them closes (see Fig. 96, *A*). Now, how are we to explain this ? In the first experiment water is absorbed into the guard-cells, and so their turgidity is increased. In the second experiment the salt solution withdraws water from the fully turgid guard-cells, and so they tend to collapse. But why do the guard-cells become more curved as their turgidity increases ? This is mainly due to the ridges by which the wall of each cell is strengthened on the side towards the pore. This side of the cell is more rigid than the other, and so offers more resistance to stretching when water is taken up. Hence the more turgid the guard-cells become, the more do their outer walls stretch as compared with their inner walls, and hence the cells become more and more convex towards the exterior, and concave towards the pore. When water is withdrawn from the cells, just the reverse happens ; the outer walls shorten more than the

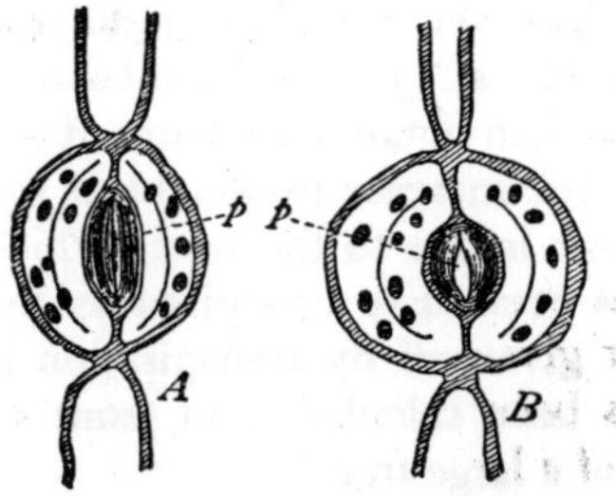

FIG. 96.—Stoma of *Amaryllis*. *p*, pore. *A*, In 2 per cent. salt solution ; the guard-cells are flaccid, and the pore closed. *B*, In distilled water ; the guard-cells are turgid, and the pore wide open. Magnified about 150 diameters. (R. S.)

inner, and so the guard-cells tend to become straight and to close the pore between them. The mechanism of the stomata is in reality much more complex than appears from the above account, but the changes described are among the most important of those which bring about the opening and closing of these organs. We have learnt, then, this much : the stomata open in light, and they open when the guard-cells are turgid. Can we, however, explain why the guard-cells become turgid in light ? Although the exact significance of the chlorophyll-granules in the guard-cells is unknown, it has been shown that, in light, the starch in the guard-cells is transformed into sugar, thereby increasing the osmotic-pressure of the cell-sap. Water is absorbed from the neighbouring cells, so increasing the turgidity of the guard-cells and causing the stomata to open. In darkness, on the other hand, some of the sugar in the guard-cells is converted into starch, resulting in a reversal of the changes previously described and leading finally to the closing of the stomata.

The aqueous vapour transpired by the stomata comes immediately from the intercellular spaces, into which it has been given out by the mesophyll-cells, including those of the spongy parenchyma. As transpiration goes on, the cells of the mesophyll constantly tend to become poorer in water, and they make good their loss by taking up fresh supplies from the water-conducting tracheids of the vascular bundles, by which every part of the leaf is traversed. We see now how appropriate is the comparison which has been drawn between the network of bundles in the leaf and the system of irrigation channels in a water meadow.

The rate at which water vapour passes out through a stoma depends upon the difference between the relative concentrations of water vapour outside and inside the stoma. When this difference is great, transpiration is rapid. Transpiration will tend to be greater with the air relatively dry. The removal of water vapour from outside a stoma by diffusion is a relatively slow process. The removal, however, is quickened by bodily movement of the air, and hence transpiration is increased by wind. Increase of transpiration consequent upon increase of wind velocity and decrease in atmospheric humidity often results in the guard-cells losing their turgidity, with the consequence that the stomata close. The transpiration may thus become even less than before. In some plants the loss of water by transpiration may be so great that the shoots lose their turgidity and wilt. It is not uncommon to see the leaves of a field of turnips wilted during a hot day. Provided that wilting is not too prolonged and the water-loss is not excessive, the leaves again become turgid as transpiration falls at night.

The more active the transpiration, the less will the supply of water by root-pressure be sufficient. The result will be that the water in the tracheæ [1] will be under tension, instead of being under pressure. Some investigators claim that the tension is so great that the continuity of the water columns is broken in some places. That the supply of water by root-pressure alone is often unable to make good the loss due to transpiration from the leaves is shown by the following experiment. If an actively transpiring plant be cut off just above the ground and placed with its cut end in water, the amount which its leaves give off as vapour may be more than ten times as much as that which is pumped up from the stump by root-pressure during the same time.

That the water ascends the stem through the wood can be well shown in any dicotyledonous tree in which the wood forms a continuous ring. Indeed, the experiment might be quite well carried out on our dicotyledonous type, the Wall-flower, but a larger woody plant is more convenient for the purpose. A ring of tissue is removed from the stem, the part stripped off reaching inwards as far as the cambium, so that

[1] A term used to include both vessels and tracheids.

the whole of the cortex and phloëm are taken away. The wound is covered up with damp cotton-wool, or protected in some other way against evaporation. No withering of the leaves takes place, they remain indefinitely fresh and turgid as before ; and if we measure the transpiration, the amount of water vapour given off is found to be undiminished. Hence it is evident that the current of water goes up just as well when the phloëm and cortex are interrupted as it does when they are complete. Only the wood and pith remain for it to pass through. It makes no difference to the result of our experiment if, as is usually the case, the pith is completely dried up and quite incapable of conducting water. Hence we can only conclude that the whole ascending current of sap passes through the wood.

In many trees, as for example the Beech, the whole thickness of wood continues to conduct water, even in trees more than a hundred years old. In others, as in the Oak, it is only the outer layers, forming the *sap-wood* or *alburnum*, by which water is conducted. In the latter case, if we repeat our ringing experiment, but make a deeper incision than before, so as to remove the outer, more lightly-coloured wood, as well as the phloëm, we get quite a different result. The leaves wither and dry up, and after a time all transpiration ceases ; the water supply has evidently been cut off from the part of the tree above the incision.

Experiments in which plants are made to absorb coloured fluids have proved that it is through the vessels and tracheids of the wood that the water ascends ; the fibres take no part in the process. Whether the living cells of the woody parenchyma and of the medullary rays do or do not help in the conduction of water is still an open question.

There is still considerable dispute as to the causes of the ascent of water in the wood. The activity of living cells of the root may have something to do with the forcing of water into the vessels, or the living cells may produce substances which pass into the vessels and draw in water osmotically. The constant loss of water vapour from the transpiring leaves is considered by some to be the most important factor in the ascent of water through the wood. This loss will cause a tension in the water in the vessels, and this will draw water through the wood, provided the continuity of the water is not completely broken under the strain. Both the above processes may be at work at the same time.

Certain physical forces can be eliminated as affording an explanation of the rise of the transpiration current. Atmospheric pressure will only maintain water to a height of about 33 feet. We know that some trees are about 300 feet high, and yet their topmost branches suffer no lack of water. Capillary action, by which water rises in narrow tubes, will not account for maintaining sap in the vessels and tracheids, for these natural tubes are not nearly narrow enough to maintain water to anything like the height required. Thus, in a tube $\frac{1}{500}$ inch in diameter, which would be rather small for a wood-vessel, water will only rise by capillarity to a height of about two feet.

A very tempting explanation, which many botanists have adopted, is to suppose that the living cells of the wood take part in the work, absorbing water from the vessels, and pumping it out again under pressure. As in certain cases, however, experiments appear to show that the sap will continue to rise in stems more than 33 feet high, which have been completely killed, this explanation is open to much doubt.

It has, however, been shown recently that a column of water will bear a considerable tensile stress without rupture. Many botanists now think that, in conjunction with the forces available in the root and the pull exercised by transpiration from the leaves, the tensile strength of the water columns in the tracheæ is sufficient to account for the ascent of sap to the tops of the tallest trees.

V. *Respiration.*—We will now leave the subject of nutrition, and say something about a chemical process of quite another kind, in which the plant instead of gaining substance loses it, but in doing so sets free energy some of which is used. This process is the *breathing* or *respiration* of plants. So long as plants are in active life, the great majority of them, like animals, are continually taking up oxygen from the air, and giving off carbon dioxide. The latter, as we know, is a product of the combustion—that is, the oxidation, of materials containing carbon. The oxygen taken up is used for the controlled oxidation of certain organic compounds. The carbon dioxide produced is given off again into the air. The breathing of plants is, in fact, just the same thing as our own breathing, or that of other animals. We must be very careful not to confuse respiration with carbon-assimilation. In both processes an exchange of gases between the plant and the atmosphere goes on. In older tissues oxygen obtains access to the inner living cells by way of the lenticels, and it must be remembered also that small amounts of oxygen are dissolved in the water absorbed by the roots. In assimilation, however, the plant gains carbon at the expense of the carbon dioxide of the air, while in respiration it loses carbon at the expense of its own carbon compounds. The former is a constructive process, by which more complex bodies are built up out of simpler ones; the latter is a destructive process, by which complex substances are broken down into simpler ones. In assimilation reduction takes place, while respiration is a process of combustion or oxidation.

In the case of green organs exposed to light, the two processes go on together. In light of ordinary intensity assimilation is much more vigorous than respiration, so that carbon dioxide is decomposed much more quickly than it is formed. In darkness, however, assimilation ceases, and then it is easy to prove that carbon dioxide is given off. The same is the case, whether in light or darkness, with any organs which are not green, such as flower-buds or seeds just beginning to germinate. In order to prove that carbon dioxide is evolved, a number of seeds are allowed to germinate under a bell-glass, which is fitted with tubes through which a current of air can be drawn. The air which enters the apparatus is deprived of all its carbon dioxide by passing through a solution of caustic potash. The same air, after it has passed over the seeds, is drawn through baryta water, in which it throws down a precipitate, owing to the formation of barium carbonate ($BaCO_3$). By drying and weighing the precipitate, we can tell how much carbon dioxide has been formed by the germinating seeds in a given time.

As a rule, the volume of carbon dioxide given off in respiration is about equal to the volume of oxygen taken in. Organs can, however, continue for some time to respire—that is, break down organic compounds with the production of carbon dioxide, even in the complete absence of oxygen. In this process, which is called *anærobic* as distinguished from *ærobic* or normal respiration, the organic compound respired is broken down into two parts, one fully oxidised, carbon dioxide, the other partly deoxidised or reduced, alcohol or some similar compound. Since these reduced compounds are often injurious, anærobic respiration can only continue for a limited time in most plants.

All living parts of plants respire. Even when the protoplasm is dormant, as in a dry seed, respiration continues, although it may be extremely slow. Respiration is, indeed, one of the fundamental properties of protoplasm. The plant, in fact, like an animal, must respire in order to live, but as in most cases the plant does not have to execute such active movements as an animal, the actual amount of oxygen which it has to consume is so much less.

Respiration being a form of combustion, heat must of necessity be set free. As a rule, however, no consequent rise of temperature is observed in the case of plants; for the plant in most cases loses heat as rapidly as it gains it. An ordinary leafy plant has an enormous surface in comparison with its bulk, and is thus exposed to very rapid cooling, both by radiation and conduction. The vaporisation of water during transpiration also uses up more heat than respiration can supply. It is only when actively breathing organs are crowded closely together in a confined space that a rise of temperature can be detected. If the bulb of a thermometer be thrust into the midst of germinating seeds, a temperature about 1·5° C. higher than that of the surrounding air may be observed, and in some plants with very crowded flower-buds, enclosed in a sheathing bract, a much greater rise of temperature takes place in consequence of respiration (as in *Arum*).

VI. *Growth.*—By growth, increase in size is usually meant. This generally involves increase in weight, although a seedling growing in darkness may actually lose weight notwithstanding the increase of the young plant in length. Normal plant growth, however, involves an increase of both size and weight.

In plant as in animal growth the protoplasm increases in amount, and one of the most remarkable properties of life is the enormous increase of protoplasm during the development of the seed into the adult plant. An oak tree, for instance, contains many thousand times as much protoplasm as the acorn from which it sprang. All this growth in the protoplasm arises from the intake into the protoplasm of the acorn of simple food substances from the soil and air, and the elaboration of these into additional protoplasm.

It has been explained in earlier chapters of this book that growth in flowering plants takes place by means of *meristems* or groups of actively dividing cells. These meristems are confined to the apical regions in most monocotyledonous stems and roots, but in Dicotyledons there are additional meristems, viz. the cambium and the phellogen (or cork cambium), both in stem and root. Any cell which retains its protoplasm is potentially capable of division, but resumption of meristematic activity in highly vacuolated cells is exceptional apart from the development of the interfascicular cambium and the phellogen in Dicotyledons.

Growth in length of the stems and roots of flowering plants involves two distinct processes. In the first place the apical meristems give rise to new cells. These also divide for a time, but most of them, even in Dicotyledons, then cease to divide. At this period vacuoles appear in the cytoplasm, and many of the cells become greatly distended, especially in a longitudinal direction. This elongation of cells no longer dividing is responsible for a large part of the increase in length of stems and roots. In the process of extension each cell at first elongates slowly, then more rapidly, and finally slowly again until it reaches its maximum length. As might be expected from the position of the elongating cells, the region of maximum elongation in stems and roots is not exactly at the apex of the stem or root, but is some distance behind it. If, for instance, the radicle of a germinating Broad Bean, when about an inch long, is marked from the apex backwards with rings of Indian ink at intervals of $\frac{1}{10}$ of an inch, and is then allowed to grow vertically downwards in a moist atmosphere, it will be found after a short time (twenty-four hours) that the rings of Indian ink are now unequally spaced. The apical zone will be slightly longer than $\frac{1}{10}$ of an inch, the second zone (from the apex) will have elongated most, the third zone will have elongated much less than the second zone, the fourth zone will have elongated only very slightly, and the fifth and subsequent zones will have elongated not at all.

Like all other physiological processes the rate of growth is dependent upon such external conditions as temperature and light, and upon certain internal conditions such as the age and degree of turgor of the cells. Other things being the same, a plant grows more rapidly the higher the temperature up to a certain point, beyond which the rate of growth decreases. In most plant stems the rate of elongation is greater in darkness than in light. The rate of elongation of a developing stem, *e.g.* that of a sunflower, is not uniform throughout the whole period of development : at first the rate of elongation is small, then it increases to a maximum, after which it becomes slower again, and finally ceases as the plant reaches its full height. In this respect the elongation of the whole shoot is comparable with that of the individual cell.

Most plants growing in temperate countries like Britain show marked periodicity of growth according to the season. Generally speaking, growth is most active in the spring and early summer, it slows down in autumn, and it is often completely in abeyance during winter. With the rise of temperature in spring the solid food reserves stored in the tissues become converted into soluble substances, sugar and the like, which pass to the growing-points of the stems and roots and also to the cambium when present. The cells of these meristematic regions then begin to divide and form new cells. In an Oak tree, for instance, most of the buds begin to expand into new shoots. Much food material is required for this comparatively sudden onset of growth in the spring. This is conveyed to the growing regions partly by the phloëm and partly by the xylem. Although the xylem generally transports chiefly water and mineral substances in solution, in the early spring it may, especially in trees, conduct upwards considerable quantities of sugar in solution, derived from the starch or fats stored in the wood during the winter. In the Sugar Maple of Canada, a large tree, the sap is commercially extracted from the wood in early spring to obtain sugar.

Some plants, however, which grow in temperate countries,

are dormant during the summer instead of winter. Tulip and Daffodil bulbs and Crocus corms, which are dormant during the summer, put out new roots in autumn, and before the winter is over green leaves appear above the ground.

VII. *Movement of Plant Organs.*—It is commonly supposed that one of the chief differences between plants and animals is that the former remain fixed in one position while the latter move freely. This belief is not strictly true, for as will be seen in Part II. of this book (*Flowerless Plants*), some of the lower plants move as freely as the simplest types of animals. Even in the flowering plants many of the plant organs move in response to changes in their surroundings although the plants as a whole remain fixed.

Let us consider in the first instance the behaviour of the young growing shoot and root of a flowering plant such as the Broad Bean (*Vicia Faba*). Normally the young root grows vertically downwards and the young shoot vertically upwards. This is true also if the plant is grown in complete darkness. If, however, still keeping the plant in darkness, it is turned on its side so as to make the root tip and the shoot tip horizontal, it will be found that movements take place in shoot and root which in a short time result in the root tip again pointing vertically downwards and the shoot tip vertically upwards.

If, for example, a seedling bean with a radicle about an inch long is pinned to a cork with the radicle horizontal, and kept under moist conditions so that growth may be continued, the portion of the young root just behind the tip will begin to show a downward curve within a few hours. In forty-eight hours or less the extremity of the root will be pointing vertically downwards while the rest will still be horizontal. The region in which the curvature takes place is a short zone just behind the root tip, and if, before putting the radicle horizontal, it is marked in the manner described on p. 206, it will be seen that the region of curvature coincides with the region of most rapid growth in length. The movement of the tip from the horizontal to the downwardly pointing vertical position is, in fact, brought about by growth, the cells on the lower side of the elongating region growing more slowly than those on the upper side. When, however, the root tip points vertically downwards again this difference in the rate of growth on the two sides of the root disappears and growth continues in a straight line.

Growth movements of this kind, whereby the organ, when displaced from its normal position with reference to some external factor, is brought back to its normal orientation, are called *tropisms*. The external factor in the present case is the gravitational attraction of the earth, acting in the direction of the earth's centre of gravity. Since the response of the radicle is one which brings the root tip to point and grow towards this centre of gravity, the radicle is said to be *positively geotropic*.

That the response of the radicle is a response to the changed direction of action of gravity on the plant organ can be shown by the following experiment. A germinating bean is fastened on to a vertical wheel so that the radicle lies at right angles *across* a radius. The vertical wheel is now kept spinning rapidly. This has two effects. In the first place it neutralises the effect of gravity on the radicle, since the radicle now passes rapidly through all possible orientations with respect to gravity. In the second place it brings into play an exactly similar mechanical force (centrifugal force) which acts, however, not downwards towards the centre of the earth, but outwards from the centre of the wheel along all radii. It will be found that in a few hours the radicle has curved so that the tip is pointing and growing outwards along a radius.

The part of the root that perceives the stimulus of gravity, *i.e.* the part that is sensitive to a change in the direction of action of gravity, is not the same as that which responds. If the root tip of a similar bean seedling is carefully cut off with a sharp razor, then, although the radicle continues for a time to grow in length, it will no longer, when laid horizontally, respond by curving downwards. From these and other experiments it is known that the root tip is the region of *perception* of the altered position.

Response is brought about by means of a special growth-regulating substance called *auxin*, which moves back from the tip to the elongating region. When the tip points vertically downwards, auxin moves from it uniformly on all sides to the growing region so that growth continues in a straight line. When the root tip is horizontal, more auxin moves out from its lower half and less from its upper half, so that the cells on the lower side of the elongating region receive more auxin than those on the upper side; and since, in the case of the root, the effect of auxin is to slow down the rate of elongation, the lower side of the radicle now extends less rapidly than the upper, and a downward curvature results.

We must now consider the behaviour of the young shoot of the bean. When placed horizontal this responds by curving so as to point and grow vertically upwards, *i.e.* away from the earth's centre of gravity. It is said therefore to be *negatively geotropic*. (Placed across a radius on a vertically revolving wheel the plumule curves so as to point towards the centre of the wheel, whereas the radicle curves so as to point outwards along a radius.)

As with the root, the perception of the altered position lies in the apex of the organ (here the plumular bud), and the response in the elongating region. Like the root tip, the stem apex is continually producing auxin and gravity has in both cases exactly the same effect on the distribution of the auxin, *i.e.* more moves out and back from the apex along the lower side and less along the upper side.

But, whereas the elongation of the root cells is *decreased* by auxin, the elongation of cells in the stem is *increased* by auxin. So the cells on the lower side of the growing stem, receiving more auxin from the horizontally placed apex, grow more rapidly and so produce the upward curvature (negative geotropic response) of the stem. As the apex is brought nearer and nearer to the vertical the flow of auxin down the two sides becomes more and more alike, until when the apex points vertically upwards the flow is uniform and growth continues in a straight line.

A word should now be said about the roots and shoots of older plants. Primary roots produced by the continued growth of the seedling radicle are almost always positively geotropic, as also are many adventitious roots. Secondary roots, which arise as branches from the primary, usually take up a position directed obliquely downwards at a more or less constant angle. They will curve back to this angle if tilted. They are said to be *plagiotropic* with respect to gravity. If, however, the primary root is injured one of the secondary roots becomes positively geotropic and grows vertically downwards.

Most of the primary shoots of plants and also the main aerial shoots from rhizomes are negatively geotropic, and if placed horizontal while they are still elongating will curve upwards again. Lateral branches usually grow out from the main stem at an angle which is characteristic of the particular plant. Like secondary roots they are plagiotropic but differ in that they are usually directed obliquely upwards instead of obliquely downwards. If the tip of the main stem is destroyed one or more of the lateral branches becomes negatively geotropic, so replacing the lost " leader."

In what has been said so far about experiments on plant shoots it has been assumed that the plants were kept during the experiments either in complete darkness or in light that was uniform on all sides of the shoot. This is very important, since growing stems respond very readily to differences in the light intensity on different sides of their apices. The young stem responds by curving towards the side having the higher light intensity and is said therefore to be *positively phototropic*. Everyone must have seen this effect on the growing stems of plants grown in a room near a window. In the open field or garden the light intensity is usually equal on all sides of the shoot so that main shoots grow vertically upwards. As with the geotropism of shoots the apex is the region of perception and the elongating region is the region of response. Auxin, too, plays the same part. When the apex is erect and equally lit on all sides auxin moves down from the apex in equal amounts on all sides and growth is continued in a straight line. But when one side of the apex is lit more strongly than another, then less auxin passes down that side (the " light " side) and more down the other side (the " shade " side). Consequently in the elongating region the " shade " side grows faster and a curvature towards the " light " side results.[1]

Leaves are usually *diaphototropic*. Their phototropic response brings them into a position in which they are transverse to the principal direction of the light, with their upper surfaces facing the brightest light. This usually takes place by a twisting of the petiole and may be well seen in privet shoots. In these shoots the leaves are borne in four rows at right-angles ; but whereas in shoots taken from the top of the bush the leaves will be found standing out equally on all sides, with their surfaces facing upwards, in shoots from the side of the bush lower down the petioles have twisted so as to make all the leaves face outwards. There appear at first sight to be only two rows of leaves on these shoots.

[1] Auxins play a part in a number of other interesting and important growth responses, including the outgrowth of lateral buds when the terminal bud is removed, and the development of new roots on cuttings

Roots which normally grow in the soil do not, as a rule, show any phototropic response. (The primary root of the mustard seedling which is markedly negatively phototropic is an exception to this general rule.)

Roots are, however, very readily influenced in their direction of growth by differences in humidity on the two sides. They respond by growing towards the region of greater humidity, *i.e.* they are *positively hydrotropic*. The actual direction of growth of any one root will be a compromise between the geotropic and the hydrotropic response, and, if the humidity difference is made great enough, considerable deviation from normal downward growth may result. If, for example, barley seeds are sown, in moist sawdust, on a narrow-meshed sieve, which has been hung obliquely in the air, the roots will at first grow vertically downward through the meshes. But, as the root tip grows down away from the oblique sieve, it comes into a region of air in which there is a very marked gradient of humidity. Since the humidity of the air on the side towards the sieve is greater than that on the side away from the sieve, the root tip begins to curve towards the sieve and the root finally grows obliquely down, along, or just inside the lower surface of the sieve.

The growth movements of stems, leaves, and roots, to which we have referred, are found in all higher plants and, as we have seen, account for the actual position these organs take up during normal growth. In climbing plants other growth movements play a part. In tendril climbers, for instance, the tip of the tendrils (usually a modified leaf) responds to contact with a solid object such as a stick or stem of another plant by growing faster on the opposite side. In consequence, the tip curls firmly round the support. This is known as *haptotropism*. In twining plants such as the hop there is no response to contact, but the shoot tip, as it grows, traces a spiral path in the air and so coils round any support which it may encounter as it spirals upwards.

In addition to growth movements (tropisms) whose direction is determined by the direction of action of some external factor, there are movements which are indeed brought about by changes in the environment but which are not related to the direction, if any, in which the external factor is operating. Many flowers, *e.g.* Tulip and Crocus, open by day, especially in warm bright weather, but close at night, or in dull cool weather. This response is largely due to change of temperature.

In some plants, *e.g.* clover, the leaflets are expanded and horizontal during the day but bend downwards at night, the diurnal movement being due to expansion or shrinkage of small swellings (pulvini) at the base of the leaflets. In one plant belonging to the same family as clover, namely, *Mimosa pudica*, "the Sensitive Plant," the leaf is also extremely sensitive to contact. When it is touched at the tip the leaflets rapidly droop in succession from the tip, and if the stimulus is great enough, the whole leaf also droops, through the response of the main pulvinus at the base of the petiole.

Rapid responses of this kind are rare in plants. The slower growth responses to gravity, light, moisture, and temperature are, however, almost universal in the higher plants and, as we have seen, they are responsible for the characteristic positions which the organs take up as they grow and develop. In general, the responses are advantageous to the plant. Under natural conditions, the negative geotropic response of the seedling shoot brings it into the air and, further, its positive phototropic response brings it from shade into brighter light. Again, the leaves on the stem normally respond so as to face the strongest light. The tropistic behaviour of the stem and leaves is thus clearly related to the assimilatory function of the foliage, and similarly the movements of the corolla of many flowers are related to pollination, especially by insects. The positive geotropism of the main root brings it down into the soil, the oblique geotropism of the laterals results in a root system permeating an increasing volume of soil as the plant above ground grows larger, while the positive hydrotropism of both main and lateral roots tends to bring the whole root system into the region of greater soil moisture. The relations between this behaviour and the root function of absorbing water and solutes is obvious.

It is, however, important to realise that a tropism is not in itself *necessarily* advantageous. Conditions can be arranged in which a response, favourable in ordinary conditions, becomes unfavourable. If, for instance, a small seed is sown at the bottom of a deep flower pot, the plumule will begin to grow upwards through the soil and may use up all its stored food without reaching the surface, whereas if it had grown down, it might quickly have reached the air and light through the hole in the bottom of the pot. In general, we may say that the responses which plants make to most commonly occurring environmental conditions are advantageous—that is, they favour the survival of the plants in their natural environment.

CHAPTER IV

TYPE III

THE SPRUCE FIR (*Picea excelsa*, Link)

I. *EXTERNAL CHARACTERS*

THE Spruce, a type which will be well known to all readers, is a coniferous forest tree, sometimes reaching as much as 150 feet in height. It is not a native of Britain, though frequently planted. Small specimens of this Fir are those generally used as our Christmas trees. In Northern Europe and in the mountainous districts of Central Europe the Spruce forms large woods, and extends almost to the extreme limits within which trees can grow. In the North it reaches Lapland and Arctic Russia, while on the Alps it is found up to a level of more than 6000 feet. There are many kinds of Spruce Fir. The species we shall describe, *Picea excelsa*, is often called the Norway Spruce.

The tree has a pyramidal form ; its main stem is straight and vertical. The principal branches grow out from it almost horizontally or slope slightly downwards ; the smaller twigs are nearly upright.

The leaves are spirally arranged and crowded together. They are needle-shaped, and are bluntly four-cornered in transverse section. Each leaf is traversed from end to end by a single vein (see Figs. 97 and 102). In the closely related Pines the leaves are borne in twos, threes, or fives on short shoots.

The branches arise in the axils of the leaves, but only a very small proportion of the axillary buds are ever developed.

The root-system consists entirely of the tap-root and its numerous ramifications. Both stem and root, like those of Dicotyledons, grow in thickness from year to year. Hence the main stem and its branches have a somewhat conical form, for they are thickest at the base and taper gradually towards the apex. The older parts of the stem are covered by a scaly bark, which peels off in thin flakes.

The tree is evergreen—that is, it does not shed all its leaves in any one season ; the same leaves remain upon it for about three years, and new leafy shoots appear each spring, so that the tree is never bare. Most of the near allies of the Spruce are evergreen, but the Larch is deciduous.

The young shoots, which will expand in the following year, pass the winter in the form of buds, which are enclosed within the *bud-scales* ; these are modified leaves of a light brown colour, and thin membranous texture.

As with all other types of trees the Spruce does not bear reproductive organs until it is several years old. The Spruce, like most, though not all of the cone-bearing trees, is *monœcious*,[1] *i.e.* the cones are of distinct sexes but are both borne on the same plant. They are extremely different from the flowers which we have considered hitherto, and indeed bear little resemblance to the flowers of any of the Angiosperms. One peculiarity which is common to the cones of both sexes, is that the floral axis is elongated and bears spirally-arranged leaves, so far resembling an ordinary vegetative branch, whereas in typical angiospermous flowers the axis is shortened, and the floral leaves are therefore brought close together at nearly the same level.

The male cones of the Spruce are borne in spring in the axils of the leaves of a shoot formed in the previous year (see Figs. 97, *B*, and 111, *A*). The cone or *strobilus* begins with a short stalk, on which are borne some bright green bracteoles.

[1] A flower, like that of the Wallflower or Lily, which contains both stamens and pistil, is called *hermaphrodite.*

There is no perianth ; above the bracteoles the axis only bears the stamens, which are numerous and arranged in a spiral. The leaf-like stamens are of a bright red colour ; in each of them the free end of the staminal leaf is bent upwards at an angle with the horizontal stalk. On its lower surface the stamen bears two pollen-sacs, lying side by side, parallel to the stalk (see Fig. 111). The details of the pollen-sacs and pollen will be described later.

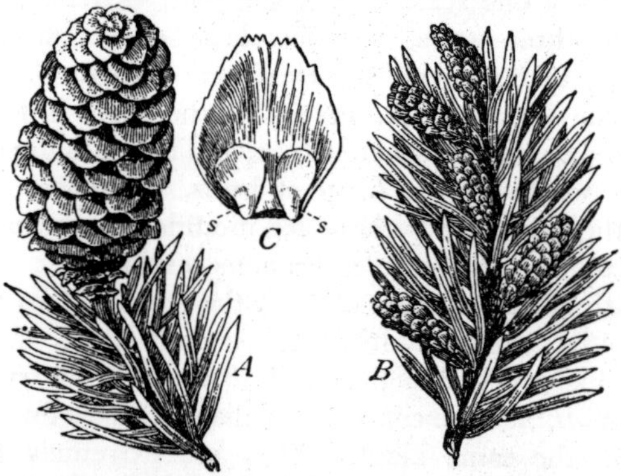

FIG. 97.—*A*, Twig of *Picea excelsa*, bearing a young female cone. *B*, Twig bearing several male cones. *C*, Ovuliferous scale from the female cone of *A*, showing two winged ovules on the upper surface. *A* and *B* slightly reduced, *C* enlarged. (After Beissner.)

The female strobili or cones grow at the ends of last year's twigs (see Fig. 97, *A*). At the end of May, when the cones open, they are from 1½ to 2 inches in length and of a beautiful red colour. At this time they are erect ; later on they become pendulous. The cone consists of a central axis, bearing crowded spirally-arranged ovuliferous scales. Each ovuliferous scale is of an obovate form (see Fig. 97, *C*), often with a blunt point at the end. These scales are not themselves the carpellary leaves, though they have all the appearance of foliar organs. More careful examination, however, shows that each ovuliferous scale stands immediately above a small narrow pointed leaf, which is called the carpellary scale. Except in the youngest stages these little leaves are very inconspicuous compared with the ovuliferous scales, but they are developed before them, and, in fact, each ovuliferous scale arises at first as an outgrowth from the upper surface of the carpellary scale and close to its base. The ovuliferous scales soon outgrow the carpellary scales which produce them, and by the time the flower opens they alone can be seen from outside, the carpellary scales being quite buried between them.

Each ovuliferous scale bears on its upper surface and near its base two ovules, with their micropyles directed inwards towards the axis of the cone. Each ovule is attached to a flat wing, much larger than itself, which is formed from the upper surface of the scale (see Fig. 97, *C*).

When the female cone matures, after its ovules have been fertilised, it undergoes but little change compared with an angiospermous flower when it becomes a fruit. The whole cone, however, grows much larger, attaining a length of from four to six inches. The scales first turn green and then brown, growing hard and woody, and the ovules, as they ripen into seeds, acquire a hard testa. Their wings remain attached to them, and become dry and membranous. The scales of the cone close firmly together, shutting in the seeds. We thus see that all the parts which constituted the young cone are still present at maturity, which is not often the case with the angiospermous flower and fruit.

The seeds are ripened in the autumn after flowering, but it is only in the following spring that the scales of the cone open out and the seeds fall. The wing to which each seed is attached offers a broad surface to the air, and so enables the seeds to be scattered more widely by the wind.

The most striking point in the whole history of the

female cone, as compared with the flower of an Angiosperm, is that there is no closed ovary at the time of pollination. When this process takes place, the scales are separated from one another sufficiently to leave an open passage down to the ovules, and it is upon the micropyle of the ovule itself that the pollen falls. There is thus no need for a stigma and a style. On this character, which is common to the whole Class of cone-bearing trees and their allies, the name *Gymnosperms*, or plants with naked seeds, is based. The group of the Gymnosperms to which the Spruce Fir belongs is the *Coniferæ* or Conifers, which also include the Pines, Larches, Cypresses, Junipers, and the Yew.

It is only after pollination that the scales close up so as to shelter the developing seeds, opening again when the latter are ripe, and so allowing them to escape. The gymnospermous structure is an obvious difference from other Flowering Plants, and will prepare us for the still greater differences which we shall find when we come to consider the development in detail.

Having now given a brief sketch of the chief points in the external morphology of our type, we will go on to examine its more minute structure.

II. *ANATOMY OF THE VEGETATIVE ORGANS*

a. Stem

The structure of the young stem in its main features is closely similar to that of a normal Dicotyledon (see Fig. 98).

The central cylinder or stele contains a single ring of vascular bundles enclosing a pith. The bundles themselves, just as in Dicotyledons, are collateral, with their xylem directed towards the centre, and their phloëm towards the exterior of the cylinder. These bundles pass out one by one into the leaves, each of which receives a single bundle only (see Fig. 98, *l*). The bundles in the cylinder are separated from one another by the primary rays, which are only two or three cells in thickness.

The primary bundles are of simple structure. The xylem consists entirely of tracheids and parenchyma associated with resin canals ; no vessels are developed at any time in the wood, whether primary or secondary, of the Fir or any other cone-bearing trees. The protoxylem, which lies at the extreme inner side of the bundle, has the usual spiral or annular tracheids ; the phloëm consists of sieve-tubes and parenchyma. There are no companion-cells, but certain cells of the parenchyma bordering on the sieve-tubes are richer in protein materials than the rest, and may discharge the function of companion-cells.

Between xylem and phloëm we find a layer of actively dividing cambium, just as in the vascular bundle of a Dicotyledon. Later on, but at an earlier stage than in the herbaceous stem of the Wallflower, divisions take place in the medullary rays, so as to form an interfascicular cambium which completes the ring. Fig. 98 shows a transverse section of a twig of the second year, when a good deal of secondary tissue has already been formed. The cambium acts precisely in the same way as in a Dicotyledon, and needs no special description.

The cortical tissue is chiefly remarkable because it contains a ring of *resin canals* (*r* in Fig. 98). The resin canal is an intercellular space, surrounded by a circle of thin-walled cells rich in protoplasm ; these cells secrete the resin, which is poured into the canal. The canal itself arises by the splitting apart of a group of cells so as to leave a space between them. The canals extend for long distances through the stem. The secreting cells lining the cavity are called the *epithelium* of the canal. The epithelium is surrounded by another ring of cells, which have rather thicker walls than those of the cortex generally. The resin may be of use to the plant by rendering it distasteful to animals, which might otherwise attack the young twigs ; it probably has other functions also. Resin-canals are also present in the wood and medullary rays. They are by no means peculiar to Gymnosperms, occurring, for example, in the Ivy.

The outline of the young stem, as shown in Fig. 98 in transverse section, is rendered irregular by the projecting

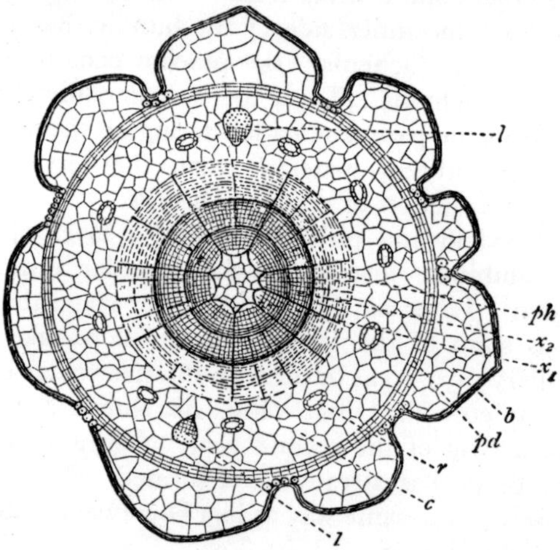

FIG. 98.—Semi-diagrammatic transverse section of a two-year-old stem of *Picea excelsa*. *b*, base of leaf ; *pd*, periderm ; *c*, cortex ; *r*, resin canal ; *ph*, phloëm ; x_2, second year's wood ; x_1, first year's wood ; *l*, leaf-trace bundle. Magnified about 40. (R. S.)

bases of the leaves. The older branches, however, are cylindrical. This change is due to the formation of *periderm*, which arises, as in the Wallflower, from a *phellogen* or cork cambium, but in the Fir the phellogen is formed by the division of a ring of cortical cells, not from the pericycle. Opposite the depressions between the leaf-bases the phellogen arises quite near the surface, but opposite the leaf-bases themselves it is deeply seated, and thus the whole layer is nearly circular, as seen in transverse section (see *pd* in Fig. 98). Cork is formed by the phellogen towards the exterior, and everything outside the layer of cork dries up, and is cast off. Thus the irregularities of the surface are removed, and the new surface formed by the cork is smooth and cylindrical. Later on, however, new layers of phellogen arise, by the division of deeper cortical layers, and so a scaly bark is formed, such as we find on the older stem or branches. The bark, as we know, consists of the cork, together with all the dried-up tissue outside it.

We will now go on at once to describe the changes which take place in the twig when it becomes a thick branch or trunk. After what we have already learned of the development of the Wallflower, it is quite easy to understand this process, though the change from a little twig, $\frac{1}{16}$ of an inch thick, to the stem of a great Fir-tree a foot or more in diameter, is certainly a remarkable one.

We have seen that there is always a layer of cambium between the xylem and phloëm of each bundle, and that by the divisions of the interfascicular tissue the cambium becomes rapidly united into a complete ring. When the cambial cells divide, they cut off cells on the inner side, which after one or more further divisions become elements of the xylem. In like manner the cells cut off from the cambium on its outer side, after subdividing, become part of the phloëm. Hence, just as in the Wallflower, the oldest part of the xylem is always that nearest the pith, while the oldest part of the phloëm is that nearest the cortex. The most recently formed part of each is found near the middle of the secondary zone next the cambium.

The secondary xylem or wood is of simple structure compared with that of most Dicotyledons. It consists entirely of tracheids and medullary rays ; there are no vessels, and there is no woody parenchyma, but the xylem is traversed by resin canals, which have their secretory epithelium. The tracheids are long, pointed elements, reaching a length of from two to four millimetres ; they are as a rule pitted on their radial walls only.

The pits are of the *bordered* kind, and afford the best examples of this structure. On each radial wall of the tracheid there is a single row of large bordered pits. Their construction is shown in Fig. 100. Seen in surface view,

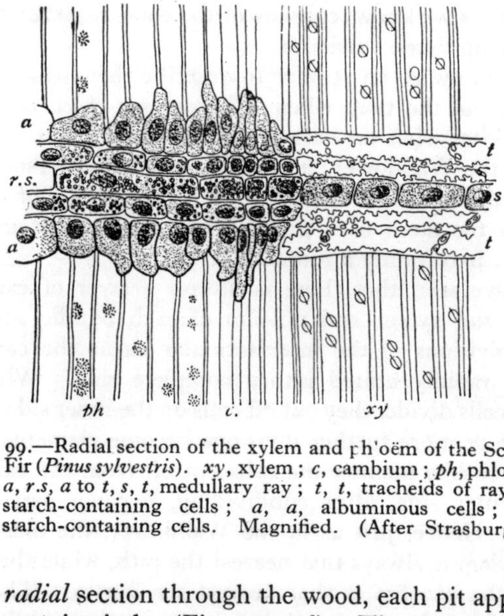

FIG. 99.—Radial section of the xylem and phloëm of the Scotch Fir (*Pinus sylvestris*). *xy*, xylem ; *c*, cambium ; *ph*, phloëm ; *a*, *r.s*, *a* to *t*, *s*, *t*, medullary ray ; *t*, *t*, tracheids of ray ; *s*, starch-containing cells ; *a*, *a*, albuminous cells ; *r.s*, starch-containing cells. Magnified. (After Strasburger.)

i.e. in a *radial* section through the wood, each pit appears as two concentric circles (Fig. 100, *A*). The meaning of the two circles only becomes clear when we examine a pit in sectional view, such as we see in either a tangential (*B*) or a transverse (*C*) section of the wood. We now find that the pit, like all other pits, is essentially a thin place in the cell-wall. The thin part of the wall or *closing membrane* is over-arched on both sides by a circular rim, which projects over it so as to leave only a narrow opening in the middle of the rim. The closing membrane itself is not of equal thickness all over, but has a thicker part in the middle, called a *torus* (*t* in Fig. 100, *B* and *C*). We now see that in the surface view (*A*) the outer circle corresponds to the external edge of the projecting border, while the little circle in the middle is the opening in the border leading to the closing membrane. The structure of all bordered pits is similar, but they are most easily studied in plants of the Fir kind on account of their large size.

We must remember that the tracheids are closed cells, so that no communication is possible between them except through their pits. The whole of the water which goes up a Fir-tree has to pass through the bordered pits thousands of times on its journey from the roots to the leaves. The thin membrane is protected from rupture by the torus.

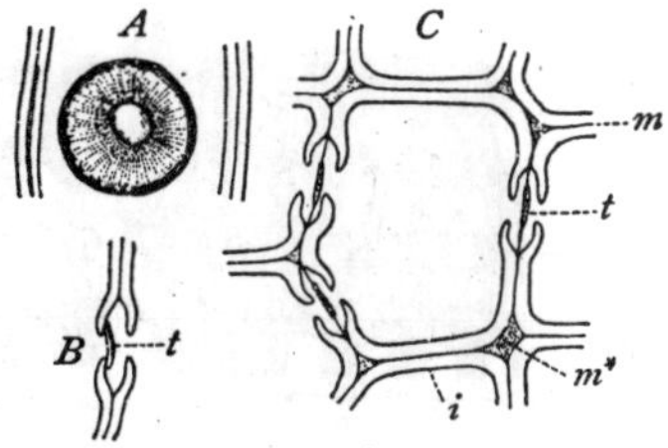

FIG. 100.—Bordered pits of a Fir. *A*, Part of tracheid with pit in radial view. *B*, Pit in tangential section: *t*, torus. *C*, Transverse section of tracheid, showing pits: *t*, torus; *m*, middle lamella; *i*, inner layer of cell-wall. Magnified about 400. (After Strasburger.)

When the pressure is much higher in one tracheid than in its neighbour, the closing membrane is bulged out, becoming convex on the side towards the lower pressure. Thus the torus is pressed closely against the opening of the pit border on that side (Fig. 100, *B*). The thin part of the membrane is thus supported by the border, and only the thickened torus is exposed to the full pressure.

The border has another important function; it serves to keep the closing membrane moist. In times when transpiration is very active, and the water-supply deficient, as in hot, dry summers, the tracheids become very nearly emptied of water. Within the minute cavities of the bordered pits, however, the water is held fast by capillary action, and so the closing membranes are prevented from drying up. Thus the entrance of air into the tracheids is guarded against.

The wood is not uniform all through. As everybody has seen, the wood of the Fir, like that of other trees in temperate climates, shows *annual rings*, each ring representing a year's growth, so that by counting them the age of the tree can be

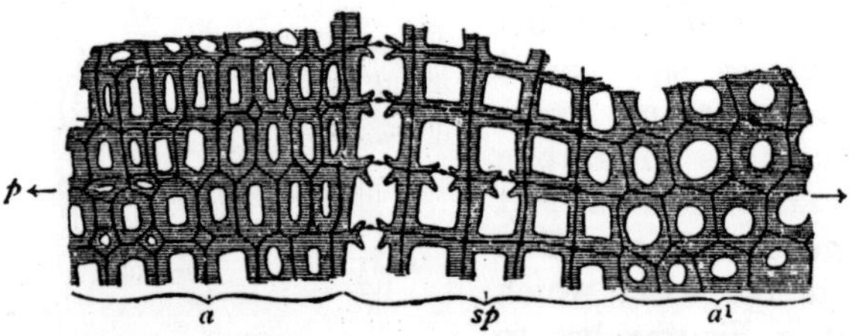

FIG. 101.—Portion of transverse section through the wood of *Picea excelsa*, showing spring and summer wood. The arrow *c* points towards the cambium, the arrow *p* towards the pith ; *a*, summer wood ; *sp*. spring wood ; a^1, beginning of summer wood of second year. Magnified 330. (R. S.)

ascertained. Each spring the cambium becomes active and produces new xylem and phloëm, but it is dormant during the winter.

The reason why we can detect the annual rings is that the wood formed in summer has a different structure from that developed in spring. Thus there is a sharp boundary between the summer wood of one year and the spring wood of the next. This difference in the case of the Fir-tree is due to the fact that in the summer wood the tracheids are much narrower radially and have much thicker walls than those formed in spring (see Fig. 101). The layers of the latest-formed wood, unlike all others, have pits on their tangential as well as on their radial walls. Through these tangential pits the water passes to the cambium, and to the spring wood as soon as it is formed in the following year.

The *medullary rays* vary very much in size. Many of the rays are only one cell thick, and some of these may be only one or two cells in height; others are several cells in thickness, and proportionately higher. The elements of which the rays are composed are of two kinds. Some are living parenchymatous cells in which starch is formed (see Fig. 99, *s*). Others, however, completely lose their living contents, and become tracheids (see Fig. 99, *t*, *t*). These have bordered pits like the long tracheids of the wood, and their walls are irregularly thickened. These tracheids in the ray serve for the passage of water in the radial direction, so that by their help communication is kept up between the different layers of the wood. The parenchymatous ray-cells have only simple pits, except where they are in contact with tracheids, in which case the pits are bordered on the side towards the tracheid. The parenchymatous cells serve for the storage of reserve food-substances, especially starch and oil. The two kinds of elements in the ray are arranged in horizontal rows. Often we find rows of tracheids at the top and bottom of the ray, and parenchymatous cells in the middle.

The resin canals which traverse the wood resemble those already described; their cavity is surrounded by a layer of thin-walled secreting cells. In each of the larger medullary rays a horizontal resin canal is usually found. These horizontal canals are in communication with the vertical ones.

Summing up the peculiarities of this example of coniferous wood, we find that its great characteristic as compared with the wood of a Dicotyledon is its greater uniformity. Vessels and woody fibres are both absent. The tracheids have to do duty for both, those of the spring wood taking the chief part in conducting water, while the thicker-walled summer tracheids serve mainly for strength. Ordinary wood-parenchyma is also absent, the only parenchyma, apart from the rays, being the secreting tissue of the resin canals. On the other hand, the rays themselves are more complex than those of most Dicotyledons. As the wood increases in thickness, new secondary rays are started by the cambium, just as we found in the Wallflower.

The secondary *phloëm* of the Fir is made up of *sieve-tubes*, variously modified *parenchyma*, and *phloëm rays* of complex structure. The sieve-tubes are long, pointed elements, resembling in form the tracheids of the wood. They have sieve-plates on their radial walls only (see Fig. 99, *ph*). The pores of the sieve-plates are excessively minute compared with those in Angiosperms. The sieve-tubes, as in other vascular plants, contain sugars and nitrogeneous substances, and transport them. The phloëm parenchyma forms tangential bands between the zones of sieve-tubes. Some of its cells contain starch or crystals of calcium-oxalate, while others become very thick-walled, forming the *stone-cells*.

The medullary rays in the phloëm consist of two kinds of cells : (1) ordinary parenchymatous cells which form starch (see Fig. 99, *r.s.*) ; (2) cells which are more elongated in the vertical direction, contain no starch, and are very rich in proteins (see Fig. 99, *a, a*). These cells (which are called the *albuminous cells*) communicate by pits with the sieve-tubes, and are believed to fulfil the same functions as the companion-cells of angiospermous sieve-tubes.

The phloëm ray is simply the outer part of the same medullary ray which passes through the xylem, and the cambium serves for the growth of both (see Fig. 99).

We see, then, that the portions of the medullary rays which pass through the phloëm show a similar differentiation to that of the same rays in the xylem. In both alike there are two kinds of elements, one kind serving for the storage of reserve materials, while the other forms part of the conducting system.

The starch which is stored in the medullary rays during the winter becomes converted into sugar in spring ; this is conveyed to the cambium and the growing regions at the extremities of shoots and roots.

b. Leaf

The leaves of the Spruce and those of most Coniferæ differ greatly from the typical leaves of Dicotyledons. They are small and needle-shaped. In our type they are very simple in structure, and the bundle-system especially is much reduced. As we have seen, a single vascular bundle enters each leaf from the stem. Sometimes this bundle divides into two, which run side by side through the leaf. In other cases, as in that figured, the bundles are united into one (see Fig. 102). The bundle forms part of a central cylinder which traverses the leaf from end to end, and is continuous with the tissues of the central cylinder of the stem. The cylinder of the leaf is surrounded by the mesophyll, which in this species shows no distinction between palisade and spongy parenchyma, though in some other Conifers this differentiation is present.

We will now consider the structure of a leaf more in detail. The form of the leaf as seen in transverse section is quadrangular. Its natural position is such that the diagonals of the square are vertical and horizontal (see Fig. 102). The epidermis has a thick outer wall, which is cuticularised. The stomata are arranged in lines along each of the four flat surfaces of the leaf. The guard-cells are sunk below the surface, and are almost covered in above by two very thick-walled epidermal cells, so that scarcely anything can be seen of the stoma itself in a surface view of the leaf. In a transverse section (see Fig. 103) there appear to be two pairs of guard-cells, one above the other, but only the lower cells belong to the stoma, and they alone take part in the opening and closing of the pore. Beneath each stoma is a large intercellular air-cavity.

Next below the epidermis comes the *hypodermis*, a layer of cells which are elongated parallel to the axis of the leaf. They have extremely thick walls, and help much in giving rigidity to the whole organ. The hypodermal layer is inter-

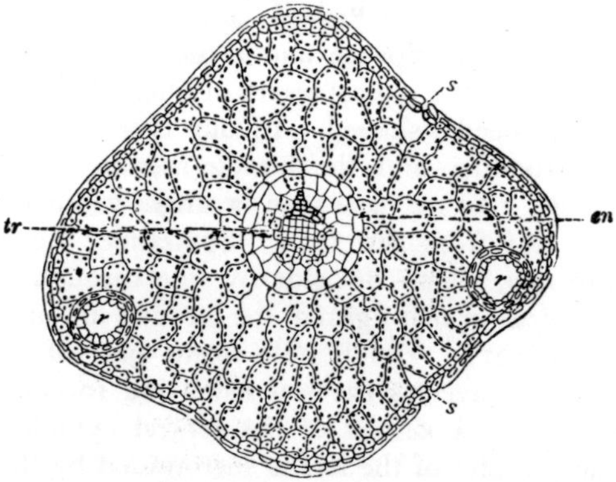

FIG. 102.—Transverse section of a leaf of *Picea excelsa*. *s*, stomata ; *r*, resin canal. In the middle observe the central cylinder with a single vascular bundle, in which the xylem is directed upwards. *en*, endodermis ; *tr*, transfusion tissue. Magnified 82. (R. S.)

rupted below the stomata, while at the corners of the leaf it may be more than one cell thick.

Within the hypodermis we find the most important part of the leaf-tissue, namely, the assimilating parenchyma, containing chlorophyll-granules. This forms a thick layer, and indeed makes up the greater part of the substance of the leaf. Its cells are thin-walled and rather irregular in shape, and their cell-walls are often somewhat folded. The cells contain innumerable chlorophyll-granules, which are embedded in the protoplasm lining the cell-wall (see Fig. 103.)

In the mesophyll we very often find two resin canals placed near the two lateral corners of the leaf (Fig. 102, *r*). The resin canals have the usual structure, namely, a layer of secreting epithelium next the cavity, and a strengthening ring of sclerenchyma surrounding the epithelium. These resin canals are not very constant in their occurrence. Some-

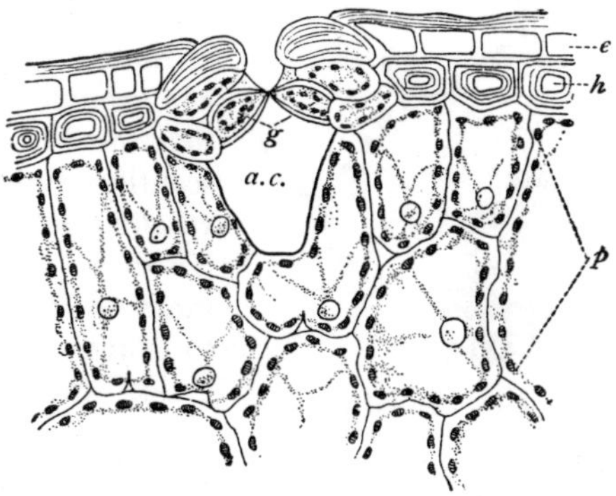

FIG. 103.—External tissues of leaf from the same section as Fig. 102; enlarged to show details. *g*, guard-cells of stoma; *a.c*, air-chamber; *e*, epidermis; *h*, hypodermis; *p*, assimilating parenchyma. Magnified 330. (R. S.)

times there is only one, sometimes there are none at all. When present, the canals end blindly at the base of the leaf, so that they do not join on to those of the stem. This shows that the leaf forms its own resin and keeps it to itself.

We now come to the central cylinder. This is marked off from the mesophyll by a regular endodermis, which has essentially the same structure as the endodermis of a root (see Fig. 104, *en*). Within the endodermis we find the conjunctive tissue of the cylinder, and in the middle of this is the vascular bundle. We will describe the latter first. It is generally a double bundle in the lower part of the leaf, but nearer the apex only one bundle can be distinguished, as in Fig. 102. The xylem consists of tracheids, with plates of parenchyma, resembling medullary rays, between them. The first-formed tracheids, on the upper side of the xylem, are spiral, the rest pitted ; the phloëm has much the same structure as in the young stem. Between xylem and phloëm is a layer of cambium. As compared with the leaves of most Dicotyledons or Monocotyledons, we see that the Fir leaf is very scantily supplied with vascular bundles. It has, however, an arrangement of its own, which to some extent makes good the deficiency. Among the cells of the conjunctive parenchyma there are a number of short tracheids with bordered pits (see Fig. 104, *t*). These tracheids are in communication both with one another and with the xylem of the bundle. They no doubt take up water from the bundle, and pass it on through the pitted endodermis to the assimilating tissue, where it is wanted. These tracheids outside the bundle form what is called the *transfusion tissue*, a very characteristic feature in the anatomy of coniferous leaves.

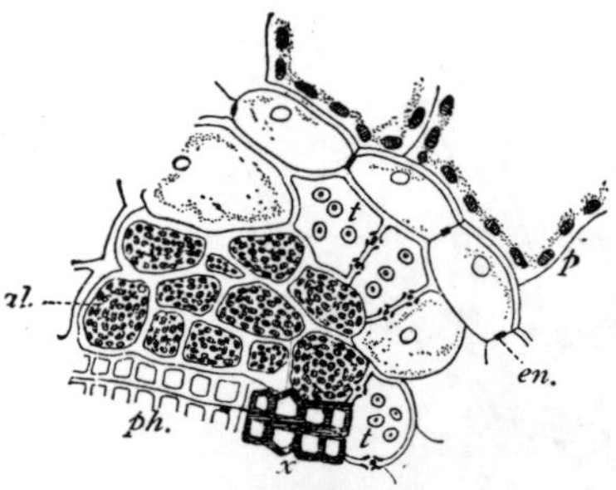

FIG. 104.—Part of the central cylinder and adjoining tissue of the leaf of *Picea excelsa*, in transverse section ; from the same section as Fig. 102, more enlarged. *p*, cells of assimilating parenchyma ; *en*, endodermis ; *t*, transfusion tissue of cylinder ; *al*, albuminous cells ; *ph*, part of the phloëm of bundle ; *x*, part of the xylem of bundle. Magnified 330. (R. S.)

The phloëm as well as the xylem enjoys the advantage of an extension into the conjunctive tissue. On each side of the bundle we find a large group of cells with dense protoplasmic contents (see Fig. 104, *al*). They join on to the phloëm of the bundle. These *albuminous cells*, as they are called, appear to answer the same purpose as the enlarged companion-cells at the ends of the finer bundles in dicotyledonous leaves. It is supposed that they collect the protein substances formed in the mesophyll, and transfer them to the phloëm for transport into the stem.

The conjunctive tissue further contains a strand of thick-walled fibres placed just outside the phloëm (see Fig. 102). We see, then, that the tissue of the cylinder is altogether rather complicated, for besides the vascular bundle and the ordinary parenchymatous cells, in which starch is formed, it contains the transfusion-tracheids, the albuminous cells, and the sclerenchymatous fibres.

The leaves remain on the tree for three or four years. During this time a certain amount of new tissue, chiefly phloëm, is formed by the cambium. Secondary growth in a leaf is not so very common, and is limited to leaves which last for a long time.

The leaf is constricted at the base, so that the mesophyll here comes to an end. Only the cylinder extends through into the stem. Hence all the food-material which goes to and fro between leaf and stem must pass through the tissues of the cylinder, which thus forms part of the conducting system of the plant.

The scale-leaves which protect the buds have a very simple structure compared with the foliage-leaves, for they have no assimilating work to do. They often contain no vascular bundle at all; two resin canals, however, are present in the mesophyll. The epidermis on the outer surface of the scale has an extremely thick cuticularised external wall, and this is no doubt the most important part of the whole scale-leaf, as it protects the young organs of the bud from the risk of drying up.

c. Root

The root-system of the Spruce, and of Conifers in general, consists of the main tap-root and its branches. Adventitious roots hardly ever occur in these plants, unless it be on the base of the hypocotyl, where they scarcely differ from the ordinary rootlets.

The general structure of the root is of the dicotyledonous

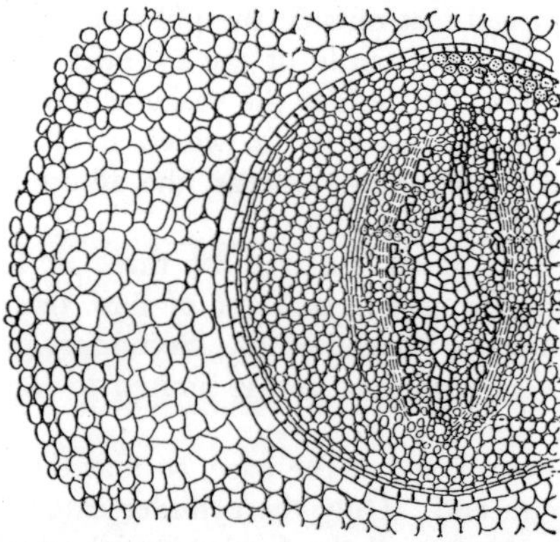

FIG. 105.—Transverse section of the root of *Picea excelsa* at the commencement of secondary thickening. *px*, *px*, the two protoxylem groups; uniting them is the primary xylem-plate. Outside each protoxylem group is a resin canal. x_2, first layers of secondary xylem; *c*, cambium; *ph*, phloëm; *p.c*, pericycle; *pd*, periderm; *en*, endodermis. Outside this is cortex. Magnified 33. (R. S.)

type—that is to say, we find as a rule a simple primary structure with a small number of strands of xylem and phloëm. This simple primary structure soon becomes completely transformed by the early occurrence of secondary growth on a great scale. The mode of development is like that of a Dicotyledon, and consequently the resulting structure is also similar. We have found many such resemblances in the vegetative organs, though, as we shall see, the two Classes are not at all closely related. The young root is clothed on the exterior by a piliferous layer of thin-walled cells. The root-hairs, however, remain short (Fig. 109, *r.h*). The Coniferæ generally are badly provided with root-hairs, and many species do not form them at all. Root-hairs increase the absorbent surface by which water and sub-

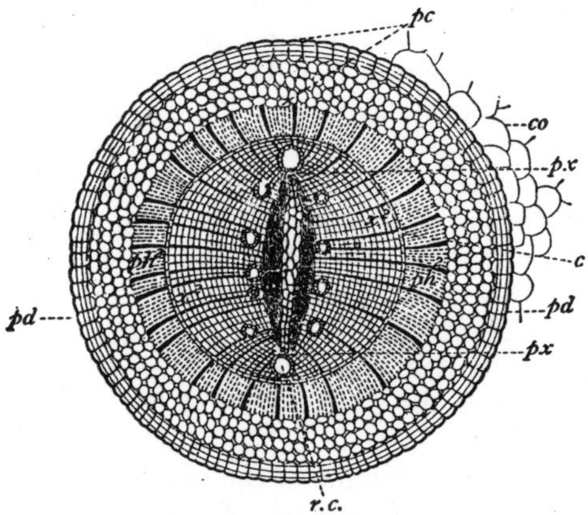

FIG. 106.—Semi-diagrammatic transverse section of a root of *Picea excelsa* at a later stage. *px*, *px*, protoxylem groups ; *r.c*, resin canal ; x^2, secondary xylem ; ph^2, phloëm ; *pc*, pericycle ; *pd*, periderm, formed from outer layer of pericycle ; *co*, remains of cortex. Magnified 22. (R. S.)

stances in solution can be taken up. The greater the transpiration, the greater is the quantity of water needed, and the more does the absorbent surface require to be increased. Now the leaves of most Coniferæ have a very thick cuticle, while the stomata are not very numerous and lie sheltered at the bottom of deep depressions. The water-conducting tissue of the leaf is also, as we have seen, but moderately developed. Hence the amount of water vapour given off in transpiration is comparatively small, and the supply of water taken up is small in proportion. Instead of root-hairs being present many coniferous roots are closely surrounded by fungal hyphæ, which absorb water and substances in solution from the soil and pass these on to the roots. Such an association between a fungus and the root-system of a plant is termed *mycorrhiza*.[1]

Within the piliferous layer is a wide zone of parenchymatous cortex, limited on the inner side by the endodermis, which immediately surrounds the central cylinder (see Fig. 105.) This endodermis has the structure with which we have already become familiar. Its radial and horizontal walls are corky and undulated, and show the characteristic dark line when seen in section (see Fig. 110, *en*).

The cylinder of the root in the Fir may be either triarch or diarch. The main root always has the former structure. We will describe the diarch type here illustrated. The triarch roots do not differ from this except in the number of strands.

The first thing that strikes us as peculiar in the structure of the cylinder is the great width of the pericycle, which forms a broad zone from six to nine cells in thickness. It is thickest opposite the middle of the two phloëm groups (see Fig. 105, *p.c*). Most of the pericyclic cells contain a quantity of starch, which is ready for use in the formation of the cells of the new tissues as they develop. Within the pericycle comes the vascular tissue, which, as seen in transverse section, occupies a somewhat elliptical area (see Fig. 105). The two protoxylem groups lie at the ends of the long axis of the ellipse, and therefore at the points where the pericycle is narrowest (Fig. 105, *px*). This first-formed xylem consists of spiral tracheids. The tissue lying between the two protoxylem groups slowly becomes converted into wood. The differentiation advances from either end towards the middle of the cylinder, until ultimately a continuous plate of primary wood is formed, extending across the middle of

[1] Mycorrhizal roots are usually profusely branched.

the cylinder (see Figs. 105 and 106). The primary wood is now complete. Its development goes on very gradually, so that the secondary growth has usually begun before the primary structure is finished. We will, however, consider the two stages separately. All the middle part of the primary

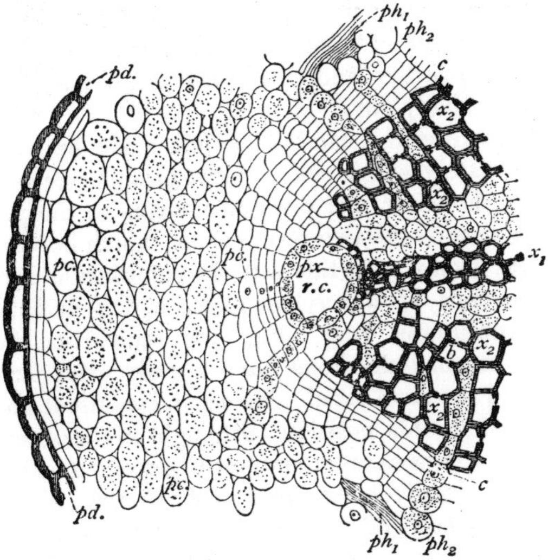

FIG. 107.—Part of a transverse section of the root of *Picea excelsa*, showing the completion of the cambium outside the protoxylem by division of pericyclic cells. *px*, protoxylem ; x_1, primary xylem-plate ; x_2, secondary xylem ; *b*, bordered pits ; *c*, cambium ; ph_1, primary phloëm (crushed) ; ph_2, secondary phloëm ; *r.c*, resin canal ; *pc*, pericycle ; *pd*, periderm. Magnified 165. (R. S.)

plate of xylem consists of pitted tracheids. Just outside each protoxylem group a resin canal is formed in the pericycle (see Figs. 105, 106, and 107).

The primary phloëm forms two bands, one on each side of the xylem-plate, from which they are separated by a few layers of conjunctive parenchyma (see Fig. 105, *ph*).

Even before the primary tissues are completely developed, cambium begins to arise by tangential divisions of the conjunctive cells lying just inside the phloëm groups. As usual, secondary xylem is formed on the inner side of the cambium and secondary phloëm on its exterior. As it is the layer of cells next the phloëm which divides to form the cambium, a band of unaltered parenchyma is always left between the primary and secondary xylem (see Figs. 105, 106, and especially 107).

For some time the cambial divisions are limited to the part immediately within the phloëm groups (see Fig. 105). Eventually, however, the division spreads to the pericyclic cells lying outside the two resin canals, which border on the protoxylem groups. In this way the cambium is completed round the ends of the xylem-plate, and henceforward it forms a continuous ring. Fig. 107 represents the stage when the cambial divisions have already extended to the cells beyond the resin canal. No secondary tissue has as yet been formed in this position, although at the sides of the xylem-plate a considerable amount of secondary xylem and phloëm has already been developed, so that the primary phloëm is crushed and obliterated. A later stage, in which the secondary tissues already form a complete ring, is shown in the diagrammatic figure 106. The structure of the secondary tissues needs no special description; it is the same as in the stem. We may, however, notice that in the first few layers of the secondary xylem the pits are formed on all surfaces of the tracheids indiscriminately (see Fig. 107, *b*). In the later-formed layers they are limited to the radial walls (see, however, p. 227).

As soon as the cambium has begun its activity, the most external layer of the pericycle also becomes the seat of new formations. The cells of this layer divide up by tangential walls, and thus give rise to a phellogen which forms cork-cells on its outer side (see Figs. 105, 106, 107, *pd*). From this time onwards the root ceases to be an absorptive organ. The corky periderm completely isolates the cortex from all communication with the conducting cylinder, and consequently the whole external tissue, from the endodermis outwards, dies away. Henceforward the root is an organ of conduction only, consisting solely of the central cylinder, which by means of its cambium retains the power of unlimited growth. In all these respects the root of a Conifer agrees closely with that of a typical Dicotyledon.

d. Growing-Points and Lateral Appendages

a. Growing-Point of the Stem

The apex of the shoot in a Fir has essentially the same structure as that of an angiospermous plant. It consists of

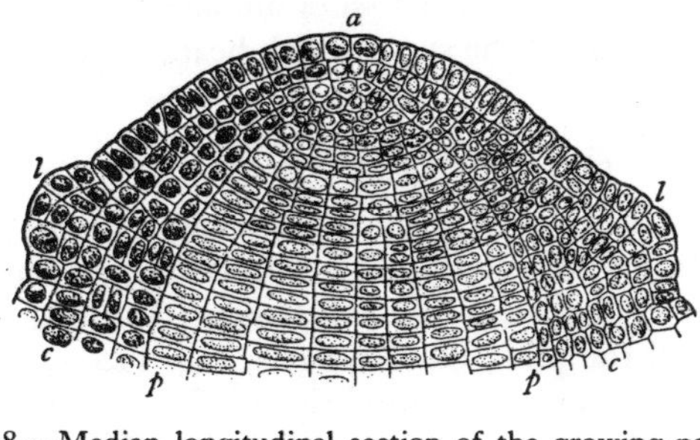

FIG. 108.—Median longitudinal section of the growing-point of the stem of *Picea excelsa*. *a*, apex ; *l*, *l*, very young leaves ; *p*, *p*, young pith ; *c*, *c*, cortex. Between *c* and *p* the vascular bundles arise. Magnified about 140. (After Sachs.)

meristem, the cells of which are small, and have the usual characteristics of embryonic tissue, large nuclei, abundant cytoplasm, and thin cell-walls (see Fig. 108).

The formative layers of the meristem are less distinct here than in Angiosperms. Whereas in the latter it is a general rule that the external layer of the apical meristem gives rise to the epidermis only, this is not the case in Conifers. Here the superficial cells of the growing-point may divide by walls parallel to the surface, and thus contribute to the formation of cortical as well as of epidermal tissue (see Fig. 108). The limit between plerome and periblem is often impossible to trace near the apex. Thus in our figure the sharp line of demarcation at *p* corresponds to the limits of the pith, not to those of the plerome. Without entering further into the question of the meristematic layers, which is a difficult and disputed one, we may safely say that their differentiation at the apex of the stem is on the whole less sharp in Gymnosperms than in Angiosperms.

The leaves and branches arise, just as in Angiosperms, exogenously, *i.e.* they are formed by the growth and division of cells which belong to the outer layers of the meristem (see Fig. 108, *l*, *l*). Their tissues are always in direct continuity with those of the shoot which bears them.

β. *Growing-Point of the Root*

The apex of the root in the Spruce and its allies grows in much the same way as that of a Dicotyledon. There is an outer layer of meristem (the *calyptrogen*), forming both the root-cap and the piliferous layer (see Fig. 110, *c*), a middle group or *periblem*, which gives rise to the cortex (*p* in Fig. 110), and within this is the *plerome*, from which the central cylinder is developed (*pl* in same figure). This, at least, is the arrangement of the apex of young roots, such as those shown in Figs. 109 and 110, soon after their first formation. It is doubtful whether the limit between the periblem and the calyptrogen can always be traced at the growing-point of the older roots.

The formation of the rootlets or branches on the parent-root also follows the dicotyledonous type, with slight modifications. The rootlets are arranged in as many vertical rows as there are protoxylem groups in the main root. Thus, in our diarch root of the Spruce there would be two opposite rows of rootlets lying in the same plane as the xylem-plate. The rootlet here, as in all Flowering Plants, arises from the pericycle of the main root. The layer of pericyclic cells which divides to form the rootlet is here the second layer from the outside. Tangential walls are first formed, which separate off the plerome, or future central cylinder of the rootlet, while the outer cells resulting from the first division again divide tangentially into periblem and calyptrogen. The more internal layers of the pericycle merely form the base of the central cylinder of the rootlet, by which it is attached to the vascular tissue of the main root.

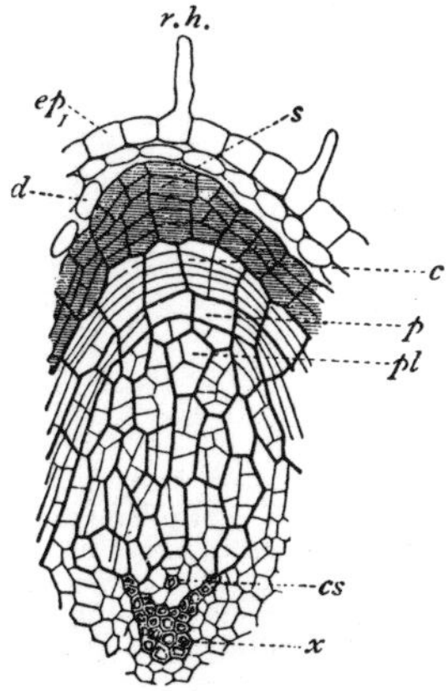

FIG. 109.—Portion of a transverse section of a root of *Pinus pinea* (allied to *Picea*), showing an early stage of development of a rootlet. *x*, protoxylem group of main root; *c.s*, resin canal; *d*, remains of cortex of main root; *ep*, its piliferous layer, in which root-hairs, *r.h*, are shown; *pl*, plerome of rootlet; *p*, periblem; *c*, calyptrogen; *s* (shaded), digestive sac, covering the apex of rootlet, and formed from the outer layer of the pericycle of the main root. Magnified about 150. (After Van Tieghem and Douliot.)

The outermost layer of the pericycle also divides, and follows the growth of the young branch, round which it forms a temporary envelope (*s* in Figs. 109 and 110) during its growth through the cortex. This envelope is the *digestive sac* (see p. 148).

The whole of the cortex, as well as the piliferous layer, has to be bored through by the rootlet before it reaches the exterior. It completely absorbs all the tissues which lie in its way.

The only peculiarities in the mode of origin of the branches of the root, in this instance, depend on the fact that the pericycle is of unusual thickness, and are not

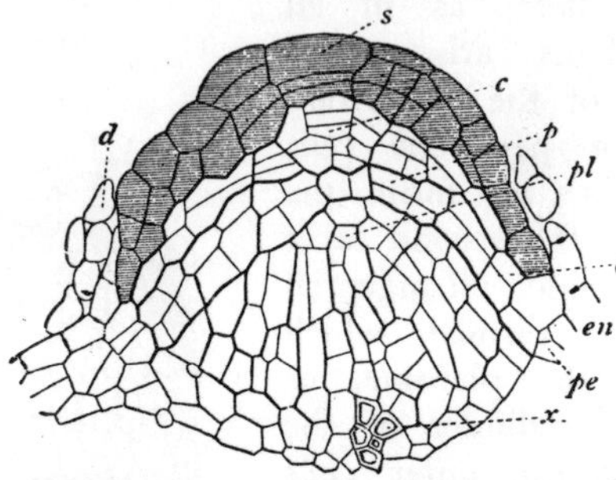

FIG. 110.—Similar section of a root of *Picea excelsa*, showing origin of a rootlet. *en*, endodermis of main root ; *pe*, outer layers of a pericycle ; *e*, cortex at base of rootlet ; other lettering as in Fig. 109. Magnified about 150. (After Van Tieghem and Douliot.)

characteristic of the class generally, but only of the Spruce and its nearer allies. In all the essential points the mode of branching of the root is identical in Gymnosperms and in Dicotyledons.

SUMMARY OF THE VEGETATIVE STRUCTURE OF CONIFERS

In the general morphology and anatomy of both stem and root the Conifers agree closely with Dicotyledons, while they bear little or no resemblance to the Monocotyledons. The distribution and longitudinal course of the vascular bundles and the mode of secondary growth are practically identical in Conifers and Dicotyledons. Yet in other respects we find very great differences. This is especially the case as regards the leaves.

Some Conifers have leaves of more complex structure than those of the Fir, but still it is characteristic of the group to have a comparatively simple bundle system in the leaf. Its simplicity is constantly compensated for by the formation of transfusion tissue and albuminous cells outside the actual bundles.

In the more minute structure there are also important differences ; the entire absence of vessels from the wood is a striking character.

As regards the phloëm, we find distinctive peculiarities in the structure of the sieve-tubes, and in the replacement of true companion-cells by elements of a different origin. These characters suggest to us that after all we are dealing with a group fundamentally distinct from Dicotyledons. It will be noticed that in the points in which Conifers differ most strikingly from Dicotyledons, they differ equally from Monocotyledons. We shall find later on that in some peculiarities of their vegetative structure the Conifers approach certain groups of the Flowerless Plants, which we have yet to consider.[1]

The study of the vegetative organs of these plants, however, throws comparatively little light on their real systematic position. We will now proceed at once to make ourselves acquainted with their reproductive phenomena.

III. *STRUCTURE OF THE REPRODUCTIVE ORGANS*

a. The Male Cone

The male cone of the Fir, as we have already learnt, is a short lateral shoot, bearing a few bracteoles on the lower, and numerous spirally arranged stamens on the upper part of its axis.

In form, the stamens are more obviously leaf-like than those of Angiosperms. The pollen-sacs are only two in number on each stamen, and are inserted side by side on its

[1] See Part II., *Flowerless Plants.*

under surface. In some other Conifers, however, they are much more numerous. While in Angiosperms the number of the pollen-sacs on the stamen is very constantly four, in the Gymnosperms their number is subject to very great variations. The dehiscence of each ripe pollen-sac takes place separately by a longitudinal slit.

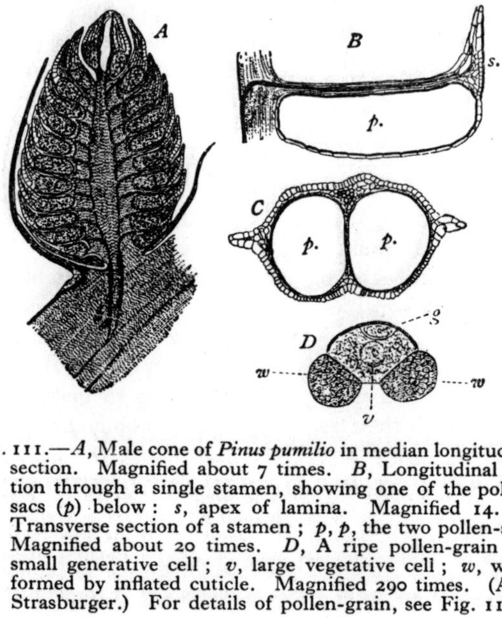

FIG. 111.—*A*, Male cone of *Pinus pumilio* in median longitudinal section. Magnified about 7 times. *B*, Longitudinal section through a single stamen, showing one of the pollen-sacs (*p*) below: *s*, apex of lamina. Magnified 14. *C*, Transverse section of a stamen; *p*, *p*, the two pollen-sacs. Magnified about 20 times. *D*, A ripe pollen-grain; *g*, small generative cell; *v*, large vegetative cell; *w*, wings formed by inflated cuticle. Magnified 290 times. (After Strasburger.) For details of pollen-grain, see Fig. 112.

Each pollen-sac arises from a small group of cells on the under side of the stamen while it is still very young. One cell of this group, lying immediately below the epidermis, becomes the *archesporium*, and after dividing up a great many times, gives rise to a large mass of pollen mother-cells. Here, as in Angiosperms, the pollen-producing tissue grows at the expense of the surrounding layers; the innermost of these layers, called the tapetum, is rich in proteins, and is the first to be digested. The outer wall of the pollen-sac is originally about three cells in thickness, but before maturity all except the epidermis have disappeared (see Fig. 111, *B*, *C*).

The division of the pollen mother-cells takes place in the early spring. Each mother-cell divides into four daughter-cells, which become the pollen-grains. The walls of the mother-cells break down. An enormous number of pollen-grains are produced altogether. If we shake a Fir-tree in May, when the male flowers are ripe, the pollen comes out in great clouds of yellow dust, which float for a long time in the air, and are carried to great distances if any wind is blowing. The Firs are *wind-pollinated* trees—that is, the transport of the pollen-grains from the stamens to the ovules is entirely dependent on currents of air. Hence a very small proportion of the pollen-grains ever reach the ovules, and in order that there may be a fair chance of pollination taking place at all, a very large number must be produced, the vast majority of which are wasted. We will now see why the pollen-grains are so light and float so long in the air. If we examine the ripe pollen-grains under the microscope, we find that each grain has a curious winged form, owing to the development of two large air-cavities in the cell-wall (see Fig. 111, D, *w*). The wall of the pollen-grain is covered on the outside by a layer of cuticle. At two places on opposite sides of the grain the cuticle becomes separated from the cellulose wall within; the spaces thus formed are at first filled with water, so that a kind of blister arises on each side of the pollen-grain. The water is absorbed and the spaces become filled with air when the pollen is ripe. This curious arrangement has the effect of reducing the weight of the pollen-grain in proportion to its surface, these lateral expansions acting as veritable wings.

Thus the grain is more easily carried to a distance by the wind.

Before the pollen-grain is ripe, important changes go on in its interior. Each pollen-grain, when first formed by

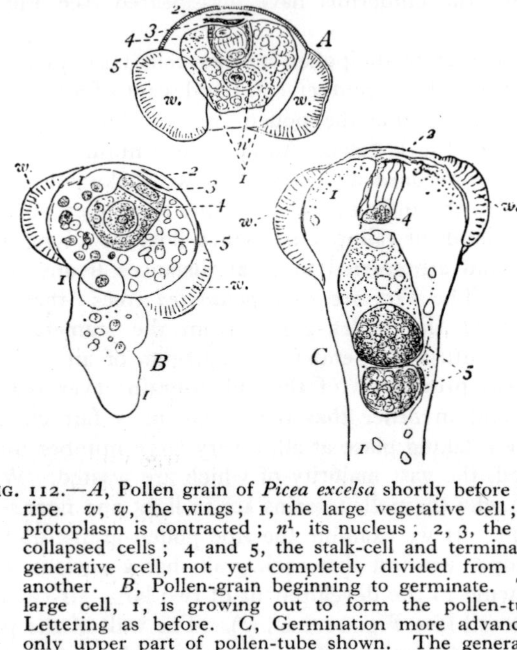

FIG. 112.—*A*, Pollen grain of *Picea excelsa* shortly before it is ripe. *w*, *w*, the wings; 1, the large vegetative cell; its protoplasm is contracted; n^1, its nucleus; 2, 3, the two collapsed cells; 4 and 5, the stalk-cell and terminal or generative cell, not yet completely divided from one another. *B*, Pollen-grain beginning to germinate. The large cell, 1, is growing out to form the pollen-tube. Lettering as before. *C*, Germination more advanced; only upper part of pollen-tube shown. The generative cell, 5, has divided into two, each with a very large nucleus. Magnified 240. (After Strasburger.)

division of its mother-cell, is merely a single cell with one nucleus. But before the dehiscence of the pollen-sac takes place, a series of divisions goes on in each pollen-grain. First, a small cell is cut off on the side of the grain farthest from the insertion of the wings. This little cell is separated from the large one by a thin wall shaped like a watch-giass.

Two more divisions take place in the large cell, cutting off two more small cells, so that there is now a row of three small cells placed one upon another, and projecting into the cavity of the large cell. The first two of the small cells collapse, the third grows bigger and divides into two, forming a little stalk-cell and a larger terminal or generative cell, which is long enough to reach quite half-way across the cavity of the grain (Fig. 112, *A* and *B*).

The pollen-grain is now ripe, and we see that within its winged membrane no less than five cells are enclosed : first, the large cell, which still forms the main part of the whole ; second and third, the two little collapsed cells next the wall; fourth, the stalk-cell ; fifth, the terminal or generative cell. We see, then, that the pollen-grain of a Fir is a much more complicated structure than that of an Angiosperm (see Fig. 112).

b. The Female Cone

The arrangement of the organs of the female cone has already been described.

We have seen that each of the conspicuous scales of the cone is an outgrowth produced on the upper surface of a much smaller carpellary scale. The much larger outgrowths which the small scales bear, and which in their turn bear the ovules, are called the ovuliferous or seminiferous scales. There is much uncertainty as to the interpretation of the parts of the female cone in Coniferæ.

In the Firs two ovules are borne on the upper surface of each ovuliferous scale. Neither the number nor the position of the ovules is at all constant in the Coniferæ generally. In the Araucarias, for example, to which the " Monkey Puzzle " belongs, there is only a single ovule to each carpellary scale, while in some of the Cypress family a great number are produced on each. The ovuliferous scale is often rudimentary or absent, as in the Araucaria family, where the ovule is borne directly on the carpellary scale. In the Yew, which is classed with Conifers, though it does not bear female cones, the ovule is not formed on a leaf at all, but is seated on the end of a short branch.

To return to the Fir: each ovule is straight (*orthotropous*), and has a single integument. The micropyle points inwards towards the axis of the cone.

It is not necessary to follow the early development of the ovule in detail, for up to a certain point it grows in very much the same way as the ovule of an Angiosperm. If the reader will refer to the development of the ovule (see pp. 102 and 156), as described in the Wallflower or the Lily, he will understand what goes on here. Here, as in Angiosperms, the important changes all go on in the nucellus or middle part of the ovule. The archesporium arises from a cell just below the epidermis of the nucellus. The archesporium is a single cell to begin with. It divides by a tangential wall; from the upper part *tapetal* cells are formed. The lower part divides up into a row of three cells, and it is usually the lowest of the three which becomes the embryo-sac. Both the epidermis and the tapetal cells undergo a great many divisions by tangential walls, so that the embryo-sac comes to be buried deep down in the nucellus, beneath a thick mass of tissue.

The embryo-sac at first, like any other cell, has a single nucleus, which soon divides up repeatedly, until there are a large number of nuclei in the protoplasm lining the cell-wall of the sac. At the same time the embryo-sac enlarges greatly. The nuclei of the embryo-sac are at first free, but after a time cell-walls are formed between them, cutting up the parietal layer of protoplasm into a layer of cellular tissue. The cells thus formed continue to grow and divide until the whole embryo-sac is filled with tissue. The growth of the sac still goes on, until by the time of fertilisation, in the latter part of June, it is about $\frac{1}{8}$ of an inch in length.

The growth and cell-division of the internal tissue keep pace with the enlargement of the embryo-sac.

We see, then, that in the Fir, *the embryo-sac becomes filled with tissue before fertilisation.* This characteristic is common not only to all Conifers, but to all Gymnosperms, and is one of the most important distinctions between this Class and all other Flowering Plants. The tissue within the embryo-sac has the same function as the endosperm of the Angiosperms, and is often called by the same name, but, as its origin is so different, it is better to use a distinct term, and call it the *prothallus* (see Part II., *Flowerless Plants*).

After the embryo-sac is filled with tissue, but before it has reached its full size, certain cells of the prothallus, at the end towards the micropyle, begin to be distinguished from their neighbours by their larger size and denser protoplasm. These cells are from three to five in number, and are always superficial cells of the prothallus, in contact with the embryo-sac wall at its upper end.[1]

While the neighbouring cells go on dividing as fast as they grow, these few special cells grow enormously without dividing. After a time, however, each of them divides near its upper end by a transverse wall. The upper and smaller daughter-cell thus formed undergoes several further divisions, both transverse and longitudinal. The lower cell remains for a long time undivided, but continues to increase greatly in size.

The stage which we have now reached is this : the embryo-sac has grown to a great size, and is filled with tissue. At its upper end are from three to five groups of cells, each consisting of a single very large oval cell below and a well-defined column of small cells above ; each group is called an *archegonium*. The large lower cell is the *venter* of the

[1] We always use the word *upper*, in describing an ovule, for the end towards the micropyle, without reference to the position which the ovule may happen to occupy.

archegonium ; the little column of cells above is the *neck.*

A general idea of the structure of the whole ovule at this period is given by Fig. 113. In this we see the pointed

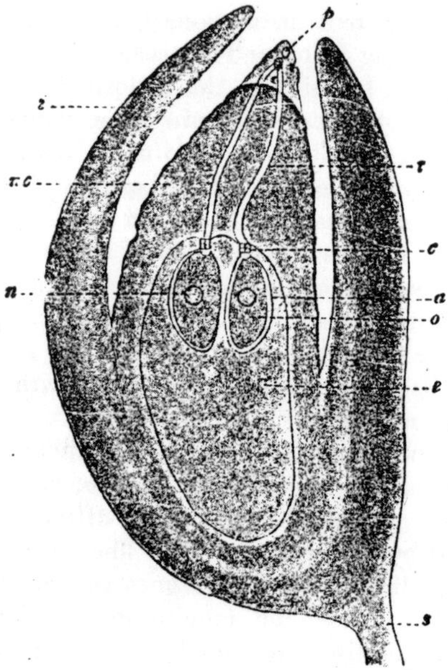

FIG. 113.—Longitudinal section of an ovule of *Picea* at the time of fertilisation. *i*, integument ; *nc*, nucellus ; *s*, part of the wing ; *p*, pollen-grains on the tip of the nucellus ; *t*, pollen tubes ; *c*, neck of an archegonium ; *a*, venter of archegonium containing the ovum, *o* ; *n*, nucleus of ovum ; *e*, prothallus. Magnified 9 times. (After Strasburger.)

nucellus, *nc*. Most of its lower part is occupied by the large oval embryo-sac, which is shaded to indicate that it is filled with prothallus, *e*. Two of the archegonia are shown in section.

The archegonia are the true female organs, in which fertilisation takes place. Just before the archegonium is ready for fertilisation, a very small lens-shaped cell is cut off from the venter at its upper end. This is called the *ventral canal cell* (see Fig. 114, *A*, *cl*). The immensely larger lower

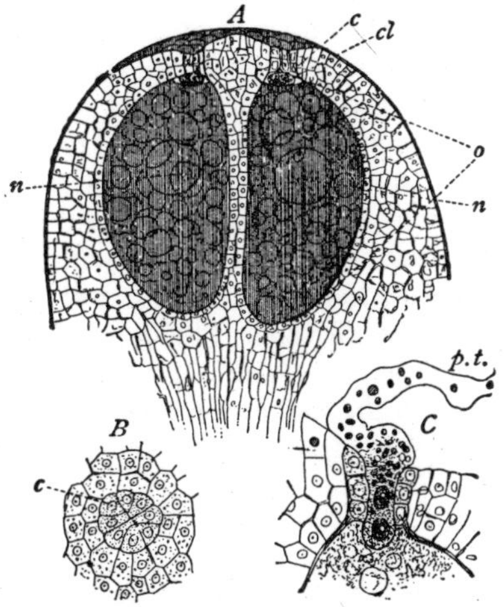

FIG. 114.—*A*, Upper part of embryo-sac in longitudinal section, showing two archegonia: *o*, ovum of an archegonium; *c*, neck; *cl*, ventral canal cell; *n*, *n* nuclei of the two ova. *B*, Portion of upper surface of prothallus, showing the neck of an archegonium, *c*. *C*, Enlarged longitudinal section passing through the neck of an archegonium at the time of fertilisation. The pollen-tube, *p.t*, has penetrated the neck. Near the lower end of the tube its two generative nuclei are seen. Magnified—*A*, about 50 diameters; *B*, about 125 diameters; *C*, about 125 diameters. (After Strasburger.)

cell which remains after this final division is the *ovum* itself, which will give rise to the embryo after fertilisation. The ova in the Fir and in most Gymnosperms are of relatively very large size, so as to be easily visible to the naked eye. The large scale of the whole reproductive apparatus at the time of fertilisation is convenient for observation, but is sometimes puzzling when we have become accustomed to the minute size of the corresponding organs in Angiosperms. Fig. 114, *A*, shows the upper part of the embryo-sac with two archegonia embedded in the prothallus.

Each ovum has a large nucleus, and is densely filled with cytoplasm, in which there are numerous vacuoles, often bearing a deceptive resemblance to nuclei.

The neck of the archegonium, as seen in longitudinal section, consists of two to four tiers of cells (Fig. 114, *A* and *C*). When seen in surface view (Fig. 114, *B*), it appears as a rosette of six or eight cells, so that the total number of cells in the neck may vary from twelve to thirty-two.

We see, then, that the arrangements within the embryo-sac at the time of fertilisation are totally different in Gymnosperms from those in all other Flowering Plants. On the other hand, they agree closely with the corresponding structures in some of the higher Cryptogams. We must now consider the processes of pollination and fertilisation.

c. Pollination and Fertilisation

Pollination, as we have already learnt, is brought about by the wind, which blows the light winged pollen-grains in all directions, so that some of them chance to alight on the female cones. This takes place in May. Just before this happens, an interesting change may be observed in the cones, which prepares them to receive the pollen. The axis of the cone elongates a little throughout its whole length, so that the ovuliferous scales are all shifted a little farther apart. Thus an open space is left between each two successive scales, through which the pollen-grains can reach the ovules.

At this time the ovule is still at an early stage of development. The embryo-sac is still small, and is not yet filled with prothallus.

The integument opens widely, leaving an open passage

through the micropyle, leading down to the apex of the nucellus. At the same time a small quantity of liquid is secreted within the micropyle on the surface of the nucellus. Some of the pollen-grains carried by the wind pass between the ovuliferous scales and come to rest on the edge of the integument of an ovule. Another change now takes place: the micropyle closes, the lips of the integument bending inwards, and bringing the pollen grains into contact with the top of the nucellus, to which they adhere owing to the liquid which has been secreted.

The germination of the pollen-grain now begins. Our Fig. 112, *B*, shows a pollen-grain, highly magnified, at the beginning of this process. The large vegetative cell sends out a tube, which penetrates the tissue of the nucellus. The vegetative nucleus passes into the tube, but becomes disorganised, and has no further part to play. So far the group of small cells within the grain has remained unchanged. Now, however, the end cell of the group, which is the generative cell, divides into two. The stalk-cell behind them breaks down; the two generative daughter-cells are set free and pass into the pollen-tube (see Fig. 112, *C*, 5). Their nuclei have now become very large and granular.

Pollination takes place in May, fertilisation not until near the end of June; the interval in the Spruce Fir is thus about six weeks. In some other Conifers, such as the Scotch Pine, *Pinus sylvestris*, and the Juniper, the interval is very much longer; the pollen falls on to the ovule in May of one year, but fertilisation is not effected until June of the next year.

To return to *Picea*: the pollen-tubes eventually grow down through the nucellus as far as the embryo-sac. They contain numerous starch-grains, which serve as food during their growth.

In the meantime the ovule has ripened. By the time that the pollen-tubes have reached the embryo-sac, it is completely filled with prothallus, and the archegonia have been formed (see Fig. 113). After the 20th of June the actual fertilisation takes place. A pollen-tube penetrates the wall of the embryo-sac just above the neck of an archegonium. It continues to grow on, passes between the cells of the neck, absorbs the ventral canal cell below, and reaches the protoplasm of the ovum itself (see Fig. 114, *c*). The two generative cells are now at the growing end of the pollen-tube. Their nuclei are shown in the figure.

The leading generative nucleus, accompanied by a small amount of cytoplasm, now passes through an opening in the membrane at the end of the pollen-tube (Fig. 115, *A*, *sn*). It traverses the intervening cytoplasm and reaches the nucleus of the ovum, with which it eventually unites [1] (Fig. 115, *B*). Fertilisation is now effected. We see thus that, different as many of the preliminary processes have been, yet the act of fertilisation in itself is essentially the same as in Angiosperms. Fertilisation is, in fact, an identical process in all plants in which it occurs at all.[2] Fertilisation involves the fusion of two sexual cells or gametes, which in all the higher plants are clearly differentiated as male and female.

d. Development of the Embryo

The fertilised nucleus, resulting from the fusion of the male and female nuclei, passes into the lower end of the ovum. It there undergoes two divisions, so that there are now four nuclei, all of which lie in the same horizontal plane (see Fig. 115, *C*, where, of course, only two of them are shown). Successive divisions now take place in a direction at right angles to the long axis of the ovum, until there are sixteen nuclei altogether, lying in four tiers one above

[1] For an account of the fertilisation of certain Gymnosperms by means of spermatozoids or motile male cells, see Part II., *Flowerless Plants*, Chapter VIII.

[2] *Double* fertilisation, however, described above in the Wallflower and Lily, is a phenomenon peculiar to Angiosperms.

another. Each nucleus of the lower three tiers is enclosed with its surrounding cytoplasm within a cell-wall; the

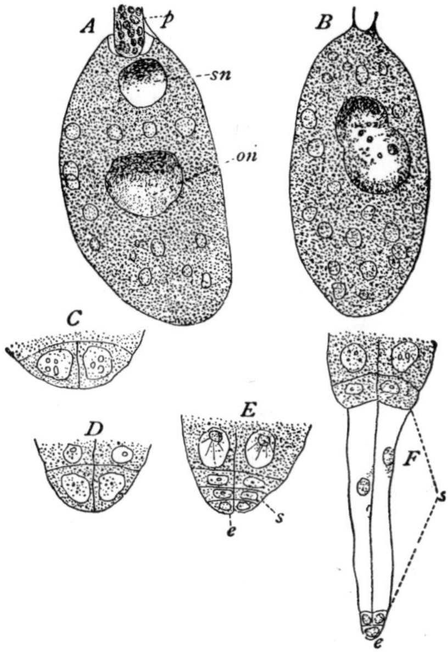

FIG. 115.—*A*, Ovum at the time of fertilisation: *p*, end of pollen-tube; *sn*, generative nucleus from the pollen-tube; *on*, nucleus of ovum. *B*, Ovum a little later: the two nuclei are in the act of fusing. *C*, Base of ovum showing first divisions after fertilisation: two of the four cells are shown. *D*, More advanced stage: four of the eight cells are shown. *E*, Still more advanced stage: *s*, cells from which suspensor will be formed; *e*, cells of the embryo. *F*, Still later stage: the suspensor *s* has grown to a great length (it is composed of four cells). The rudimentary embryo, *e*, has been pushed downward by the growth of the suspensor. Magnified about 70 diameters. (After Strasburger.)

uppermost four are free nuclei (see Fig. 115, *D*, *E*). It is to be noticed that only a part of the ovum is concerned in these divisions; much the greater portion remains undivided and has nothing to do directly with the development of the young plant. It only serves as a store of food for the embryo.

The four free nuclei now become disorganised ; the tier of four cells next below them undergoes little change, and remains in its original position. The third layer, however, *s* in Fig. 115, *E* and *F*, elongates enormously, forming the ***suspensor***, which here consists of four parallel cells. The lowest group of four cells is pushed forward by the suspensor deep into the middle of the prothallus. It is this lowest group, *e* in Fig. 115, *E* and *F*, which alone forms the actual embryo.

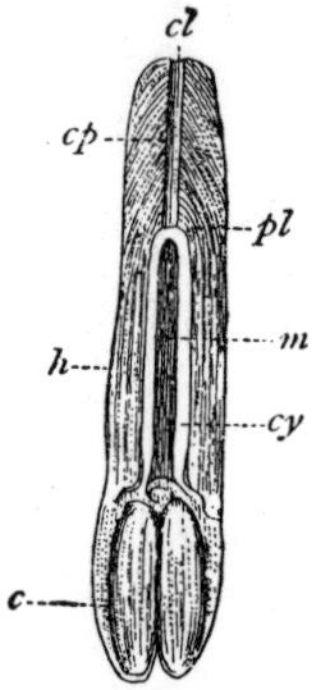

FIG. 116.—Longitudinal section of the embryo of *Picea* taken from a ripe seed. *cp*, root-cap ; *cl*, central tissue of root-cap ; *pl*, growing-point of root ; *m*, pith ; *cy*, vascular tissue of cylinder ; *h*, hypocotyl ; *c*, cotyledons, of which only two are shown. Magnified 10 diameters. (After Strasburger.)

In the Spruce Fir only one embryo can be formed from each archegonium. All the archegonia, however, may be fertilised ; so, for a time, from three to five developing embryos may be present side by side in the prothallus of the same ovule. One, however, always gains the upper hand, and in the ripe seed only this one remains. In some other Conifers, such as the Scotch Pine, the four cells of the suspensor separate from each other, and each bears a separate embryo at its end. In this case, therefore, each archegonium gives rise to four embryos, and altogether there may be from twelve to twenty in the young seed. Here also, however, only one out of the whole number comes to perfection.

We will not follow the further development of the embryo in *Picea* in detail. Its growth, with which the cell-division keeps pace, goes on from June until October, by which time the seed is ripe. The suspensor at first serves to supply the young embryo with food, but by the time the seed is ripe its function is at an end. The embryo then extends through the whole length of the embryo-sac, but is still enveloped in a mantle of prothallus; the middle part of the prothallus has been absorbed by the embryo for its own nutrition. The structure of the embryo in the ripe seed is shown in Fig. 116, which represents it in longitudinal section. The root is directed towards the micropyle. The apex of the root is protected by an enormously thick root-cap. The root itself is still but little developed, while the hypocotyl occupies nearly half the length of the embryo. The cotyledons are numerous, generally numbering about eight. This is not, however, a character common to Conifers generally, for many of them, such as the Yew, have only two. The eight cotyledons of the Firs probably represent the subdivisions of two.

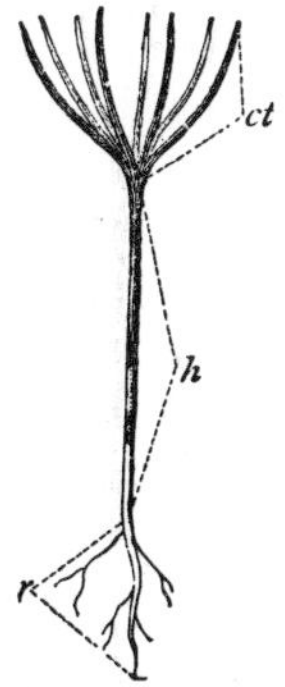

FIG. 117.—Young seedling of *Picea*. *ct*, cotyledons; *h*, hypocotyl; *r*, roots. Natural size. (After Kerner.)

The cone opens in the spring, and the winged seeds are scattered by the wind, as already described.

When the seed germinates, the root grows out through the micropyle, and turns downward into the ground. The cotyledons during germination absorb the remainder of the prothallus. The apex of the stem directs itself upwards; the seed-coat is at first lifted up upon the cotyledons, but is soon cast off as they expand (Fig. 117). The cotyledons become green, and the young plant is now able to assimilate its own food. In some Conifers the cotyledons within the seed become green before germination. This is one of the few examples of chlorophyll developing in darkness.

Summary

We will now pass rapidly in review those facts in the reproduction and development of the Fir which are typical of gymnospermous plants generally, and which indicate their relations, on the one hand to the Angiosperms, and on the other hand to the Flowerless Plants.

Apart from the peculiarities in the general morphology of the male cone, which have been sufficiently dealt with already, we find a characteristic feature in the considerable number of cells which are formed within each pollen-grain. This, as we shall see later on, is a very significant fact. The production of all these intermediate cells, before the actual generative cells are set apart, is really a cryptogamic character, and is one indication, among many, that the Gymnosperms are on a different level from all other Flowering Plants.

The gymnospermy, or direct pollination of the ovule without the intervention of a stigma and style, is itself a point of great importance. Fertilisation by means of a pollen-tube is one of the chief characters of Phanerogams. In Gymnosperms we still have the pollen-tube, but its work is more limited than in Angiosperms, for its function only consists in conveying the generative cells through the tissues of the nucellus.

On the side of the female organs, the differences as compared with Angiosperms decidedly outweigh the resemblances.

The filling of the embryo-sac with an extensive prothallus before fertilisation is exactly what we shall find occurring in the highest Cryptogams, but is quite different from anything in Angiosperms, where only a few cell-divisions intervene between the formation of the embryo-sac and that of the ovum.

The archegonia, with their multicellular necks, are quite foreign to the Angiosperms, while they agree in every detail with the female organs of the higher Cryptogams. In fact, we may say that from the formation of the embryo-sac up to the act of fertilisation the whole development is cryptogamic.

The embryology, on the other hand, though highly peculiar, is not of such far-reaching significance, for the mode of development varies greatly among the Gymnosperms themselves.

The seed is still a marked phanerogamic character ; after fertilisation, the differences from Angiosperms become much less striking than before. The seeds of Gymnosperms, however, are naked whereas those of Angiosperms are enclosed in the ovary or fruit.

The Gymnosperms are known, from geological evidence, to be enormously more ancient than any other flowering Plants ; they still retain many of the characters of their yet more primitive cryptogamic progenitors.[1]

[1] See Part II., *Flowerless Plants*, Chapter I.

GLOSSARY

Adventitious.—Not according to rule ; as in the case of roots which spring from the stem and not from the main root.

Albuminous.—1. Containing a nutritive tissue (endosperm) when ripe ; used of seeds.

2. Containing protein ; used of certain cells in Gymnosperms.

Alburnum.—The sap-wood.

Anatropous.—Bent back parallel to the funicle, so that the micropyle faces the placenta, the nucellus remaining straight ; used of ovules.

Andrœcium.—The stamens of a flower, collectively.

Annual.—A plant which lives for only one year or a single season, producing seeds before it dies.

Annual Rings.—The rings of wood in shrubs and trees, formed annually from the cambium.

Anterior.—Turned away from the main axis.

Anther.—The part of a stamen which contains the pollen.

Anthocyanin.—A class of pigments commonly present in the cell sap of petals and other parts of plants.

Antipodal Cells.—The cells in the basal part of the embryo-sac, away from the micropyle.

Archegonium.—The organ in which the female cell is produced in the Gymnosperms and Higher Cryptogams.

Auxin.—A special growth-promoting substance in plants.

Axil.—The angle between stem and leaf, on the upper side of the latter.

Axillary.—Arising in the axil.

Axis.—The stem or main root, as distinguished from its appendages, such as leaves or rootlets.

Bast.—Phloëm ; the tissue which conducts the food-substances formed in the leaf ; it may include fibres and parenchyma.

Biennial.—A plant which lives for two years, flowering only in the second year.

Bract.—A small leaf in the axil of which a flower is borne.

Bracteole.—A small leaf on the flower-stalk.

Bulb.—An underground storage organ, serving for perennation, in which swollen leaves or leaf bases are the chief storage tissue,

Calyptrogen.—The layer at the growing-point of a root, which forms the root-cap.

Calyx.—The outermost set of leaves in the flower.

Cambium.—The layer of actively dividing cells, which forms the secondary xylem and phloëm.

CAMPYLOTROPOUS.—With the nucellus curved back so that the micropyle faces the placenta ; used of ovules.

CAPSULE.—A dry, dehiscent fruit, formed of two or more united carpels.

CARPEL.—One of the innermost floral structures which bears or encloses the ovules.

CARPELLARY SCALE.—The lower of each pair of scales in a female Fir-cone, bearing the ovuliferous scale on its upper surface.

CASPARIAN STRIP.—The thickened part of certain walls of the endodermal cells.

CELL.—The structural unit in living organisms, consisting of cytoplasm, nucleus, and usually cell-wall.

CENTRAL CYLINDER.—The inner region of a stem or root, including xylem, phloëm, and conjunctive tissue.

CENTRIFUGAL.—Developing from within outwards.

CENTRIPETAL.—Developing from without inwards.

CHALAZA.—The base of the nucellus of an ovule, where it joins the funicle and where the integuments are attached.

CHLOROPHYLL.—The green colouring matter of plants necessary for carbon-assimilation in sunlight.

CHLOROPHYLL-GRANULE.—A cytoplasmic body (plastid) in which chlorophyll is contained.

CHLOROPLAST.—A plastid containing chlorophyll.

CHROMOPLAST.—A plastid containing colouring matter other than chlorophyll.

CHROMOSOMES.—The essential thread-like constituents of the nucleus, which divide longitudinally during mitosis.

COLLOIDAL.—Matter is said to be colloidal when it consists of a mixture of two phases. These may be both liquid, *e.g.* an emulsion of oil and water, or one may be solid and the other liquid, *e.g.* a suspension of clay particles in water. In each of these examples one of the phases (oil or clay) is discontinuous. A jelly (*e.g.* gelatine), however, consists of a continuous framework of gelatine with a continuous water phase permeating it (like the air spaces in a sponge). If the particles of the discontinuous phase are *extremely* small, the " system " is a true solution.

COMPANION-CELL.—A cell associated with a sieve-tube and formed from the same mother-cell.

CONJUNCTIVE TISSUE.—The tissue of the central cylinder, in which the xylem and phloëm are embedded ; it includes pith, medullary rays, and pericycle.

CONNECTIVE.—The part of the anther to which the pollen-sacs are attached.

CORK-CAMBIUM.—The layer of actively dividing cells forming the cork and phelloderm.

CORM.—An underground storage organ resembling a bulb, but in which the storage tissue is chiefly stem.

COROLLA.—The set of leaves in a flower, next within the calyx ; usually of a bright colour.

CORTEX.—The zone of tissue lying between the central cylinder and the epidermis.

COTYLEDON.—A seed-leaf ; the first leaf or one of the first two leaves of the seedling.

CUTICLE.—The external layer of the outer cell-walls of the epidermis.

CYTOPLASM.—The clear, homogenous part of the protoplasm of a cell, excluding the nucleus.

DEHISCENCE.—The opening of a fruit to emit the seeds, or of an anther to emit the pollen-grains.

DERMATOGEN.—The layer at the growing-point which forms the epidermis.

DIARCH.—Having two primary groups of xylem and two of phloëm ; used of roots. So also TRIARCH (3), TETRARCH (4), and POLYARCH (numerous).

DIGESTIVE SAC.—The temporary cap of a young rootlet, while still within the parent root.

ECTOPLASM.—The outer, clear layer of the cytoplasm.

EGG-APPARATUS.—The three cells (ovum and synergidæ) at the end of the Angiosperm embryo-sac towards the micropyle.

EMBRYO.—The young plant produced from the fertilised ovum.

EMBRYO-SAC.—The large cell of the ovule, containing, in Angiosperms, the egg-apparatus, antipodal cells, and polar nuclei, and, in Gymnosperms, the prothallus with archegonia.

ENDARCH.—The position of the protoxylem when on the inner edge of the metaxylem.

ENDODERMIS.—The innermost layer of the cortex, adjoining the central cylinder.

ENDOGENOUS.—Arising from deep-seated tissues.

ENDOPLASM.—The inner, granular layer of the cytoplasm.

ENDOSPERM.—A tissue formed in the embryo-sac, which feeds the embryo.

ENZYMES.—Complex substances, formed by living cells, in the presence of which certain chemical changes are rapidly effected.

EPHEMERAL.—A plant which lives for only a few months, producing seeds before it dies.

EPIDERMIS.—The skin of plants ; the superficial layer of tissue in stems and leaves.

EPITHELIUM.—A layer of cells secreting resin or other substances.

ETIOLATION.—The condition of a plant grown in darkness, in which the shoots are yellow, the internodes are elongated, and the leaves are small.

EXALBUMINOUS.—Containing little or no endosperm when ripe ; used of seeds.

EXARCH.—The position of the protoxylem when on the outer edge of the metaxylem.

EXODERMIS.—The outermost layer of cortex in a root, just below the piliferous layer.

EXOGENOUS.—Arising from the external tissues.

EXTRORSE.—Opening outwards ; used of anthers.

FASCICULAR.—Belonging to a vascular bundle.

FERTILISATION.—The union of the male cell with the female cell or ovum.

FIBROUS CELLS.—Long, thick-walled cells of the wood, serving for strength, sometimes retaining their living contents.

FIBROUS LAYER.—The thickened layer in an anther, which brings about its dehiscence.

FILAMENT.—The stalk of a stamen.

FRUIT.—The ripened ovary, containing the seeds.

FUNICLE.—The stalk of an ovule.

GAMETES.—Sexual cells, differentiated in the higher plants as male and female.

GENERATIVE CELLS.—The male cells or gametes, which effect fertilisation.

GEOTROPISM.—The response of plant organs due to the stimulus of gravity.

GROWING-POINT.—The part (usually the apex) of a stem or root, where the new tissues and organs are formed.

GUARD-CELLS.—The two cells of a stoma, enclosing the pore.

GYNŒCIUM.—The carpels of a flower, collectively.

HAIR.—An outgrowth of the epidermis.

HEART-WOOD.—The part of the wood which no longer conducts the transpiration stream, serving only for mechanical support.

HERMAPHRODITE.—Containing stamens and pistil in the same flower.

HILUM.—1. The scar on a seed, where it was attached to the funicle.
2. The first-formed part of a starch-granule, round which the successive layers are deposited.

HYDROTROPISM.—The response of plant organs due to a difference in relative humidity on different sides.

HYPOCOTYL.—The part of the stem below the cotyledons.

HYPODERMIS.—An outer zone of the cortex, next below the epidermis.

HYPOPHYSIS.—The cell in an embryo which joins the suspensor to the embryonic cell.

INFLORESCENCE.—A group of flowers (or a single flower) on a stem.

INTEGUMENT.—A coat or envelope of an ovule or seed.

INTERFASCICULAR.—Lying between the vascular bundles.

INTERNODE.—The portion of stem lying between two nodes.

INTRORSE.—Opening inwards ; used of anthers.

Ions.—The electrically charged parts into which a molecule of an acid, base, or salt breaks up when the substance is dissolved in water.

Leaf-trace.—A vascular bundle, or group of bundles, passing out from the stele into a leaf.

Lenticels.—Gaps in corky tissues, which allow of aeration of the internal living cells.

Leucoplast.—A colourless plastid, forming a starch-granule.

Lignin.—A substance of complex composition which impregnates the walls of woody cells.

Medulla.—The pith.

Medullary Ray.—A vertical plate of parenchyma running radially through xylem and phloëm.

Meristem.—A tissue of actively dividing cells, from which the permanent tissues are developed.

Mesophyll.—The internal tissue of the lamina of a leaf, including both palisade and spongy parenchyma.

Metaxylem.—The later-formed part of the primary wood.

Micropyle.—The opening in the coats of an ovule, through which the pollen-tube or pollen-grain enters.

Mitosis.—The complicated division of the nucleus, which involves the longitudinal fission of each chromosome, and the separation of the daughter chromosomes into the daughter nuclei.

Monœcious.—Having the stamens and pistils in different flowers on the same plant.

Mother-cell.—A cell which produces other cells by division, as the mother-cell of the four pollen-grains.

Mycorrhiza.—The intimate association of fungal hyphæ with the subterranean organs of certain plants.

Nectary.—A gland secreting honey.

Node.—The part of the stem to which a leaf is attached.

Nucellus.—The central body of the ovule, surrounded by the integuments and containing the embryo-sac.

Nuceolus.—A dense granule usually present in the resting nucleus.

Nucleus.—A definite organ of the cell, forming part of its living substance and controlling its development and multiplication.

Octants.—The 8 cells into which the embryonic cell first divides.

Orthotropous.—Straight, not in any way bent or curved ; used of ovules.

Ovary.—The part of the pistil in which the ovules are contained.

Ovule.—The young seed before fertilisation, consisting of nucellus, one or more integuments, and usually a stalk or funicle.

Ovuliferous Scale.—The upper scale of each pair in a Fir-cone, borne on the upper surface of the carpellary scale and itself bearing the ovules.

OVUM.—The female cell or gamete, which, after fertilisation, gives rise to the embryo.

PALISADE PARENCHYMA.—The part of the internal tissue of the lamina of a leaf which is turned towards the light; consisting of vertically elongated cells, rich in chlorophyll.

PAPILLA.—A small outgrowth or elevation of the surface.

PARENCHYMA.—Cellular tissue, with the cells not much elongated and usually having square or rounded ends.

PEDICEL.—The stalk of a flower.

PERENNIAL.—Lasting for several years ; used of herbaceous and woody plants.

PERIANTH.—The envelopes of a flower, including calyx and corolla when both are present.

PERIBLEM.—The layer at the growing-point, from which the cortex is formed.

PERICYCLE.—The outermost layer of the central cylinder, adjoining the cortex.

PERIDERM.—The secondary skin of plants, including cork on the outside and phelloderm on the inside, with cork-cambium between.

PERISPERM.—A tissue formed from the nucellus which feeds the embryo ; occurring in a few seeds, *e.g.* Water Lily.

PETAL.—A leaf of the corolla.

PHANEROGAMS.—Flowering Plants, including Angiosperms and Gymnosperms.

PHELLOGEN.—Cork-cambium, the layer of actively dividing cells forming the periderm.

PHLOËM.—Bast ; the tissue which conducts the food-substances formed in the leaf ; it may include fibres and parenchyma.

PHOTOSYNTHESIS.—Carbon-assimilation ; the decomposition of carbon dioxide in sunlight, by the aid of chlorophyll, with the resulting formation of new organic material.

PHOTOTROPISM.—The response of plant organs due to unequal intensities of illumination on different sides.

PHYLLOTAXIS.—The arrangement of the leaves on the stem.

PILIFEROUS LAYER.—The superficial layer, bearing the root-hairs in young roots.

PISTIL.—The central organ of a flower, composed of one or more carpels, and enclosing the ovules.

PITH.—The internal tissue of the central cylinder, lying inside the zone of vascular strands.

PITS.—The thin spaces in the walls of vessels, tracheids, and other cells.

PLACENTA.—The part of the ovary to which the ovules are attached.

PLASTID.—A cytoplasmic granule, usually forming starch, and often containing chlorophyll or other colouring matter ; includes chloroplasts, chromoplasts, and leucoplasts.

PLEROME.—The middle part of the tissue of the growing-point, forming the central cylinder.

PLUMULE.—The young stem and leaf-bud of the embryo.

POLAR NUCLEI.—The two nuclei, one from each pole of the Angiosperm embryo-sac, which unite, to form, after fertilisation, the nuclei of the endosperm.

POLLEN.—The cells discharged from the stamens ; they produce the male cells by which fertilisation is effected.

POLLEN-SAC.—The sac in which the pollen-grains are produced ; usually there are four such sacs in the stamen of an Angiosperm and two or more in that of a Gymnosperm.

POLLEN-TUBE.—The tube which grows out from a pollen-grain and conveys the male cells to the ovule.

POLLINATION.—The depositing of the pollen on the stigma or on the ovule.

POLYPETALOUS.—With the petals separate.

POLYSEPALOUS.—With the sepals separate.

POSTERIOR.—Turned towards the main axis.

PRIMORDIAL UTRICLE.—The cytoplasm lining the cell-wall and enclosing the central vacuole.

PROCAMBIUM.—A strand of cells in the meristem of the growing-point, from which a primary vascular strand will be formed.

PRO-EMBRYO.—The early stage of the embryo, when it consists of the embryonic cell and suspensor.

PROSENCHYMA.—A tissue consisting of elongated cells with pointed ends.

PROTHALLUS.—The tissue which bears the sexual cells.

PROTOPHLOËM.—The first-formed elements of the phloëm or bast.

PROTOPLASM.—The living substance of plants and animals.

PROTOXYLEM.—The first-formed elements of the xylem or wood.

RACEME.—An inflorescence consisting of a long main axis, bearing stalked lateral flowers, the lowest opening first.

RADICLE.—The young main root of the embryo.

RAPHE.—The part of the funicle which is adherent to the nucellus in an anatropous ovule.

RECEPTACLE.—The axis of a flower, on which the perianth, stamens, and pistil are borne.

REPLUM.—The persistent placenta in Cruciferæ, to which the seeds are attached.

RESPIRATION.—The breathing of plants, involving oxidation and the giving off of carbon dioxide.

RHIZOME.—A horizontal underground stem.

ROOT-CAP.—The sheath of protective tissue which envelops the tip of a growing root.

ROOT-HAIR.—A unicellular outgrowth of the young root, by which water and salts are absorbed from the soil.

ROOT-PRESSURE.—The osmotic pressure exerted by the cells of the root, resulting in the pumping up of water into the stem.

SAP-WOOD.—The part of the wood which conducts the transpiration stream.

SEED.—The sexually produced reproductive body of the Flowering Plants, consisting of the fertilised and matured ovule, containing an embryo, with or without a nutritive tissue (endosperm).

SEPAL.—A leaf of the calyx.

SEPTUM.—1.—A partition in the ovary or fruit, separating two carpels.
2.—A cell-wall separating one cell from another.

SIEVE-PLATE.—The perforated cell-wall through which the cells of a sieve-tube communicate with each other. Sieve-plates also sometimes occur on other walls of the sieve tubes.

SIEVE-TUBE.—A vessel of the phloëm, having perforated cell-walls between its cells.

SPERMATOZOID.—A male cell which has the power of active movement, as in most Cryptogams and in Cycads among Gymnosperms.

SPONGY PARENCHYMA.—The part of the internal tissue of the lamina of a leaf which is turned away from the light ; consisting of chlorophyll-containing cells, with large intercellular spaces.

STAMEN.—One of the organs in a flower or male cone, which bears the pollen-sacs.

STARCH-GRANULE.—A body consisting of starch $(C_6H_{10}O_5)n$ and formed by a plastid.

STELE.—The central cylinder, containing the vascular tissue.

STIGMA.—The part of the pistil which receives the pollen.

STOMA.—A pore, serving for the interchange of gases, and consisting of two guard-cells with an opening between them.

STROBILUS.—A cone, such as the male or female fructification of a Fir-tree.

STYLE.—The part of the pistil which connects the stigma with the ovary.

SUSPENSOR.—A filament of cells, forming a temporary organ of attachment for the embryo.

SYNCARPOUS.—Consisting of two or more united carpels.

SYNERGIDÆ.—The two cells accompanying the ovum in Angiosperms, and assisting in fertilisation.

TAPETUM.—The cells lining the inside of a pollen-sac, or adjoining the embryo-sac mother-cell in an ovule.

TAP-ROOT.—The main root of the plant, developed from the radicle of the embryo.

TESTA.—The seed-coat.

THALAMUS.—The receptacle.

TORUS.—The thickened portion of the closing membrane of a pit.

TRACHEA.—A water-conducting element of the wood ; either a vessel (cell-fusion) or a tracheid (single cell).

TRACHEID.—A water-conducting element of the wood, formed from a single cell, not undergoing fusion.

Transfusion Tissue.—The network of short tracheids, lying outside the vascular bundle in the leaf of a Conifer.

Transpiration.—The giving off of water vapour through the stomata, from the surface of the plant, especially from that of the leaves.

Tropism.—Curvature of a plant organ in response to an external stimulus.

Tuber.—A short and thick underground branch, serving for propagation.

Vacuole.—A clear space in the cytoplasm, filled with cell-sap.

Valve.—One of the segments into which the wall of a capsule splits up, on dehiscence.

Vascular Bundle.—A strand consisting of xylem and phloëm associated together.

Venter.—The lower part of an archegonium, containing the ovum.

Ventral Canal-cell.—The small cell lying immediately above the ovum, in Gymnosperms and the Higher Cryptogams.

Versatile.—Turning freely on its support, as the anther on the filament in the lily, etc.

Vessel.—A water-conducting element of the wood, formed by the fusion of a row of cells.

Whorl.—A ring of leaves or other organs, all inserted at the same level on the axis.

Xylem.—The wood ; that part of the vascular tissue which serves principally for the conduction of water, though also for food-storage and mechanical support.

INDEX

FUNDAMENTALS OF MODERN BOTANY

FUNDAMENTALS OF MODERN BOTANY

By

DUNKINFIELD HENRY SCOTT

Re-Edited by

F.T. BROOKS, M.A., F.R.S.

WITH 127 ILLUSTRATIONS
AND A GLOSSARY

Vol. II

ANMOL PUBLICATIONS PVT. LTD.
NEW DELHI - 110 002 (INDIA)

ANMOL PUBLICATIONS PVT. LTD.
4374/4B, Ansari Road, Daryaganj
New Delhi - 110 002
Ph.: 23261597, 23278000
Visit us at: www.anmolpublications.com

Fundamentals of Modern Botany

ISBN 81-261-1470-3 (Set)

PRINTED IN INDIA

Published by J.L. Kumar for Anmol Publications Pvt. Ltd., New Delhi - 110 002 and Printed at Mehra Offset Press, Delhi.

PREFACE TO THE TENTH EDITION

Having assisted in the revision of the last edition of *Structural Botany*, Part I., Dr. D. H. Scott requested me to re-edit *Structural Botany*, Part II.

In consultation with him it was decided to introduce certain new types in the present edition, namely, *Peronospora parasitica*, *Saccharomyces*, and *Euglena viridis*, and to exclude *Cladothrix dichotoma*.

Since the Ninth Edition of this book was prepared, a good deal of research has been done in connection with some of the types of Algæ and Fungi described therein, notably on *Ulothrix zonata*, on the life-history of *Ectocarpus siliculosus*, on the function of the spermatia of *Puccinia graminis*, and on the organisation of the hymenium of the Mushroom. As far as possible the results of these recent investigations have been incorporated; further information about the general biology of Fungi and Bacteria has also been included.

In preparing this new edition I have been greatly assisted by botanical friends. Professor F. E. Fritsch and Dr. Margery Knight have spent much time and trouble in giving me the benefit of their wide knowledge of the Algæ. Professor O. V. Darbishire has made valuable suggestions concerning the structure and mode of life of the Lichen, *Xanthoria parietina*, and he has most generously provided Figs. 99, 99*, 101, 101*, and

102. To all these and to my colleague, Mr. G. E. Briggs, I offer my best thanks for help freely rendered. I am also indebted to Mr. R. W. Marsh for Figs. 95*, 104*, and 117.

For the alterations which have been made in the present edition I am wholly responsible. I have tried—imperfectly, I fear—to retain the style of the original author.

F. T. BROOKS.

NOTES ON RECENT EDITIONS

THE chief alterations in the Seventh Edition were the introduction of a new type, *Chlamydomonas*, and the addition of a Glossary. A new figure was also added to illustrate further the spermatozoids of *Cycas*. I was indebted to Professor Dame Helen Gwynne-Vaughan, Professor F. E. Fritsch, and Professor G. S. West for kind advice and help on various points.

In the Eighth Edition, parts of the concluding chapter were modified in so far as the sketch of the relationships between the main groups was affected by remarkable discoveries (especially those of Drs. Kidston and Lang) in the older Devonian Flora. An account of these results will be found in the third edition of *Studies in Fossil Botany*.

In the Ninth Edition the wording was altered at several places, often in order to indicate a change of view as to the affinities of the plants discussed. Greater caution in questions of descent has proved necessary. In particular, the once popular belief in the origin of the Fungi from Algæ has been given up, chiefly on the ground of arguments adduced by Mr. F. T. Brooks in his Presidential Address to the British Mycological Society, 1923.

D. H. SCOTT.

AUTHOR'S PREFACE

Owing to the immense variety of organisation among the Cryptogams, it has been necessary to describe a much larger number of types in the present volume than in Part I. While it was possible to give some idea of the main outlines of structure in Flowering Plants by the full description of three representatives, it has seemed desirable to select no less than twenty-three types for the illustration of Cryptogams, and even then many important groups have been left out. The increased number of types has involved a curtness of treatment, in most cases, which only the relative simplicity of many of the forms has rendered possible.

It is hoped, however, that the essential morphological points have been brought out, and that a certain continuity has been maintained throughout the book, so that the study of the selected examples may serve to give a connected idea—however elementary—of the great groups of plants. In order to afford a general view of the whole field, the concluding summary has been added.

When theoretical points are touched on, the great aim has been to avoid dogmatism, and, so far as space permitted, to put the reader in possession of the evidence as a whole. This applies especially to the question of alternation of generations, as to the nature of which such different views are held.

As regards the fundamental homologies between Cryptogams and Phanerogams, an attempt has been made to demonstrate, and not merely to state them. Unless the student be taught to follow the reasoning by which such conclusions are arrived at, morphology loses at once its interest and its educational value.

It may be well to state again explicitly that the use of this book requires to be accompanied from the first commencement onwards—(1) by the study of living plants in the field, without which all botanical teaching is dull and barren; and (2) by practical work in the laboratory.

The author is indebted to the Trustees of the British Museum for permission to make use of the cuts, reproduced in Figs. 113–117, from Mr. Arthur Lister's Monograph of the Mycetozoa. As in Part I., the figures signed R. S. have been drawn from nature by Mrs. D. H. Scott. Figs. 5 and 44 are the work of Mr. W. C. Worsdell. The source of all figures not original is acknowledged in the descriptions. Special thanks are due to Professor J. Bretland Farmer for his kind help in connection with the Liverworts and the Fucaceæ.

D. H. SCOTT.

CONTENTS

TYPE V

TYPE VI

B. THE MOSSES

TYPE VIII

CHAPTER III

A. THE CHLOROPHYCEÆ

TYPE IX

TYPE X

TYPE XI

TYPE XII

TYPE XIII

TYPE XIV

TYPE XV

TYPE XVI

TYPE XVII

TYPE XVIII

CHAPTER IV

AN INTRODUCTION TO STRUCTURAL BOTANY

CHAPTER I

THE VASCULAR CRYPTOGAMS

TYPE IV

Selaginella Kraussiana

Selaginella, from which our first type of Cryptogams is taken, is a large genus, containing about five hundred species, most of which inhabit the damp forests of tropical countries. A few are natives of Europe, and one, *Selaginella spinosa,* grows in our own country, on boggy moors, or in mountainous districts. Some of the tropical species are universally grown in hothouses, and are often popularly called Lycopodium, but the real *Lycopodium*, or Club Moss, is quite a different, though an allied, genus.

In general appearance the *Selaginellas* resemble large Mosses, for they have long, usually creeping, stems, thickly clothed with numerous small leaves. With the true Mosses, however, which we shall describe later on, they have nothing whatever to do.

Selaginella is chosen as our first flowerless or Cryptogamic type, because in its reproduction and general course of development the genus comes as near to

Flowering Plants as any other Cryptogam [1] now living. In other respects, such as its vegetative anatomy, the structure of *Selaginella* is peculiar to itself. We shall therefore pass rapidly over this part of its organisation, and give most of our attention to those reproductive processes which illustrate the relation between Cryptogams and Phanerogams. We will, however, begin by examining the external characters of one or two of the species.

I. *EXTERNAL CHARACTERS*

A. Vegetative Organs

Selaginella Kraussiana, A. Br.,[2] a native of S. Africa, Madeira, and the Azores, and the commonest species cultivated in greenhouses, has a creeping stem, which, however, rises a little above the surface of the ground. The main stem is repeatedly forked, and the two branches arising at each bifurcation are alike. From the principal shoots other smaller branches are given off laterally, and these again bear still finer ramifications. The origin of the branches is really the same all through the plant, for all branches are really lateral, but they are formed so near the growing-point that the latter seems to give rise to two equal shoots. In the earlier ramifications, both shoots develop similarly, so that we cannot distinguish between the main axis and the branch. This

[1] The word *Cryptogams*, constantly used for Flowerless Plants, dates from Linnæus, who lived in the eighteenth century. It implies that in these plants the process of fertilisation is hidden, while in Flowering Plants (*Phanerogams*) it is manifest. This distinction no longer holds goods, for, with the help of the microscope, fertilisation is at least as easy to observe in Cryptogams as in Phanerogams. The names, however, are still kept up.

[2] Alexander Braun, the authority for the name.

is not the case with the later-formed branches, which are evidently different from the axis which bears them. When a growing-point gives rise to two perfectly equal shoots, the branching is said to be *dichotomous*. In *Selaginella*, the branching is not really dichotomous, but it comes very near to being so.

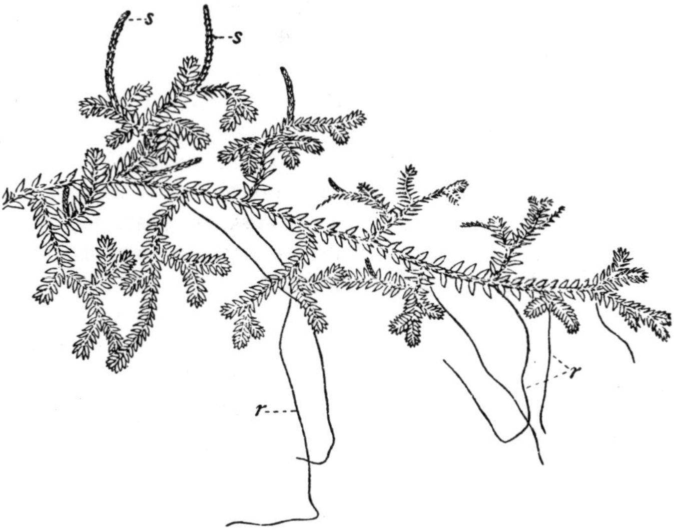

FIG. 1.—*Selaginella Kraussiana* ; general view. *r*. rhizophores ; *s*, spikes or cones. (Reduced.)

The stem bears very numerous small leaves, which are separated by distinct internodes on the older parts, but are crowded together towards the growing-points. The leaves are arranged in four rows, two of which spring from the lower and two from the upper side of the stem. The leaves on the lower side are much larger than those on the upper (see Fig. 2).

The arrangement, if carefully examined, is found to

be in pairs, each pair consisting of one of the large lower leaves and one of the small upper ones, which are exactly opposite each other.

Each leaf bears on its upper surface and close to the base, a small membranous outgrowth, *the ligule*, which is best observed on the very young leaves (see Figs. 10 and 16), as it soon withers and disappears. This ligule is characteristic of the whole genus *Selaginella*, and one other living genus, *Isoëtes*, and, unimportant as it seems, is a very ancient character, for it is found in a large family of fossil plants of the coal period (*Lepidodendreæ*).[1]

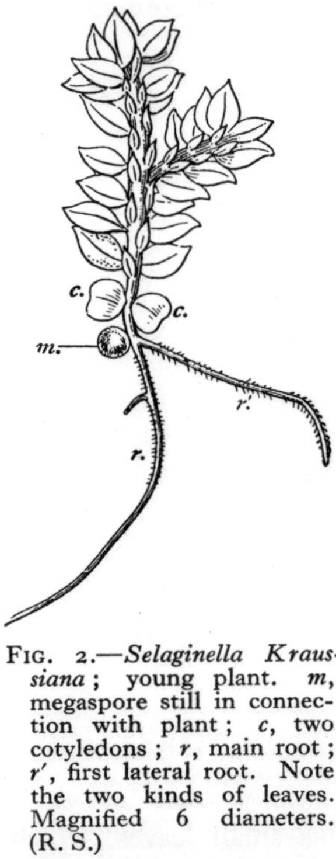

FIG. 2.—*Selaginella Kraussiana*; young plant. *m*, megaspore still in connection with plant; *c*, two cotyledons; *r*, main root; *r′*, first lateral root. Note the two kinds of leaves. Magnified 6 diameters. (R. S.)

At each ramification of the stem, a root-like organ is given off, which arises at the side of the stem, just below the fork (see Figs. 1 and 3). These organs, *the rhizophores*, are colourless and destitute of leaves; they grow straight down to the soil and resemble roots, but have no root-caps. On coming into contact with the ground they branch, giving rise to subterranean rootlets, which have root-caps as usual.

If the plant which we examine is fruiting, we shall find that some of the branches, instead of creeping along near the ground, grow straight upwards; it is these

[1] See *Studies in Fossil Botany*, 3rd ed.

vertical branches which form the terminal spikes or cones. The cones bear the reproductive organs; they differ from the vegetative branches in the fact that all their leaves are of the same size (see Figs. 1 and 3).

Other species of *Selaginella* differ very widely from that just described.

Some are minute creeping plants of almost microscopic dimensions, with unbranched stems (*S. simplex*); others have climbing stems, which ascend tall trees, and may attain a length of 60 feet (*S. exaltata*); while in others again the stem is stiff and erect, rising vertically to a height of three feet from the ground (*S. grandis*).

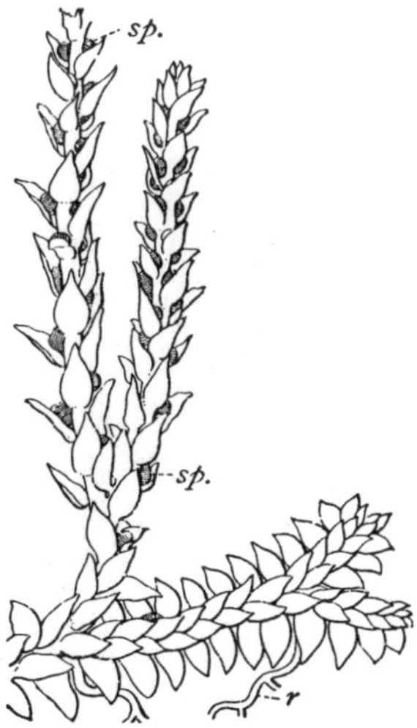

FIG. 3.—*Selaginella helvetica*, showing procumbent stem and two fertile spikes. *r*, rhizophore; *sp*, sporangia. Slightly magnified. (After Dodel-Port.)

A still more important variation is in the arrangement of the leaves. The majority of species agree with *S. Kraussiana* and *S. helvetica* (see Fig. 3), in having four rows of leaves, two large and two small, in the vegetative region, while in the spikes all the leaves are alike. In another group, however, to which our native species (*S. spinosa*) belongs (see Fig. 5), the leaves are all similar, and are arranged spirally, both on the ordinary stem and on the spike. In certain foreign species again, the case is just opposite, for the leaves of the spike, like those of the vegetative stem, are of two kinds.

B. Reproductive Organs

The reproductive organs of a *Selaginella* plant are the ***sporangia***, containing the spores. Each sporangium is a stalked sac, reaching a diameter of about a millimetre ($\frac{1}{25}$ inch), and is borne in the axil of one of the leaves of the cone (see Figs. 4 and 5).

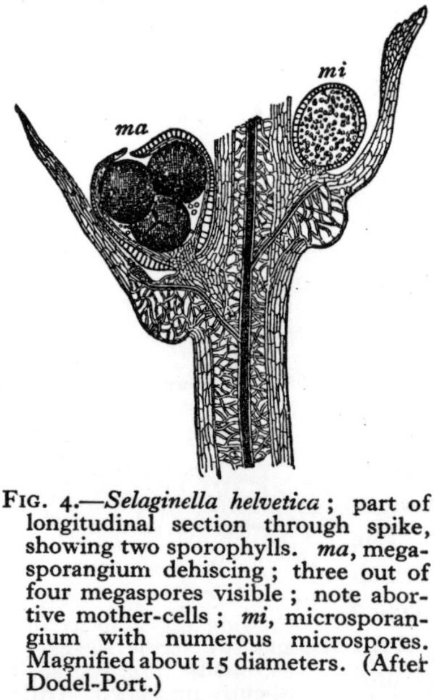

Fig. 4.—*Selaginella helvetica*; part of longitudinal section through spike, showing two sporophylls. *ma*, megasporangium dehiscing; three out of four megaspores visible; note abortive mother-cells; *mi*, microsporangium with numerous microspores. Magnified about 15 diameters. (After Dodel-Port.)

The sporangia are of two kinds: the one kind (the ***microsporangium***) contains very numerous small ***spores***[1] (***microspores***), comparable in size to pollen-grains. The other kind (the ***megasporangium*** or ***macrosporangium***) contains only four spores, ***megaspores*** or ***macrospores***, but these spores are so large that the sporangium which contains them has to be much larger than that which holds the innumerable microspores (see Fig. 4).

Both kinds of sporangia are borne on the same cone; generally the microsporangia are the more numerous, and occupy the axils of all the upper leaves of the cone, while the few megasporangia are found at the base of the cone only. The arrangement, however,

[1] The word *spore* is applied to any single cell which becomes isolated from the parent plant for reproductive purposes; cf. Part I. p. 113.

varies in different species. The development and structure of the sporangia will be further described below.

II. *INTERNAL CHARACTERS OF THE VEGETATIVE ORGANS*

Among the Flowerless Plants we find a very great variety in characters which in the Phanerogams remain fairly constant throughout whole Classes.

This holds good especially for the internal structure. A description of the anatomy of the Wallflower was sufficient to give a fair general idea of the chief anatomical features of the Dicotyledons generally, and so it was with our other types of flowering plants. With the Cryptogams the case is quite different. Not only is the anatomy of *Selaginella* peculiar to that one genus among plants now living, but the variation of structure among the species is so great that a general description, even of the genus as a whole, is impossible. In an elementary book, we cannot enter into all these variations; we can only give a short description of two or three forms, which may serve to give some idea of the peculiarities of the genus and of the range of variation among its species.

FIG. 5.—*Selaginella spinosa*; fertile spike. Magnified $1\frac{1}{2}$ diameters. (W. C. W.)

a. The Stem

In each of our types of Flowering Plants we found that the stem was traversed by one central cylinder, consisting of the vascular bundles and conjunctive tissue (see Part I. pp. 47, 152, 247). We learnt further that the bundles of the stem are directly continuous with those of the leaves. These facts hold good, with certain exceptions, for the Phanerogams generally.

In the Selaginellas the arrangement is totally different. The number of cylinders or *steles* varies from one up to five or more, not only in different species, but sometimes even in different parts of the same plant. The conjunctive tissue is very little developed, and pith is almost always absent, the whole interior of the cylinder being occupied by a solid strand of wood. Consequently it is generally impossible to distinguish the limits of the individual vascular bundles in the stele, or, to be more accurate, the stele in the stem is not differentiated into distinct bundles. Lastly, the vascular system of the stem is not built up entirely of leaf-trace bundles. The greater part of the xylem and phloëm can be traced continuously through the whole stem, and only certain portions of the vascular tissue are directly connected with the bundles of the leaves.

It will be convenient to begin with a short description of the anatomy of our native species, *S. spinosa*, which, though exceptional in the genus, illustrates several points of importance.

In the upper part of the ascending branches the stem has the structure shown in transverse section in Fig. 6.

There is a single central cylinder traversing the middle of the stem. This is surrounded by a wide intercellular space, which is bridged over at intervals by long radiating cells connecting the stele with the cortex. The latter is thick and of ordinary parenchymatous structure, and is bounded externally by a large-celled epidermis without stomata. Now, returning to the stele, we find the structure quite different from anything which we have previously met with in a stem. There is no pith whatever; the whole interior of the cylinder is occupied by solid wood, which consists entirely of tracheids. The development of this central mass of wood is also peculiar, for the first-formed elements or protoxylem-groups lie at the outside of the wood; in this particular case there are seven such groups, and it is

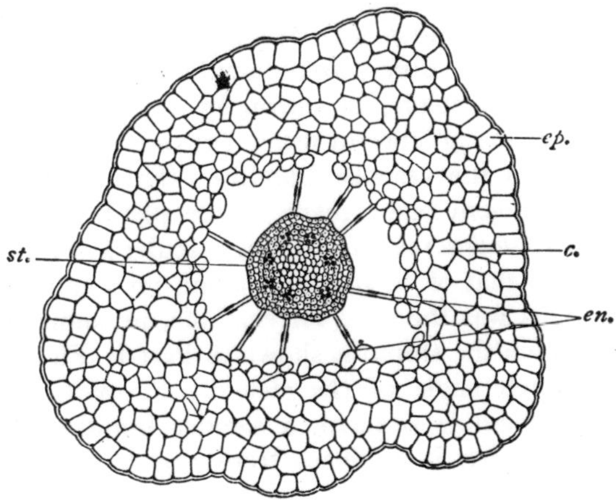

FIG. 6.—*Selaginella spinosa*; transverse section of stem. *ep*, epidermis; *en*, trabeculæ representing endodermis; *c*, external cortex; *st*, stele; the seven dark groups are protoxylem. Magnified about 35 diameters. (After Harvey Gibson.)

from these points that the development of the xylem starts; so we see that in this stem the wood develops centripetally, just as it does in the root of other plants. This is a very important difference from flowering plants.

This centripetal development of the xylem holds good as a general rule[1] for the stems of the Selaginellas and their allies.

[1] In the trailing part of the stem of *S. spinosa* it appears that the protoxylem is central. See also p. 13.

Surrounding the xylem is a ring of phloëm, consisting of parenchyma and sieve-tubes, but with no companion-cells. The sieve-tubes, like those of the Conifers, have their sieve-plates on the lateral walls. The whole stele is bordered by a layer of cells containing starch. Outside this layer is the intercellular space. Each of the cells, which stretch across the space, has a cuticularised band; these cells represent the endodermis.

We see, then, that we have a vascular structure in this plant which differs from anything which we have seen before in stems, as shown by (1) the centripetal xylem; (2) the absence of pith; (3) the want of separation between the vascular bundles. This type of stele is a very ancient one: many of the plants of the coal period (*Lepidodendron*, etc.) had a vascular system almost exactly like that of *S. spinosa*, though on a much larger scale. This was the case, for example, in stems such as that of which the stump is shown in Part I., Fig. 5.

From each angle of the stele in *S. spinosa*, where the protoxylem is situated, a slender bundle runs out to a leaf, which it traverses from end to end without branching.

As we have already mentioned, the structure of this species is exceptional in the genus; its interest lies chiefly in the resemblance to so many fossil forms, from which we may probably infer that it is a very primitive type of structure. A great many Selaginellas, like *S. spinosa*, have only a single vascular cylinder, or, in other words, are *monostelic*; but most commonly the single stele has a simpler structure.

If we now return to the species, *S. Kraussiana*, with which we started, we find a totally different arrangement. In this species, the stem is traversed by two parallel steles, each of which has a single protoxylem-group. The structure of these steles, their course through the stem, and their relation to the leaves, are sufficiently indicated in Figs. 7 and 8. In other species the steles are more numerous and are sometimes fused together in a complicated manner. The anatomical peculiarities of the stem of the genus *Selaginella* may be summed up as follows :

(1) The stele usually contains no pith.

(2) The vascular tissue of the stele is not divided into distinct bundles.

(3) The xylem is usually developed centripetally.

(4) In many species there is more than one stele.

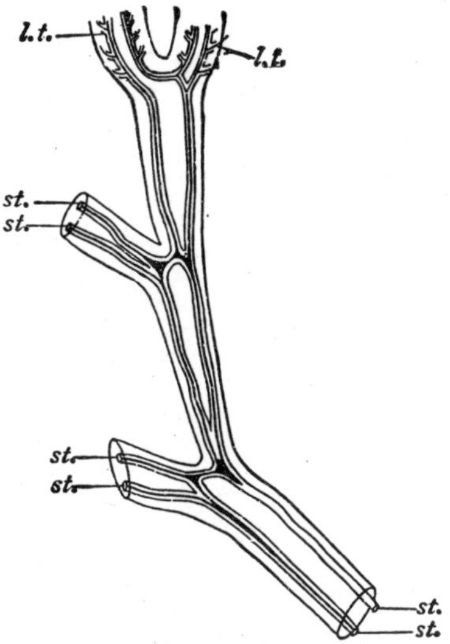

FIG. 7.—*Selaginella Kraussiana* ; diagrammatic transparent view of stem. *st*, the two steles, anastomosing at base of branches ; *l.t*, leaf-trace bundles, only shown in upper part. (After Harvey Gibson.)

As regards the details of the tissues, it is only necessary to add that the tracheids of the protoxylem are annular or spiral, as is usually the case. The other tracheids usually have long transverse pits, and are hence called scalariform (see Fig. 23, p. 45), from the ladder-like appearance which these pits give to their walls. We shall find this form of tracheid very general among the higher Cryptogams, and shall study it more fully in the Ferns. In one or two species of *Selaginella* true vessels, arising by cell-fusion, occur in the wood.

b. The Leaves

The leaves of *Selaginella* are of excessively simple structure; each leaf, as we have seen, receives a single vascular bundle from the stem. The bundle traverses the leaf from end to end, forming the midrib; it has no branches, neither is there any transfusion-tissue, which in Conifers takes the place of the branched veins. The bundle consists of a slender strand of tracheids surrounded by a thin layer of phloëm. Around the whole is a bundle-sheath.

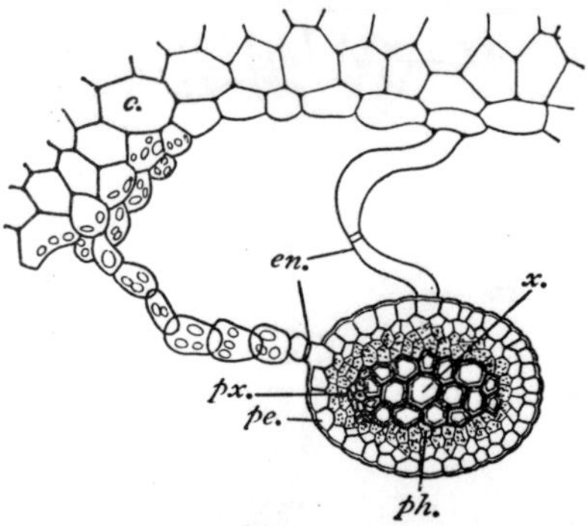

FIG. 8.—*Selaginella Kraussiana*; part of transverse section of stem showing one stele. *x*, the wood; *px*, protoxylem; *ph*, phloëm; *pe*, pericycle; *en*, endodermal cells forming the whole or part of trabeculæ; *c*, inner layers of cortex. Magnified about 100 diameters. (After Harvey Gibson.)

The mesophyll of the leaf is very slightly differentiated, the intercellular spaces being a little larger towards the lower surface. The epidermis, like the mesophyll, contains chlorophyll; the chlorophyll bodies in each cell are few and unusually large.

The stomata, which have the ordinary structure, are usually found on the under-side of the leaf only, and especially in the neighbourhood of the midrib. The membranous ligule at the base of the leaf on its upper surface has already been mentioned (p. 4; Figs. 10 and 16, pp. 17 and 28).

This is the simplest type of leaf that we have yet met with.

c. The Rhizophores and the Roots

These organs are generally similar to one another in structure ; the rhizophores in fact may be regarded as roots which have not yet begun to form a root-cap. The anatomical structure is simple, but unlike that in most other roots. There is a single stele, which contains only one group of xylem and one of phloëm. This structure, which may be called *monarch*, is pretty general in *Selaginella* and its allies. It is a very ancient character, for the rootlets of the fossil relations of *Selaginella*, which lived in the Carboniferous epoch, had an almost identical structure. The rhizophore, as distinguished from the root, of *S. Kraussiana* is peculiar in having central protoxylem.

d. Growing-Points and Mode of Branching

a. The Stem

The growing-point of the stem in *Selaginella* differs from that in Flowering Plants in the fact that the meristem at the apex shows no trace of stratification, *i.e.* of any arrangement in distinct layers. In many species, among which is *S. Kraussiana*, we find at the apex a single cell, larger than its neighbours, from the divisions of which all the new tissues and organs are ultimately produced. The presence of this single *apical cell*, as it is called, is very general, though not universal, among Cryptogams, and contrasts sharply with the small-celled meristem characteristic of the Flowering Plants. We shall, however, have better opportunities of studying growth by an apical cell when we come to other groups of Cryptogams, so we will defer the further consideration of this subject.

The *branches* arise laterally by the growth of a group

of cells just below the apex; the main axis and its lateral branch, however, often develop so equally that the stem appears to be forked. The branching of *Selaginella* may be described as an apparent bifurcation or *dichotomy*. True dichotomy only exists when the growing-point itself is equally divided. The branches of *Selaginella*, unlike those of Flowering Plants, are not axillary.

The development of the leaves proceeds in the same way as in the Phanerogams (see Part I. pp. 84, 169, 270).

β. *Rhizophores and Roots*

In those species which have special rhizophores, the latter may either grow by means of an apical cell, or may have a small-celled meristem more like that of the higher plants. As soon as the rhizophore reaches the ground, roots begin to grow out at its end. The growing-points of these roots arise *endogenously*, *i.e.* from the interior of the tissue at the end of the rhizophore. The roots themselves, which grow by means of an apical cell, differ from most other roots in their manner of branching. The rootlets arise quite close to the growing-point, where they are only covered by the root-cap of the parent root, so that they are not really endogenous in origin like other roots, which arise beneath the cortex. This is a very rare case, and does not hold good for most other Vascular Cryptogams.

We have found great differences between *Selaginella* and the Flowering Plants, even as regards the vegetative organs. The structure of both stem and root, the apical growth, and the mode of branching are all quite distinct from anything we met with in our Phanerogamic types. These differences, however, are unimportant compared with those which we shall find in the reproductive organs and life-history. We will now go on to describe the reproductive processes. We shall find that they at first seem quite distinct from those in Flowering Plants, but a careful comparison will enable us to see that there is, after all, a close relation between the two methods of reproduction. Among the flowerless plants *Selaginella* is one of those which in its course of development approaches most nearly to the Phanerogams. It is this fact which makes it of special interest to the student of botany.

III. *REPRODUCTION AND LIFE-HISTORY*

a. The Sporangia and Spores

We have already learnt that *Selaginella* is reproduced by spores, which are of two kinds. The organ in which the spores are immediately produced is the *sporangium*. In *Selaginella* a single sporangium is borne in the axil of each fertile leaf or *sporophyll* of the cone. The cone is a vertical shoot differing but little from the ordinary vegetative shoots of the plant, and bearing many sporophylls (see Figs. 3, 4, and 5).

We will now trace the development of a single sporangium. It does not matter whether we take a microsporangium or a megasporangium, for up to a certain point they develop in the same way.

Each sporangium arises just below the growing-point of the cone, from the outgrowth of a little group of meristematic cells, situated either exactly in the axil between leaf and stem or rather higher up, on the stem itself. A little ridge of tissue is thus produced, which at first consists of uniform cells. Very soon, however, a few cells in the middle of the young sporangium, lying immediately below its epidermis, begin to be distinguished by their more abundant protoplasm (see Fig. 9, *B*). Only one or two such cells are visible in a radial section such as that shown in the figure. This little group of cells is called the *archesporium* (see Part I. pp. 113, 275), for ultimately, after much growth and numerous cell divisions, it produces the spores. The archesporium soon becomes surrounded by a well-marked layer of cells, the *tapetum*, formed partly from the surrounding tissue, and partly from the archesporium itself (see Fig. 9, *A* and *B*, *t*). The whole sporangium continues to grow and its cells to divide. At the stage shown in Fig. 9, *A*, it already has a short stalk, and the archesporium has given rise to a many-celled tissue.

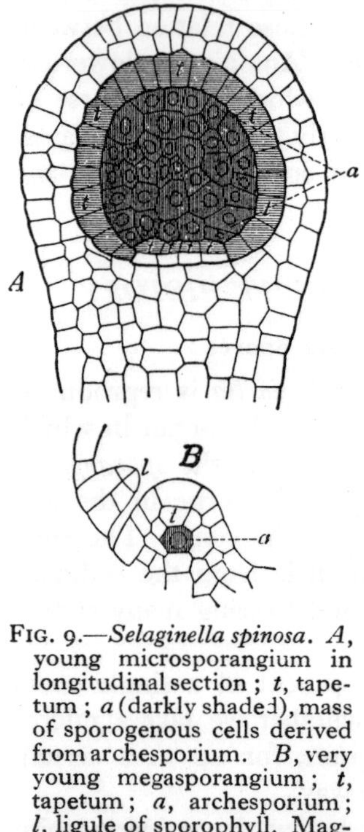

FIG. 9.—*Selaginella spinosa*. *A*, young microsporangium in longitudinal section; *t*, tapetum; *a* (darkly shaded), mass of sporogenous cells derived from archesporium. *B*, very young megasporangium; *t*, tapetum; *a*, archesporium; *l*, ligule of sporophyll. Magnified about 300 diameters. (After Goebel.)

Up to this point both kinds of sporangia behave exactly alike. The reader will at once see that, thus far, the development of the young sporangium is in all respects similar to that of a pollen-sac (compare Fig. 9, *A*, with one of the four pollen-sacs shown in Fig. 39, A, in Part I. p. 110. It is also evident that in the youngest stages there is a considerable resemblance to an ovule at its first origin (cf. Fig. 9, *B*, with Fig. 44, 1 or 2, in Part I. p. 121).

Henceforth it will be necessary to distinguish between

a microsporangium and a megasporangium. We will first describe the former.

The divisions of the archesporial cells of a microsporangium give rise to a mass of spore-producing tissue, each cell of which now rounds itself off and becomes a *spore mother-cell*. Each of the spore mother-cells, lying free within the cavity of the sporangium, next divides into four, the division taking place exactly in the same way as in the pollen mother-cells of Dicotyledons (see Part I. Fig. 39, B, p. 110). The four daughter-cells become the *microspores*. Each group of four is tetrahedrally arranged. The microspore acquires a cell-wall of its own, the outer layer of which is thickened and cuticularised.

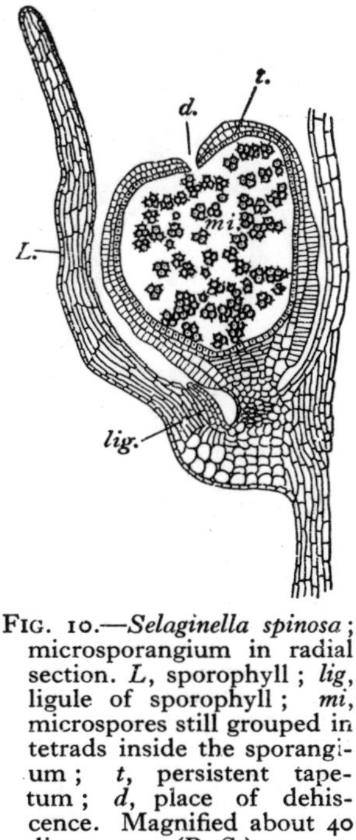

FIG. 10.—*Selaginella spinosa*; microsporangium in radial section. *L*, sporophyll; *lig*, ligule of sporophyll; *mi*, microspores still grouped in tetrads inside the sporangium; *t*, persistent tapetum; *d*, place of dehiscence. Magnified about 40 diameters. (R. S.)

The microsporangium is now ripe, and the space within the sporangium-wall is filled with thousands of microspores (see Figs. 4 and 10).

We see that in every respect the *microsporangium*, throughout its whole development, closely resembles a *pollen-sac*, while the *microspores*, in their structure and mode of origin, precisely correspond to the *pollen-grains*.

We will consider the further destiny of the microspores later on, and will now pass to the megasporangium.

Up to the time when the spore mother-cells become

isolated from one another, the development of the megasporangium goes on in just the same way as that of the microsporangium. Now, however, a striking difference manifests itself. In the megasporangium, out of all the numerous mother-cells, only a single one undergoes division ; all the rest remain undivided and are abortive (see Fig. 4).

The one favoured mother-cell divides into four tetrahedrally arranged daughter-cells ; each daughter-cell becomes a *megaspore*. The four megaspores develop enormously, and gradually displace and absorb all the remaining mother-cells, which, however, can be seen for a long time lying inert in the sporangial cavity. The four megaspores, as they grow, gradually take possession of the whole interior of the sporangium, which itself grows to a greater size than the microsporangium (see Figs. 4 and 11). The megaspores acquire greatly thickened cell-walls, the outer layers of which are cuticularised, and often have a rough and warty surface.

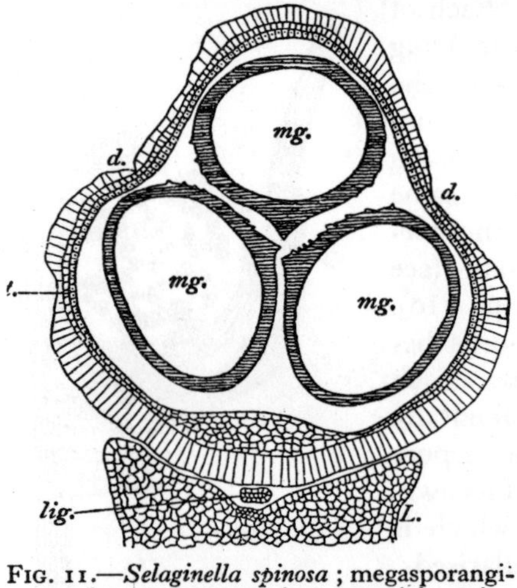

FIG. 11.—*Selaginella spinosa* ; megasporangium from same cone in tangential section (transverse to sporophyll); *mg*, megaspores, three out of the four are visible ; *t*, persistent tapetum ; *d*, line of dehiscence ; *lig*, ligule ; *L*, sporophyll. Magnified about 4c diameters. (R. S.)

We see then that a megasporangium differs from a

microsporangium in the fact that only one mother-cell divides, and that its daughter-cells occupy the whole sporangium, which thus contains four spores only.

The megaspores, the diameter of which is about twenty times that of the microspores, attain their great dimensions at the expense of the abortive mother-cells.

We cannot understand the relation of the microspores and megaspores to each other, or to the reproductive cells of Flowering Plants, until we are made acquainted with their further history. We will therefore now go on to describe the changes which take place in the spores, on their germination.

b. Germination of the Microspores

The microsporangium, when ripe, opens by transverse dehiscence, the walls splitting along a line parallel to the surface of the adjoining leaf (see Fig. 10). The microspores are thus set free, and if they fall on damp earth germination takes place.

The first change that happens is that the spore divides into two cells of very unequal size. A very small cell is cut off on one side of the spore (see Fig. 12, *A*, *p*). This little cell takes no further part in the development, and may be called the *prothallus-cell*. The large cell now divides into two equal parts (see Fig. 12, *A*); each half undergoes several further divisions. The final result is that we find the spore divided up into about a dozen cells. One of these is the little prothallus-cell first formed, which remains unaltered. Of the remainder, eight form an external layer, within which the central cells are enclosed, their number being either two or four according to the species. In *S. Kraussiana* there are four (Fig. 12, *B*, *c*). While these divisions are in progress, the spore grows a little, becomes more spherical, and bursts its outer hard membrane, so that its contents are now enclosed only by its inner cellulose wall. The central cells next undergo several successive divisions ; the small cells thus formed become rounded and lie freely in the space enclosed by the external layer (Fig. 12, *C*). Subsequently the external cells become disorganised, their contents contributing to the nutrition of the central group. The prothallus-cell, however, persists all through. The round central cells are alone concerned in the further development.

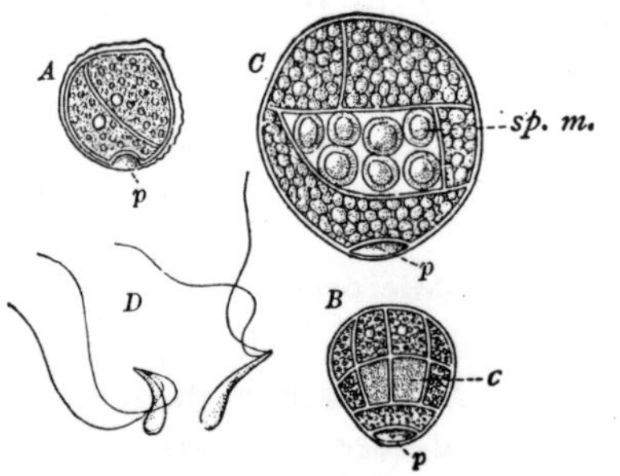

FIG. 12.—*Selaginella* ; germinating microspores. *A*, young stage ; *p*, prothallus-cell ; first wall formed in antheridium. *B*, more advanced stage ; *c*, central cells which will form spermatozoids. *C*, mature stage ; *sp. m*, spermatozoid mother-cells, surrounded by cells of wall of antheridium. *D*, free spermatozoids, each with two cilia. *A*, magnified 290 diameters ; *B*, magnified 290 diameters ; *C*, magnified 640 diameters ; *D*, magnified 780 diameters. (After Belayeff.)

When the external layer becomes disorganised, its cell-walls disappear and the contents flow together into a structureless mass, in which the round cells are embedded. In the meantime important changes go on in the contents of these cells. The large nucleus, which each cell contains, becomes converted into a long, rather club-shaped body, which has a spiral twist. At the thin end of this body there is a little cytoplasm, and at this point two excessively fine cytoplasmic threads are attached (see Fig. 12, *D*). The whole body now constitutes a *spermatozoid*, and the cytoplasmic threads are its *cilia*. The cell in which a spermatozoid is formed is called its *mother-cell*. The surrounding cells having completely broken down, the spermatozoid mother-cells are let loose into the water, for the whole process of the germination of the microspores can only go on in water. The microspores are, however, so small that a very little water is sufficient, such as we should find on the surface of the ground after rain or heavy dew.

The spermatozoids next become free from their mother-cells, the walls of which dissolve. As soon as the spermatozoids are at liberty, or even sooner, their cilia begin to lash about in the water, and when free the spermatozoid sets off in active locomotion, exactly like some water-animalcule.

The movement is a double one: the spermatozoid travels through the water with its narrow ciliated end foremost, and at the same time it rotates about its own axis. Its motion, in fact, is just like that of a rifle bullet through the air, or that of the screw of a steamer through the water.

We must remember that the spermatozoids are of a very minute size; the body is about 0·0125 mm. ($\frac{1}{2000}$ of an inch) long, the cilia about twice that length.

The spermatozoids are the bodies which perform the act of fertilisation. Each spermatozoid corresponds to one of the generative cells in the pollen-tube of Flowering Plants (see Part I. pp. 185 and 286). We know that the generative cell chiefly consists of a very large nucleus, with only a little cytoplasm. This is also true of the spermatozoid, which is all nucleus, except the small part at the pointed end and the cilia, which are cytoplasmic. The spermatozoid is an actively moving cell, which swims off on its own account, and may eventually find its way to an ovum. It thus differs from the generative cell of the higher plants, which, except in some of the Gymnosperms (see p. 311), is conveyed passively to its destination by the growth of the pollen-tube.

It may seem strange to us at first to find a cell belonging to a plant swimming actively about, as if it were an animal. When the first examples of such moving vegetable cells were observed, about the year 1830, the discoverer was so much astonished that he thought he had caught the plant at the very moment of its turning into an animal! Now, we know better. Actively moving cells are produced by most cryptogamic plants; sometimes they are male cells, as in *Selaginella*; sometimes they are sexless spores (see below, p. 144). Movement, in fact, is not specially characteristic of animals as distinguished from plants, for all protoplasm is capable of spontaneous motion. We have seen this already in the case of *Elodea* (Part I. p. 42), only there the movements go on within a closed cell-wall. Wherever movement is of advantage to the plant, we find that its protoplasm can show itself just as active as that of animal cells. In plants, however, owing to their different mode of nutrition, the necessity for locomotion arises less often.

We cannot follow the fate of the spermatozoids to the end, until we have seen how the megaspores germinate. Before we go on to this, however, we will carry the comparison between a microspore and a pollen-grain rather further than we have done hitherto.

If we refer back to the account given in Part I. (pp. 276 and 286) of the maturation and germination of the pollen-grain in the Spruce Fir, we shall recall the fact that several cell-divisions take place before the generative cells are formed. In like manner we have found several cell-divisions in the microspore of *Selaginella* before the spermatozoid mother-cells are formed.

The little prothallus-cell, which is first cut off, probably corresponds to the first two cells cut off in the pollen-grain of *Picea* (see Part I. Fig. 111, cells marked 2 and 3, p. 278).

In neither case do these cells take any part in the further development. The succeeding divisions are more numerous in *Selaginella* than in our gymnospermous type. The whole resulting group, including both enveloping and central cells, constitutes an ***antheridium***, the characteristic male organ of the Cryptogams. In the higher Cryptogams, this organ always consists of an enveloping layer of cells enclosing the central group from which the spermatozoids are derived. In the Gymnosperms the antheridium is only represented by the " stalk-cell " and the two generative cells (see Part I. Fig. 111, 4 and 5, p. 278). The whole is enclosed in the undifferentiated vegetative cell, which forms the bulk of the pollen-grain. In *Selaginella* the vegetative cell ceases to exist ; it is all used up in forming the antheridium, while in Gymnosperms it persists in order to produce the pollen-tube.

We see, then, that microspores and pollen-grains, which agree exactly in their mode of origin, agree also up to a certain point in their mode of germination. The differences between them are connected with the different means by which fertilisation is effected.

c. Germination of the Megaspores

Unlike the microspores, the megaspores of *Selaginella* begin to germinate while still in the sporangium. Each of the four megaspores is tetrahedral in shape, like a microspore. It contains at first a single nucleus and abundant cytoplasm, in which is a large vacuole containing oil. The nucleus lies near the angle, where the megaspore joins its three sister-cells. We will call this angle the *apex* of the megaspore. The first sign of germination is the division of the nucleus into two. The divisions are repeated many times, and soon cell-walls begin to appear in the cytoplasm, between the daughter-nuclei.

The cell-formation is at first limited to the apical part of the spore, but it gradually spreads downwards and inwards. In Fig. 13 a megaspore is shown which is already nearly half filled with tissue. As a rule, the cell-division extends so far that the tissue fills the whole cavity of the spore. In some cases this process is com-

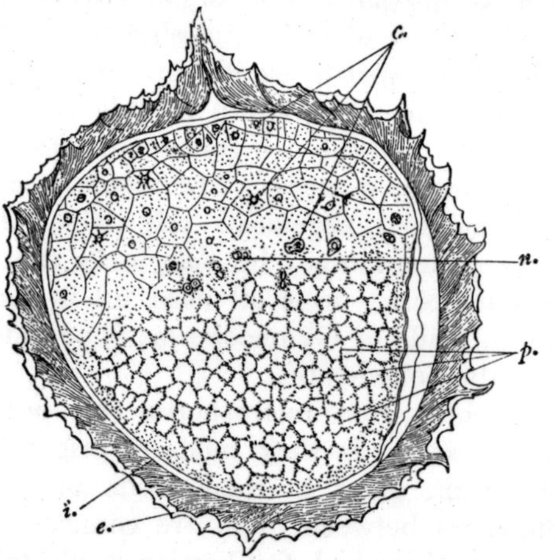

FIG. 13.—*Selaginella Martensii*; germinating megaspore. *c*, cellular tissue of prothallus, only complete in upper part of spore; *n*, free nuclei; *p*, undivided protoplasm of spore forming a network; *i*, inner, *e*, outer, layer of cell-wall of spore. Magnified 335 diameters. (After Heinsen.)

pleted even before the megaspores are set free from the sporangium, while in other cases the lower part of the tissue is developed after the spores have fallen on the ground.

The tissue which fills the megaspore is called the *prothallus*. The prothallus gives rise to the *archegonia*, or female organs. A cell at the apical end of the prothallus grows larger than the rest, and divides into two by a wall parallel to the outer surface. The upper cell divides by two longitudinal walls, crossing each other at right angles, into four, and each of these four cells divides by a transverse wall into two. Thus a neck is formed, consisting of eight cells arranged in two tiers (see Fig. 14, *n*, *n*). (Of course only four of the cells can be seen in longitudinal section.)

In the meantime the lower cell, which has so far remained undivided, forms an outgrowth which penetrates between the cells of the neck. This outgrowth is cut off as a distinct cell, called the *neck canal-cell.* Another smaller cell, the ventral canal-cell, is cut off below it; the remaining lower portion of the original central cell is the *ovum* (cf. Fern-archegonium, p. 66, Fig. 35). We see, then, that the development is just like that in *Picea*, except that in *Selaginella* we find two canal-cells instead of one. In *Selaginella* a real canal is formed, for the canal-cells break down, and an open passage, containing only mucilage, is left between the cells of the neck leading down to the ovum within (see Fig. 14).

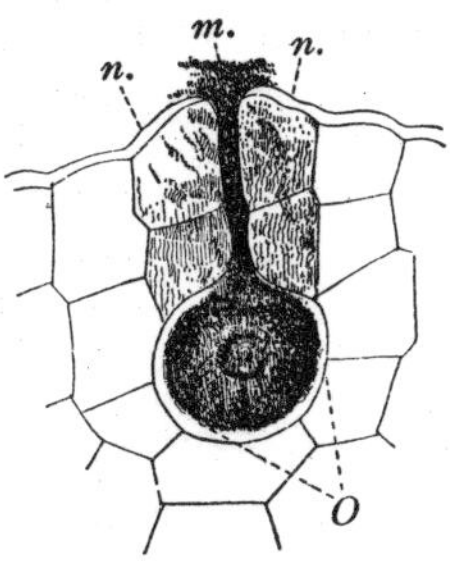

FIG. 14.—*Selaginella*; archegonium ready for fertilisation. *O*, ovum; *n*, cells of neck; *m*, mucilage in canal. Magnified about 500 diameters. (After Pfeffer.)

The first archegonium is sometimes formed before the megaspore is shed.[1] After the dehiscence of the megasporangium, which takes place as in the microsporangium by a transverse slit, the megaspores are shed on the ground. The growth of the prothallus

[1] It has occasionally been observed that fertilisation takes place and the embryos develop while the megaspores are still retained in their sporangium.

continues, the coats of the megaspore are ruptured at its apical end, and so the upper part of the prothallus becomes exposed, turns green, and grows a little beyond the limits of the megaspore (see Fig. 15).

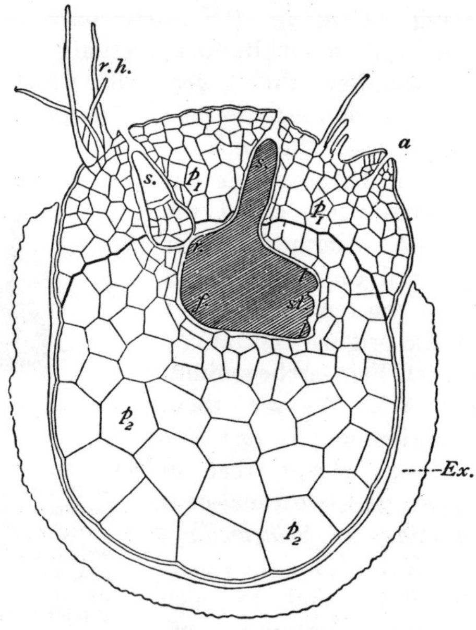

FIG. 15.—*Selaginella Martensii*; longitudinal section through old prothallus showing two embryos. p_1 and p_2, prothallus; *a*, unfertilised archegonium; *r.h*, rhizoids; *s*, suspensors of embryos; larger embryo shaded, no cells shown; *r*, root; *f*, foot; *st*, stem; *l*, *l*, cotyledons; *Ex*, wall of megaspore. Magnified 165 diameters. (After Pfeffer.)

Other archegonia are formed around the first one, and a few root-hairs (*rhizoids*) grow out from the prothallus. The mode of development of the prothallus bears a striking resemblance to that of the "endosperm" of the Spruce Fir or other Gymnosperms, so that we are justified in calling both by the one name of prothallus (see Part I. p. 281). The archegonia are formed in the same way in both. The prothallus of *Picea* is developed within the embryo-sac, that of *Selaginella* within the megaspore.

d. Fertilisation and Embryology

The archegonia are fertilised by spermatozoids; this takes place under water. The spermatozoids, when liberated from the ruptured antheridia, swim actively through the film of water covering the damp earth, and some of them are attracted to the archegonia of any female prothallus which lies near enough. The probable nature of the attraction will be considered when we come to the Ferns (see p. 68). At this time the mucilage formed from the disorganised canal-cells not only fills the canal of the archegonium, but spreads a little beyond its opening (see Fig. 14, *m*). The details of fertilisation are not so well known in *Selaginella* as in the Ferns, but there is no doubt that the spermatozoid becomes caught in the mucilaginous drop and then passes down through the canal to the ovum below. In other plants it has been proved that the spermatozoid unites with the ovum, the nucleus of the former fusing with that of the latter.

The really important distinction, then, between the fertilisation of a Cryptogam, such as *Selaginella*, and that of an ordinary Flowering Plant, consists in the mode in which the male cell is conveyed to the ovum. In the Cryptogams, the journey is accomplished by the active locomotion of the male cell itself; the spermatozoid moves by means of its cilia, and this can only take place under water. In most Phanerogams the generative cell is carried from the pollen-grain to the ovum by the growth of the pollen-tube; the final result, the union of the nuclei of the two cells, and, no doubt, of certain portions of their cytoplasm also, is the same in both Sub-kingdoms. It is now ascertained that spermatozoids are formed in certain Gymnosperms. (See p. 311.)

After fertilisation the ovum surrounds itself with a cell-wall of its own, and soon divides by a transverse septum into two cells. The upper cell, *i.e.* that lying next to the neck of the archegonium, becomes the sus-

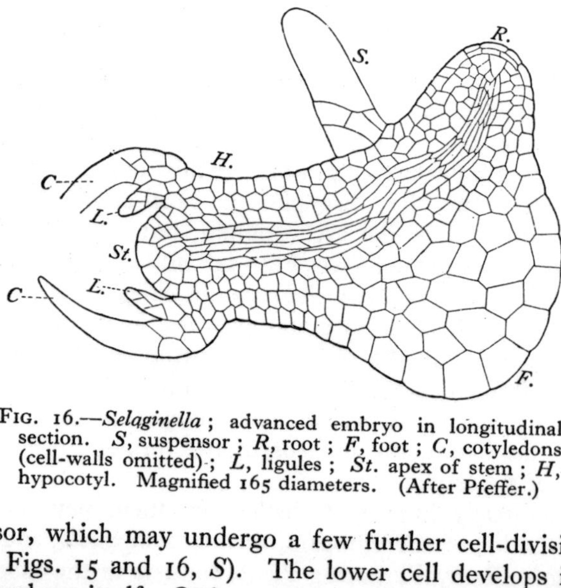

FIG. 16.—*Selaginella* ; advanced embryo in longitudinal section. *S*, suspensor ; *R*, root ; *F*, foot ; *C*, cotyledons (cell-walls omitted) ; *L*, ligules ; *St.* apex of stem ; *H*, hypocotyl. Magnified 165 diameters. (After Pfeffer.)

pensor, which may undergo a few further cell-divisions (see Figs. 15 and 16, *S*). The lower cell develops into the embryo itself. Owing to the growth of the suspensor in length, the embryo is carried deep down into the tissue of the prothallus (see Fig. 15). The embryonic cell undergoes segmentation, and very soon the first organs of the embryo are marked out. The apex of the stem lies at the end opposite the suspensor, but not quite in the middle ; the two cotyledons lie on either side of it. One side of the hypocotyl grows out into a temporary organ, the *foot*, which grows so rapidly as to force the apex of the stem to one side (see Figs. 15 and 16). This organ serves to absorb food for the young plant from the tissues of the prothallus. The first root appears a little later, between the foot and the suspensor. Fig. 15 gives a general idea of the position of these organs, and their relation to the prothallus. In Fig. 16 a slightly more advanced embryo is shown in greater detail. At this stage the young stele, consisting of procambial tissue, can already be traced from the apex of the stem to that of the root. It will be noticed that the cotyledons, like all the other leaves of the plant, possess ligules.

The stem begins to branch while the embryo is still enclosed in the prothallus. A young plant of *Selaginella Kraussiana* after germination is shown in Fig. 2. It is still attached by its foot to the megaspore, or rather to the prothallus inside it.

In possessing a suspensor, *Selaginella* resembles the Gymnosperms and most other Flowering Plants. The position of the organs of the embryo, however, is different. Although the embryo is dicotyledonous, like that of many Gymnosperms as well as of the Dicotyledons, properly so called, the apex of the stem is displaced towards one side, and the first root, though it arises near the suspensor, is not in a line with it (cf. Part I. Figs. 114, *F*, and 115, pp. 288, 289). The peculiarities of the embryo are connected with the presence of the absorbing organ called the foot, which in *Selaginella* and most of the higher Cryptogams performs the function of absorbing food from the prothallus, while in most Flowering Plants the corresponding function is discharged by the cotyledon or cotyledons.

Comparison between Selaginella and the Gymnosperms

At first sight the differences between even the highest Flowerless Plants, such as *Selaginella*, and the Flowering Plants, seem so great that we see little in common between them. In the Cryptogams we find no obvious flowers and no seeds, while fertilisation takes place in a totally different way from that which prevails among typical Phanerogams. If, however, we carefully compare the development of *Selaginella* with that of a *gymnospermous* Flowering Plant, we shall find it quite possible to trace the corresponding stages in their life-history ; and, having accomplished this for the lower Flowering Plants and the higher Cryptogams, it will not be impossible to extend the comparison further so as to include the Angiosperms on the one hand, and the simpler Cryptogams on the other. The relations between Gymnosperms and Angiosperms have already been briefly discussed in Part I. Chap. IV.

As regards the male organs, the comparison has been drawn above (p. 17). We need only recapitulate the results here. We found that the development and structure of the microsporangium of *Selaginella* agreed very closely with that of a pollen-sac in the Flowering Plants. The microspores produced in the former are developed precisely in the same way, and have just the same structure as the pollen-grains formed in the latter. When the microspore germinates it begins by cutting off a little inactive cell (the prothallus-cell), and the pollen-grain of a Gymnosperm does the same. The subsequent divisions lead to the formation of the spermatozoids in the Cryptogam, and of the generative cells in the Gymnosperm. In both cases these are the bodies which effect fertilisation. That the generative cells are homologous with spermatozoids has long been recognised.

Two Japanese botanists were the first to make the brilliant discovery that in some of the Gymnosperms the generative cells become converted into actively moving spermatozoids comparable to those of the Ferns (see p. 311). The antheridium, *i.e.* the organ in which the male cells are formed, is more complex in *Selaginella* than in the Gymnosperms, for in the latter it has almost become reduced to its most essential part, the generative cells themselves. The pollen-tube of the Gymnosperm is not represented in *Selaginella*, for in the latter the whole contents of the microspore are used up to form the prothallus-cell and antheridium. The comparison of the development gives us then the following chief results :

Selaginella.		*Gymnosperm.*
1. Spermatozoids	=	Generative cells.
2. Prothallus and antheridium	=	Cell-group in pollen-grain.
3. Microspore	=	Pollen-grain.
4. Microsporangium	=	Pollen-sac.

The student, however, must clearly understand that it is quite useless to learn up the names of the equivalent organs, unless he thoroughly grasps the developmental facts on which their comparison is based.

We will now compare the female organs in the two Classes. It is best to start with the ovum, which is beyond question the same thing in both. The ovum in *Selaginella* is produced in an archegonium, which is almost exactly like that of a Gymnosperm, such as *Picea*, and which develops in just the same way. The archegonia of *Selaginella* are formed from superficial cells of the prothallus, exactly as those of the Fir are formed from superficial cells of the " endosperm." The development, moreover, of the prothallus itself is just like that of the " endosperm." In both cases free nuclear division first takes place, then cell-formation begins, and the tissue thus formed goes on growing until it has filled all the available space. We can have no doubt, then, that the female prothallus of *Selaginella* corresponds to the " endosperm " of the Fir, which we may, if we like, call by the same name. The only difference is that, in *Selaginella*, the prothallus grows a little way out from the megaspore, becomes partly green, and forms a few rhizoids. It is, in fact, a more independent structure in the case of the Cryptogam, developing freely on the ground, instead of within the closed tissues of the ovule.

Now the cell in which the prothallus of *Selaginella* develops is the *megaspore*, while that in which the " endosperm " of *Picea* arises is the *embryo-sac*. We therefore arrive at a new term in the comparison ; the megaspore is the equivalent of the embryo-sac.

There are some differences, however, to be dealt with at this point : in normal cases the megaspore is set free and completes its development on the ground, while the embryo-sac remains always enclosed in the ovule or seed. We must remember, however, that the megaspore itself begins its germination while still in the megasporangium, and, in exceptional cases, even the embryo-plant may be developed in this position. The thick cuticularised wall of the megaspore is obviously a necessity for its protection when it becomes freely exposed. It is interesting, however, to know that in some Gymnosperms the membrane of the embryo-sac likewise becomes cuticularised. It has also been found that in some plants of this class (Cycadeæ) the development of the " endosperm " is only completed while the seeds are lying on the ground, and in a few cases the " endosperm " has been observed to burst through the embryo-sac and seed-coats and to become green, just like the prothallus of *Selaginella*.

A more serious difficulty is that there are four megaspores in *Selaginella*, and only one embryo-sac in the

Fir, though there are Gymnosperms which have more than one, as *Gnetum*. In most Gymnosperms the sister-cells of the embryo-sac become abortive at an early stage of their development, as is the case also with the sister-cells of the fertile megaspore in some fossil relations of *Selaginella*, exceptionally in species of *Selaginella* itself, and constantly in the heterosporous Water-Ferns.

The organ in which the megaspores are produced is a megasporangium; that in which the embryo-sac develops is the ovule. Both organs arise in the same way from a group of cells near the growing-point. The similarity of their development has already been pointed out (see p. 15). We infer, then, that the megasporangium corresponds to the ovule, or more strictly to the nucellus of the ovule, for the megasporangium has no integument.

We have found, however, that the megasporangium and microsporangium are just alike in the earlier stages of their growth; the former, as we have seen, corresponds to the nucellus of an ovule, the latter to a pollen-sac. Hence we must draw the conclusion that a pollen-sac and the nucellus of an ovule are equivalent structures—a result which could only have been arrived at by a comparison with Cryptogams. We may sum up our inferences as to the relations of the female organs in the two types thus:

Selaginella.		*Gymnosperm.*
1. Ovum	=	1. Ovum.
2. Archegonium	=	2. Archegonium.
3. Prothallus	=	3. Endosperm.
4. Megaspore	=	4. Embryo-sac.
5. Megasporangium	=	5. Nucellus of ovule.

If we try to carry this comparison further, and to find the equivalent in the Cryptogam of the carpels and stamens of the Flowering Plant, we meet with some difficulty. In *Selaginella* neither kind of sporangium is

3

borne actually on a leaf, but in its axil. Similar cases, however, are known among Flowering Plants. We may regard the leaves, in the axils of which the sporangia of *Selaginella* are situated, as representing stamens or carpels, according as the adjoining sporangium is a micro- or megasporangium. There is, however, no differentiation between carpels and stamens in *Selaginella* or in any Cryptogam. We may compare the whole spike of *Selaginella* to a hermaphrodite [1] flower with no perianth, and with stamens and carpels resembling each other. In some Gymnosperms also, as in certain members of the group Cycadeæ,[2] the stamens and carpels are just alike in their vegetative parts.

We see, then, that although Cryptogams and Phanerogams appear to differ so completely from one another, we can yet successfully compare them together and determine the relations between their organs.

Organs which resemble each other in their development and their place in the life-history, so that we regard them as morphologically the same organ, are said to be *homologous* one with another. Organs, on the other hand, which are morphologically different, but are adapted to the same physiological function, are said to be *analogous*. To go back to our old illustration in the introduction to Part I. (p. 4), the tuber of a potato is *homologous* with a branch of the stem, but *analogous* with a fleshy root such as that of a carrot. In our comparison between

[1] Flowers are called *hermaphrodite* when, as in those of the Wallflower and the Lily, the stamens and carpels are both contained in the same flower.

[2] As the Cycadeæ have been referred to more than once, it may be worth while to mention that they are a group of Gymnosperms, of great geological antiquity, now represented by a few tropical genera, with a palm-like habit. They all have pinnate leaves of great size. The student will find a magnificent collection of living Cycadeæ in the Palm-house at Kew.

Selaginella and a Gymnosperm we have aimed at establishing the *homologies* of the various organs; such comparisons are essential in order to determine the relationships of different groups of plants, for it is only by tracing the homology or morphological equivalence of organs that we can form any idea of the probable modifications which may have taken place during the course of descent.

The proof that the reproductive organs in Flowering and in Flowerless Plants are homologous was due to a German botanist named Hofmeister, and is one of the greatest discoveries ever made in morphology. We have every reason to believe that the Flowering Plants are descended from Cryptogams, which resembled *Selaginella* in having two kinds of spores. Their actual ancestors, however, which have been extinct for millions of years, were no doubt totally different in all other respects from any Cryptogams now living (see p. 311).

One point remains: we have not yet considered the homologies of the seed. One part of the seed, the testa, is not represented in *Selaginella*, for the megasporangium is without integuments. The megasporangium itself corresponds to the nucellus of an ovule, as we have already seen, but it never develops into anything of the nature of a seed. This is because the megaspores are normally set free from the megasporangium before fertilisation takes place, so that the seed-stage is never reached. The megaspore, when it is filled with prothallus and contains an embryo, bears a certain resemblance to a seed, but there is no complete homology; for, as we have already seen, the development shows that the megaspore is homologous with the embryo-sac only.

The typical seed, such as we find in the higher Flowering Plants, represents a persistent, integumented megasporangium, containing a single megaspore, which produces a prothallus, and, after fertilisation, an embryo, while still *in situ*. The shedding of the seed thus corresponds to the detachment of the entire megasporangium, together with its integument and contents. Bodies closely analogous with seeds are found in some fossil Lycopods allied to *Selaginella*, but the true seed of the higher plants appears to have been evolved among the Pteridosperms, Palæozoic Seed-plants resembling Ferns. Some primitive seeds, however, may have only developed an embryo when germination set in, as sometimes happens in Cycads at the present day.

TYPE V

The Male Fern (*Aspidium Filix-mas*, L.)

The Ferns are a vast group, enormously outnumbering all the other Vascular Cryptogams put together. The Class in the widest sense includes about 150 genera and over 6000 species. In our own native Flora seventeen genera and about forty species are represented. The limits of both genera and species are, however, very indefinite among the majority of Ferns. The Ferns are the only Class of Vascular Cryptogams which has successfully held its own down to the present day, while the other groups are represented by comparatively few surviving forms.

It need hardly be said that among this immense group of plants every possible variety of habit is to be found, while in structure the differences are scarcely less great. In size, we find every gradation, from the Tree Ferns of the tropics and New Zealand, which may reach 60 feet in height, down to minute Filmy Ferns hardly larger than Mosses. The main outlines of the life-history are, however, with few exceptions, fairly uniform throughout.

The particular Fern which we have chosen as a type serves well to illustrate the chief points in the development and morphology, but we cannot expect any single representative to give a fair conception of the Class as a whole.

I. *EXTERNAL CHARACTERS*

A. Vegetative Organs

The Male Fern, one of the commonest British Ferns, grows abundantly in woods and hedgerows. The short, stout stem grows obliquely upwards, and rises but little above the surface of the ground. It often reaches a length of about eight inches and a diameter of about one inch ; but it appears much thicker than it really is, because it is completely covered with the bases of the old leaves.

The stem at first has the form of a cone with the thin end downwards ; for it grows thicker for a time towards the top, till a constant diameter is attained (Fig. 17). This is commonly the case in plants which have no secondary growth in thickness, and in which, therefore, the increase in bulk must depend entirely on gradual strengthening of the growing-point. In this respect, though in no other, Ferns resemble the Monocotyledons (see Part I. p. 173).

The leaves, often called *Fronds*, are of very large size, one to three feet long, and much subdivided (see Figs. 17 and 18).

This is the first example of a *compound leaf* we have had. A compound leaf is one in which the lamina or blade is completely subdivided, so that its several parts, called *leaflets*, resemble distinct leaves. The leaves of the Male Fern are *pinnate*, that is, the main stalk, or *rachis*, bears two rows of leaflets, or *pinnæ*, one row on each side (see Fig. 18). The pinnæ are often subdivided themselves in the same way, and then the whole leaf is

FIG. 17.—General view of the Male Fern. *a*, apex; *b*, base of stem, which is covered with the remains of old leaves, and bears numerous adventitious roots; *c*, *c*, young leaves, showing circinate vernation. Greatly reduced.

said to be *bi-pinnate*. In the specimen figured, the pinnæ are deeply lobed, but not completely subdivided. Each lobe, like that shown singly in Fig. 18, B, may be called a *segment*. The segment is traversed by a midrib, springing from the rachis, and giving rise to lateral veins, which fork, and end near the edge of the leaf. This type of venation is a common one in Ferns (see Fig. 18, B).

The petiole and rachis, and sometimes also the larger veins of the leaflets, are clothed with brown chaffy hairs, the *paleæ* or *ramenta*, which are characteristic of the Class (see Fig. 18).

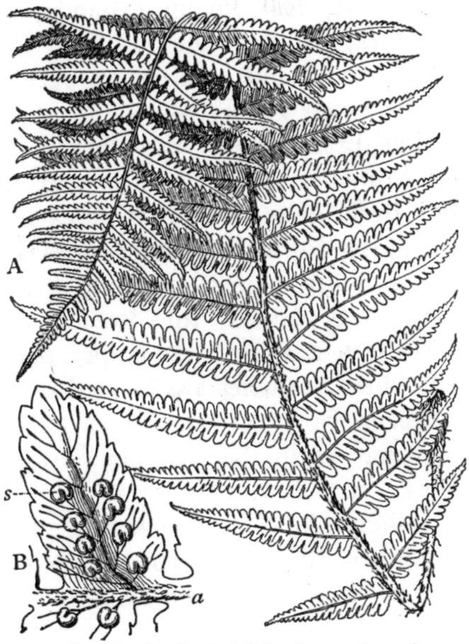

FIG. 18.—A, leaf of Male Fern (much reduced). B, part of a fertile pinna seen from below; *a*, rachis; *s*, sorus. Magnified. (After Luerssen.)

The leaves develop very slowly, arising in the bud two years before they unfold. The growth of the leaf goes on at the apex like that of the stem. Thus the leaf-stalk is the first part to be formed, and is generally the only part developed in the first year of growth. The blade is formed later, and this also grows from the base upwards. The blade of the young leaf is rolled up in such a way that the rachis or midrib forms a spiral like a watch-spring, the apex of the leaf being at the centre of the spiral (see Fig. 17, *c*). Everyone who has ever watched a Fern coming up in spring must have noticed the form of the young leaves. The curvature is due to the greater growth of the under-side of the leaf, which is external in the bud. Each leaflet is coiled up in a similar way. This mode of folding of the young leaf is called *circinate* or crosier-like *vernation*, vernation being a general word for the folding of a leaf in the bud. When the leaf finally expands, the inner side grows more rapidly than the outer, so that the curves become straightened out. Circinate vernation is characteristic of the Ferns generally.

The branching of the stem in this Fern is peculiar; no branches at all are formed at the growing-point, but buds arise on the petioles of some of the leaves, springing from their outer sides a little above the base. These buds, though their first origin takes place very early, only develop into branches at a much later time, and often not until the upper part of the leaf has died off. Few branches are formed in this particular Fern.

Ferns vary very much as regards their branching; in some, as in the Bracken Fern, the stem forks at the apex; in a few, as in some Filmy Ferns, the branching is axillary like that of flowering plants, while in others, as in some of the Tree Ferns, the stem does not branch at all.

The *roots* which we find on an ordinary full-grown plant are all adventitious, for the original main root of the embryo dies away very early. The adventitious roots, which arise at the bases of the leaves, usually three below each leaf, are very slender and much branched. An old stem is densely clothed with a matted growth of adventitious roots (see Fig. 17).

B. Reproductive Organs

The ordinary Fern plant, such as we have described, is purely asexual. Like *Selaginella*, it bears, at this stage, sporangia only, but, unlike that genus, the sporangia and the spores which they contain are all of one kind. The sporangia of the Male Fern and of most other Ferns are borne on the lower surface of the ordinary foliage leaves, so that here there is no difference between vegetative leaves and sporophylls. In this respect such Ferns are on a lower level, as regards the physiological division of labour, than any plants which we have yet considered.

If we examine one of the fertile leaves in summer, the clusters of sporangia, or *sori*, as they are called, are very conspicuous on the under-surface. They are usually absent from the basal part of the leaf. On the larger segments, the sori are arranged in two short rows, one on each side of the midrib (see Fig. 18, B), while on the smaller segments there may be only one or two sori altogether. Each sorus is covered by a kidney-shaped membranous envelope called the *indusium*, and is seated just over one of the lateral veins. The individual sporangia, which cannot be distinguished without the aid of a lens, are very numerous in each sorus, and every sporangium contains a large number of spores, so that the reproduction of the plant is extremely well provided for.

On germination, each spore gives rise to a *prothallus*, which is a much larger structure here than in *Selaginella*, and leads quite an independent existence (see Fig. 31, p. 61). The prothallus is a flat, green, heart-shaped body, sometimes as much as half an inch in diameter, attached to the soil by the root-hairs or rhizoids, on its under-surface. Prothalli may be found in abundance covering the damp ground where Ferns are growing. In Ferns the same prothallus usually bears both kinds of sexual organs, the antheridia and archegonia. After fertilisation the ovum formed in one of the archegonia becomes an embryo, which eventually grows up to be a new Fern plant.

In Ferns, then, we have, in normal cases, a sharp *alternation of generations*. The Fern plant is the asexual generation, or *sporophyte*, producing the sporangia, and ultimately the spores. The prothallus is the sexual generation, or *gametophyte*,[1] producing the antheridia and archegonia, in which the sexual cells are developed.

II. *INTERNAL STRUCTURE OF THE SPOROPHYTE*

A. The Vegetative Organs

1. The Stem

a. The Vascular System

In all the plants with which we have had to do so far (except some of the Selaginellas), the vascular system of the stem evidently constitutes a single cylinder, or *stele*. In the Male Fern and most other Ferns this is only the case while the stem is still quite young. As the growing-point strengthens, a more complex structure is gradually built up, and the stem soon becomes *polystelic*, *i.e.*, in all the later-formed part of the stem the original *stele* has broken up into a network of vascular strands or steles ; in fact the simple *monostelic* structure is only found at the very base or oldest portion of the stem ; this part soon dies away altogether, so the whole stem of a Male Fern, as soon as it is once fairly established in the ground, is polystelic. This is the case with the majority of Ferns, but a few, such as the Filmy Ferns, retain a monostelic structure throughout life.

If we examine with the naked eye, or with the aid of a

[1] The word *gamete* means a sexual cell.

lens, the transverse section of the mature stem of the Male Fern, we at once see the cut ends of the steles, arranged in a ring and embedded in ground tissue (see Fig. 21).

FIG. 19.—Longitudinal section of small stem of Male Fern. *a*, apex; *l*, *l*, bases of leaves; *st*, stem; *g*, leaf gap. Natural size. (After Sachs.)

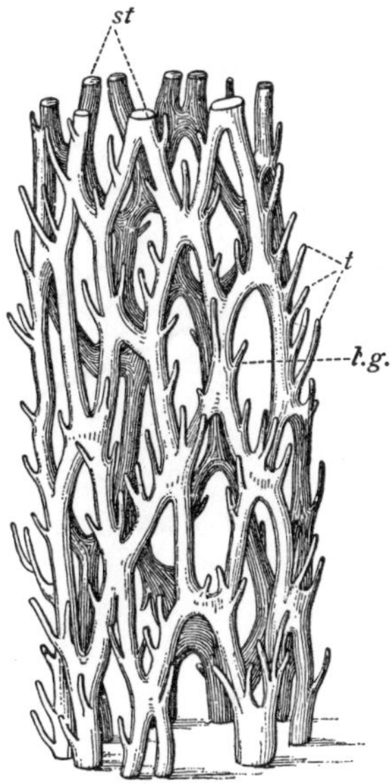

FIG. 20.—Bundle system of the Male Fern dissected out. *st*, principal steles of the stem; *l.g.* leaf gap corresponding to the insertion of a leaf; *t*, steles passing out into the leaf. Magnified. (After Reinke.)

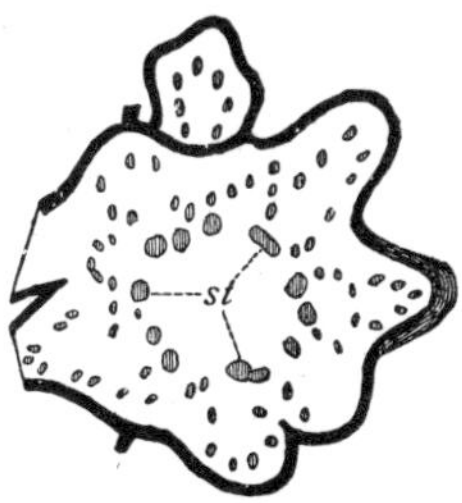

FIG. 21.—Transverse section of stem of Male Fern, showing the bases of leaves. *st*, principal steles of the stem. (After De Bary.)

In order to understand the arrangement it is necessary to make a dissection, carefully removing the parenchyma and leaving behind the vascular skeleton only. Such a preparation is shown in Fig. 20. We now see that the steles form a hollow network, with large diamond-shaped meshes. Each mesh corresponds to the base of a leaf; the steles bordering the mesh give off branches, which enter the petiole (see Figs. 19, 20). As the leaf-bases of the Male Fern completely cover the surface of the stem, and no internodes are developed, every transverse

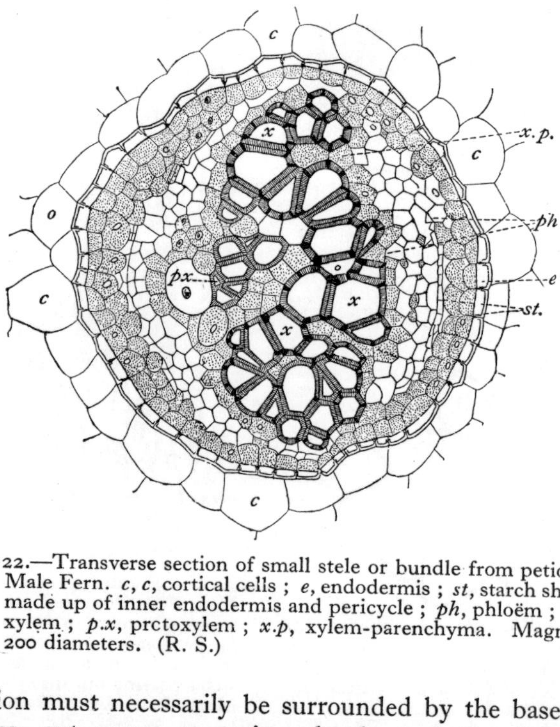

FIG. 22.—Transverse section of small stele or bundle from petiole of Male Fern. *c, c*, cortical cells; *e*, endodermis; *st*, starch sheath, made up of inner endodermis and pericycle; *ph*, phloëm; *x, x*, xylem; *p.x*, prctoxylem; *x.p*, xylem-parenchyma. Magnified 200 diameters. (R. S.)

section must necessarily be surrounded by the bases of leaves, cut across at various levels, and showing the steles entering them obliquely from the stem (see Fig. 21).

Every stele consists essentially of a central mass of wood surrounded by a ring of phloëm (see Fig. 22). This arrangement is usual in Ferns, and vascular strands of this kind are often called concentric bundles. As, however, in many cases each of them almost exactly repeats the structure of the whole vascular cylinder of the embryonic stem, it is more convenient to describe them as steles or cylinders, but no sharp distinction can be drawn.

The wood consists of tracheæ and parenchyma ; the tracheæ of Ferns are generally of the kind called *scalariform*, or ladder-like, from the peculiar structure of their walls, shown in Fig. 23. This structure depends on the form of the pits, which are slightly bordered and much elongated in the transverse direction, so that the thickened ridges between them resemble the rungs of a ladder. It was stated a few years ago that the elongated elements of the wood, commonly called tracheids, are everywhere in open communication with each other, the closing membranes of the pits being absorbed, but the most recent investigations confirm the older view, that in Ferns, with very rare exceptions, the water-conducting elements remain permanently closed, and are not of the nature of vessels.

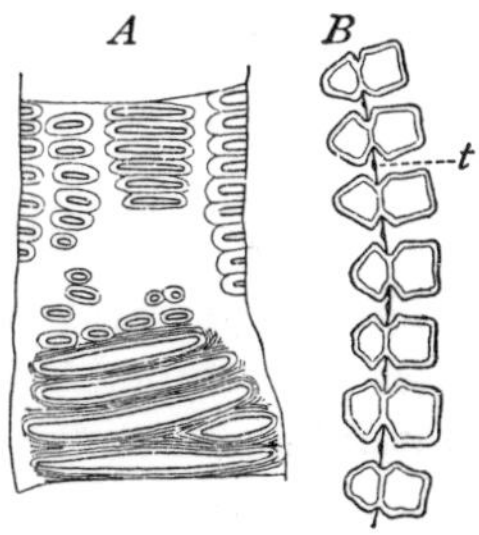

FIG. 23.—Portions of scalariform tracheæ. *A*, part of wall in surface view. Magnified 187 diameters. *B*, part of wall in section, showing bordered pits. *t*, torus on closing membrane. Magnified 375 diameters. (After De Bary.)

In the stele figured (see Fig. 22) there is only one group of protoxylem (*px*) lying on one side of the wood. In the larger steles of the stem there are usually two or three such groups. Spiral tracheids occur at these points, but usually become destroyed very early as the stem grows in length. Surrounding the wood is a layer of parenchyma containing starch, and then we come to the phloëm-zone, consisting of sieve-tubes and parenchyma. The former have their sieve-plates on the lateral as well as on the oblique terminal walls. They are not unlike those which we observed in *Pinus*.

The phloëm again is surrounded by a belt of parenchyma very rich in starch, beyond which we come to the endodermis. The endodermis is really two cells thick, but its inner layer cannot be distinguished from the pericycle except by the fact that its cells fit on exactly to those of the outer endodermal layer. This outer layer alone has the usual structure of an endodermis (see p. 75, Part I.) and becomes thick-walled.

β. *Other Tissues of the Stem*

The great mass of the ground-tissue, in which the steles are embedded, consists of ordinary parenchyma containing abundant starch. The outer cells have thicker walls, and those nearest the epidermis are narrow and fibrous. They serve to give mechanical stiffness to the stem. The epidermis itself has a thick brown outer wall, and otherwise presents no peculiarities. It bears flat chaffy scales, or *ramenta,* which are very characteristic of the plant, and indeed of almost all Ferns. They are sometimes of large size, reaching half an inch in length, and consist of a plate of tissue one cell thick, attached to the epidermis at one end ; they arise each from the growth and division of a single epidermal cell.

2. The Leaf

As we have already seen, each leaf, at least in the mature plant, receives several steles from the stem (see Figs. 19, 20, and 21). The structure of the petiole is simple enough. The steles (see Fig. 22), as seen in transverse section, are arranged in a horseshoe, embedded in ground-tissue, the outer layers of which consist of very thick-walled cells. The basal part of the petiole is densely clothed with chaffy ramenta, which are more scattered higher up on the leaf.

A bundle enters each pinna of the leaf, branching off from one of the two larger steles which are situated near the upper surface of the leaf-stalk. This bundle gives off branches to the right and left, which enter the successive segments of the lamina, and by their further ramifications supply its vascular system (see Fig. 18, B). As we trace the bundles into the finer veins of the leaf, we find that the upper part of the phloëm gradually dies out, so that the ultimate branches of the bundle system come to be collateral instead of concentric. This is very generally the case in Ferns.

If we now endeavour to sum up what we have learnt of the vascular system of the Male Fern, we see that its most striking peculiarity consists in the polystely of the stem, where each strand of wood and phloëm resembles an entire central cylinder rather than a single vascular bundle. As we follow the leaf-traces outwards, however, we find that the steles assume more and more the character of simple vascular bundles, until in the lamina they have the same collateral structure as in the leaves of flowering plants. It is evident that no sharp line can be drawn between stele and bundle.

Returning to the lamina of the leaf, we find that its structure is distinctly bifacial. The mesophyll towards the upper surface consists of closely-packed squarish cells, forming a kind of palisade-parenchyma, though the palisade form is not well marked. The lower portion of the mesophyll, on the other hand, is made up of irregularly branched cells, attached to each other by only small parts of their surface, so that large intercellular spaces are left between them. This tissue is thus a typical spongy parenchyma. All the cells of the mesophyll contain abundant chlorophyll granules (see Fig. 28, *A*).

The epidermis of the lower surface alone bears the stomata, which are very numerous (see Fig. 24).

The stomata are characteristic : each pair of guard-cells is half surrounded by a subsidiary cell, shaped like a horseshoe. The subsidiary cell and guard-cells are ultimately derived from a single mother-cell, which is cut out from one of the epidermal cells by a curved wall.

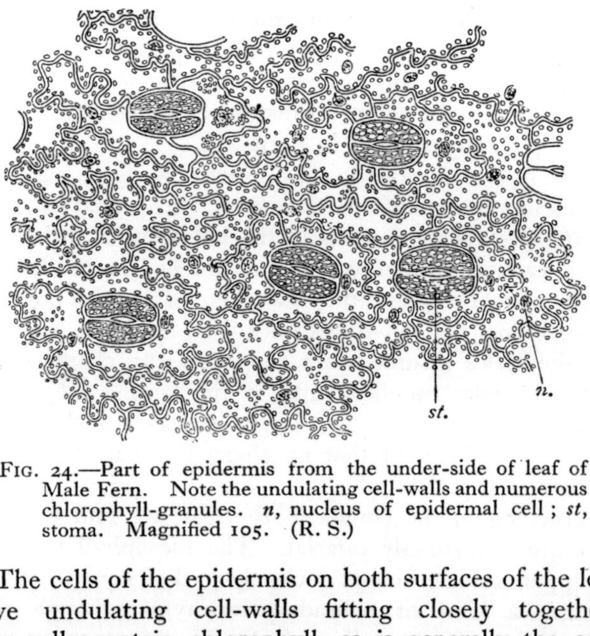

FIG. 24.—Part of epidermis from the under-side of leaf of Male Fern. Note the undulating cell-walls and numerous chlorophyll-granules. *n*, nucleus of epidermal cell ; *st*, stoma. Magnified 105. (R. S.)

The cells of the epidermis on both surfaces of the leaf have undulating cell-walls fitting closely together. The cells contain chlorophyll, as is generally the case in the epidermis of Ferns, though less usual among flowering plants.

3. The Root

The adventitious roots of the Male Fern arise, as we have already seen, at the bases of the leaves, though they are in direct connection with the principal steles of the stem.

The structure of the root in Ferns is, with one or two exceptions, essentially similar to that of the root in flowering plants. In the Male Fern and in many other Ferns the vascular cylinder of the root is diarch (see Fig. 25). The first-formed elements of the wood, protoxylem, lie at the two ends of the xylem-plate, exactly as in the Wallflower (see Part I. p. 73), and the development of the wood advances from these two points in centripetal direction to the middle of the cylinder.

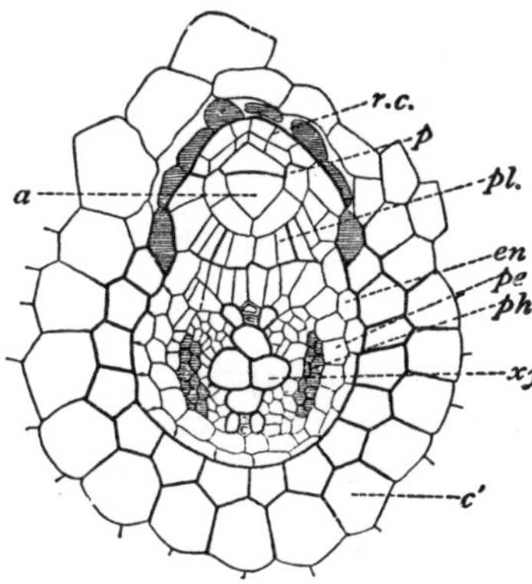

FIG. 25.—Transverse section of central part of young root of a Fern, showing origin of rootlet. *xy*, diarch xylem; *ph*, phloëm; *pe*, pericycle; *en*, endodermis; *c'*, cortical cells; *a*, apical cell of rootlet; *r.c*, root-cap; *p*, cortex of rootlet; *pl*, stele of rootlet. Magnified about 150 diameters. (After Van Tieghem.)

The small first-formed tracheids are spirally thickened, the larger elements, developed later, are scalariform. On either side of the xylem-plate, and therefore alternating with the protoxylem-groups, are two strands of phloëm. The whole is surrounded by a single layer of pericycle, and this again by the endodermis, which has the usual cuticularised bands on its radial cell-walls. The cortex consists of two zones—an inner thick-walled region forming a firm sheath round the cylinder, and an outer portion in which the cells have thinner walls. We often find that the cell-walls of the inner zone are not uniformly thickened; at the points opposite the two ends of the xylem-plate, the cells remain comparatively thin-walled so as to leave a free passage, through which the water absorbed from the soil can reach the wood, and thus pass upwards to the stem and leaves. At the exterior of the whole root is the piliferous layer, which bears numerous unicellular root-hairs. We see that, except for minute details, such a root resembles a young root of the Wallflower (see Part I. p. 73), but in the case of the Ferns there is no secondary growth of thickness. When we come to consider the development, we shall find considerable differences from any of the previous types.

4. The Growing-points

a. The Stem

With rare exceptions, the development of both stem and root in Ferns can be referred to a single apical cell, from the divisions of which all tissues and organs arise. This important cell can be easily distinguished from its neighbours, which are derived from it, by its larger size and characteristic form. In the Male Fern and most other members of the class, the apical cell of the stem has the form of an inverted three-sided pyramid or tetrahedron, with its curved base directed outwards (see Fig. 26, which is taken from a simpler Fern-stem, but illustrates the essential features).

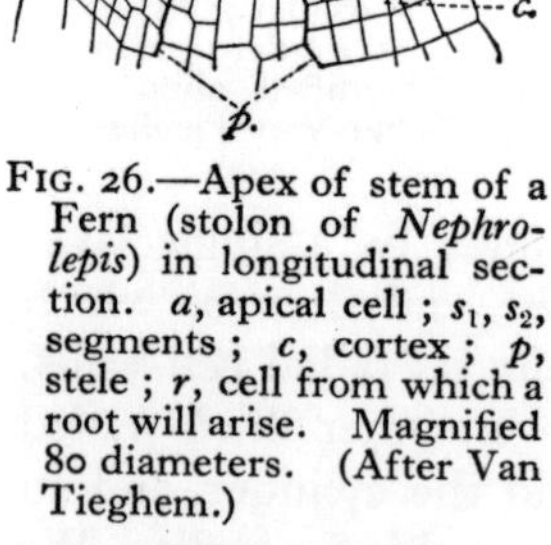

FIG. 26.—Apex of stem of a Fern (stolon of *Nephrolepis*) in longitudinal section. *a*, apical cell ; s_1, s_2, segments ; *c*, cortex ; *p*, stele ; *r*, cell from which a root will arise. Magnified 80 diameters. (After Van Tieghem.)

In longitudinal section, therefore, the cell appears triangular ; its three sides are in contact with the adjacent tissue, while the curved base is free and faces upwards (assuming the stem to be erect). The apical cell divides in regular order by walls successively parallel to each of its three sides. The cells thus cut off are called segments (see Fig. 26). By the growth and repeated subdivision of the three rows of segments all the tissues of the stem are produced. The stem figured is monostelic ; here the first tangential walls formed in the segments mark the limit between the central cylinder and the surrounding cortex. In a polystelic stem, such as that of the Male Fern, the steles are not marked out until after more numerous divisions have taken place. It is probable that each leaf owes its origin to the outgrowth of cells derived from a single segment.

β. *The Root*

The root, like the stem, carries on its apical growth by means of a single cell, which here also has the form of an inverted three-sided pyramid. The essential difference between the divisions in the apical cell of the root and in that of the stem, is that, in the former, cell-walls are not only formed parallel to the three sides, but also parallel to the base of the pyramid. The segments thus cut off from the outer end of the apical cell (see Fig. 27) go to form the root-cap ; those cut off laterally build up the tissues of the root itself, in much the same way as in the case of a monostelic stem.

The mode of branching of the root in Ferns differs in one important respect from that in the higher plants. In Ferns each rootlet arises, not from the pericycle, but from the endodermis, and, in fact, its origin can always be traced to a single endodermal cell, lying opposite one of the groups of protoxylem. The cells destined to give rise to rootlets can be distinguished by their larger size. The cell in question divides up by inclined walls, so as to form at once a pyramidal apical cell, by means of which the further development of the rootlet is carried on (see Fig. 25). The young root, as it makes its way through the tissues of the parent organ, is at first enveloped in a digestive sac, derived from an inner layer

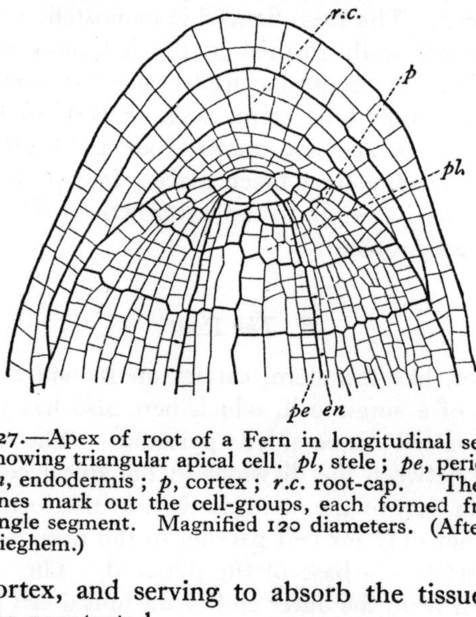

FIG. 27.—Apex of root of a Fern in longitudinal section, showing triangular apical cell. *pl*, stele ; *pe*, pericycle ; *en*, endodermis ; *p*, cortex ; *r.c.* root-cap. The dark lines mark out the cell-groups, each formed from a single segment. Magnified 120 diameters. (After Van Tieghem.)

of the cortex, and serving to absorb the tissues which have to be penetrated.

In the Ferns, the pericycle has nothing to do with the development of the rootlet, beyond forming a pedicel by which it is connected with the vascular tissues of the main root.

The origin of the adventitious roots, which play so important a part in the organisation of Ferns, follows the same rule which holds good for the rootlets. Every adventitious root arises from an endodermal cell bordering on one of the steles of the stem. In Fig. 26, for example, the shaded cell marked *r* is destined to produce a root. We see from this that the first differentiation of the root-forming cells in the stem takes place very early.

γ. *The Leaf*

The development of the leaf in Ferns, like that of the stem and root, goes on at the apex, whereas in most Flowering Plants the growth of the leaf chiefly takes place at the base. It is only when still very young that a Fern leaf grows by means of a single apical cell. This cell soon divides up so as to form a row of marginal cells, all of which take equal parts in the subsequent cell-formation.

B. Reproductive Organs of the Sporophyte

We have already seen that the sporangia of the Male Fern are grouped in sori, and that the sori are seated on the back of the leaf, over the lateral veins of a segment or pinnule (see Fig. 18, B).

Beneath each sorus is a prominent mass of tissue, which we may call the *placenta* (see Fig. 28, *r*). This receives a short branch from the vascular bundle immediately below it. The placenta grows out at its summit into the kidney-shaped indusium, which consists of a membrane, one cell in thickness, attached to the placenta by a massive stalk (see Fig. 28, *i*). The sporangia spring from the sides of the placenta, and are all roofed in by the indusium. Each sporangium consists of a long slender stalk, made up of two or three rows of cells, bearing the terminal spore-case or capsule (see Fig. 28, *B*, *C*, *E*). A club-shaped hair, secreting resin, is usually borne on the stalk. The capsule is not spherical but much flattened, resembling the case of a watch in form ; its wall when mature consists of a single layer of cells ; its interior is occupied by the spores, forty-eight to sixty-four in number, which are of a brown colour when ripe.

The sides of the capsule are formed of cells with thin membranes, but around its edge runs a single row of larger cells with peculiarly thickened walls of a rich brown colour, forming a very conspicuous feature under the microscope, when the sporangium is ripe. This special row of cells is called the ring or *annulus* (see Fig. 28, *B*, *C*, *E*). The annulus starts from the stalk at one side, passes over the crest of the capsule, and extends about half-way down on the other side. Here it suddenly comes to an end. The cells of the wall immediately below the termination of the annulus are broad and flat; this is the place where the capsule ultimately opens (see Fig. 28, *B* and *C*, *st*). In the annulus both the inner and the radial cell-walls are much thickened; the free outer walls of the cells, however, remain thin. The function of the annulus is to cause the dehiscence of the sporangium when ripe.

Each sporangium arises from *a single superficial cell* of the placenta; in this respect it differs from the sporangia hitherto considered, namely, the pollen-sacs and ovules of Flowering Plants, and the two kinds of sporangia in *Selaginella*. The great majority of Ferns are distinguished by the unicellular origin of their sporangia. A single cell, then, grows out from the surface of the placenta and soon begins to divide. One or two basal cells are often cut off, to begin with, by transverse walls, but they are of no great importance. The terminal cell next undergoes division by inclined walls (see Fig. 28, *G*), three of which are formed in succession, inclined to each other at an angle of 120°, as seen from above. In side view, as shown in the figure, only two of these walls can be seen, and they join each other at an acute angle. The result of these three divisions is to carve out a

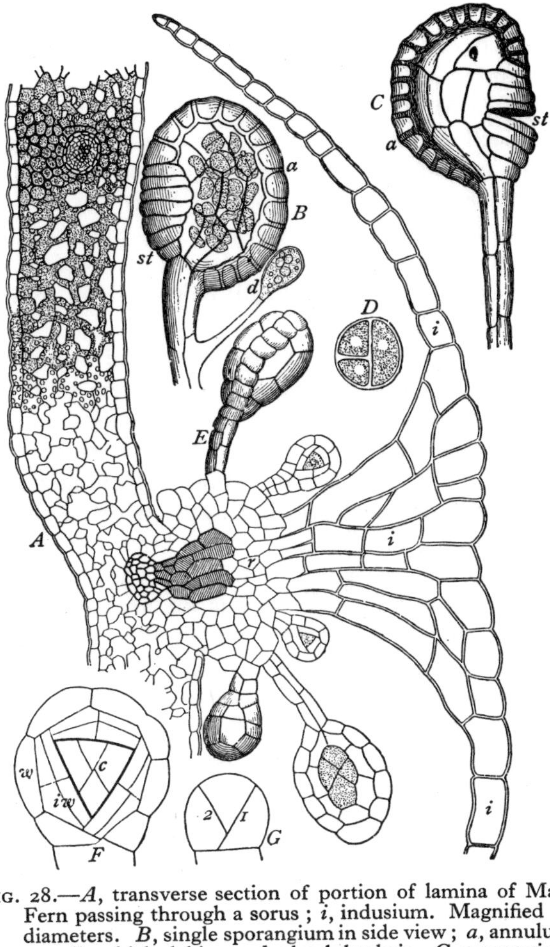

FIG. 28.—*A*, transverse section of portion of lamina of Male Fern passing through a sorus; *i*, indusium. Magnified 80 diameters. *B*, single sporangium in side view; *a*, annulus; *st*, cells which dehisce; *d*, glandular hair. *C*, sporangium dehiscing at *st*; *a*, annulus. *B* and *C* magnified 100 diameters. *D*, spore mother-cell divided (only three spores visible). Magnified 350 diameters. *E*, nearly ripe sporangium. *F*, young sporangium; *c*, sporogenous cells; *iw*, tapetum; *w*, wall. *G*, very young sporangium, showing first divisions. Magnified 260 diameters. (After Luerssen.)

three-sided pyramidal cell with a free base, quite like the apical cell which we have already described in the stem and root. The next wall formed runs parallel to the free base of this pyramidal cell, so now we have a central cell surrounded on all four sides by the segments which have been cut off from it. The segments undergo a great many more divisions, and form the wall of the capsule, which remains only one cell thick, as all the cell-divisions are at right angles to its surface. The lateral segments also cut off cells below, which go to build up the stalk.

In the meantime the pyramidal central cell has itself divided by walls parallel to its four sides, so that it is now surrounded by an inner layer of cells separating it from the wall of the capsule. These intermediate cells undergo further divisions in various directions and form the *tapetum*, the ultimate destiny of which is to afford food material to the developing spores (see Fig. 28, *F*, *iw*).

The central cell which remains is the essential part of the whole structure, for this is the *archesporium*, from which the spores themselves are produced (see Fig. 28, *F*, *c*). We see, then, that in this case the archesporium begins as a single cell. It undergoes several cell-divisions (see Fig. 28, *F*, *c*). The cells thus produced round themselves off and become the mother-cells of the spores. In many Ferns there are sixteen or more mother-cells in each sporangium, but in the Male Fern there are usually not quite so many.

The spore mother-cells are spherical; as the sporangium has grown more rapidly than they have, they do not fill the whole interior, but float freely in a half-liquid mass derived from the disorganised tapetal cells. Each mother-cell now divides twice so as to form four cells, each of which has at first the shape of a quadrant of a sphere (see Fig. 28, *D*). These four daughter-cells are the spores. As they ripen they become kidney-shaped, the convex side corresponding to the free outer surface of the mother-cell, while the concave edge of each spore represents the line of junction with its sister-cells. The spore membrane becomes much thickened, and consists of two layers, the outer of which is strongly cuticularised, and assumes a dark-brown colour. We have now seen how the myriads of microscopic dust-like spores which we find on the back of a Fern-frond are produced. It remains for us to learn how they are scattered.

It is the annulus which causes the sporangium to open; dehiscence takes place when the wall of the ripe sporangium has begun to dry up. The cells of the annulus lose water, and consequently contract, the thin outer walls of the cells becoming concave instead of convex (see Fig. 28, *C*). The water in the annulus cells is in a state of tension. The whole annulus tends to straighten, and this finally causes the rupture of the sporangium at the broad, thin-walled cells at the end of the annulus (see Fig. 28, *C*, *st*). The annulus not only straightens, but bends back on itself in the opposite direction. This primary movement takes place quite slowly. The annulus cells continue to lose water: their thin outer walls are drawn in yet more, and the annulus bends back still further. This goes on until the tension in the water is so great that the water comes away from the walls of the cells. This causes a sudden release of the tension, and the walls of the cells and the annulus as a whole recover with a violent jerk. It is this secondary movement that causes the ejection of the spores, which may then be carried to a great distance by the wind.

On the island of Krakatoa (Malay Archipelago), the vegetation of which was completely destroyed by the volcanic eruption of 1883, and which lies about eleven miles from the nearest land, Ferns were among the very first plants to reappear after the catastrophe.

It is a very general rule that the dehiscence of sporangia is so contrived as to take place in dry weather; the advantage of this to the plant is obvious. When the air is dry the spores form a powdery dust, which is easily scattered by the wind, whereas in wet weather they hang together in damp clusters, and could never be properly disseminated.

We have now traced the history of the reproductive process in the asexual generation. The most important points in which the Male Fern differs from *Selaginella* are the totally different arrangement of the sporangia, the origin of each sporangium from a single cell, and the fact that sporangia and spores are all of one kind.

As regards the two latter points, however, all Ferns do not agree with the Male Fern, for in some members of the class the sporangia have a multicellular origin, while others are heterosporous. It now remains for us to follow the germination of the spores, to see how the prothalli are produced from them, to learn how fertilisation is effected, and finally to study the origin of the embryo, which develops once more into the asexual Fern-plant, and thus completes the cycle of life.

III. *THE GAMETOPHYTE OR SEXUAL GENERATION*

A. Development and Structure of the Prothallus

Fern spores can be sown successfully on ordinary garden earth, on peat, on sand, or even on pieces of tile. The last-mentioned material has the advantage that very clean cultures can thus be obtained. It is well to heat the soil, or whatever else is used, up to at least 100° C. (the boiling-point of water) before sowing the spores, so as to destroy the germs of other organisms, which are sure to be present, and might compete only too successfully with the young prothalli. It is important not to sow the spores too thickly, or else when they germinate the prothalli overcrowd each other. The cultures must, of course, be kept moist. It is best to cover them with a bell glass, and to water from below.

After about a week, the beginning of germination may be observed ; the spore starts growing and bursts its brown outer membrane. By this time the spores, which in their resting condition are without chlorophyll, will have begun to turn green. Fern spores which contain no chlorophyll when ripe keep their power of germination for a long time. In a few kinds, such as the Royal Fern, *Osmunda*, the spores are green, and will only germinate if sown at once. The first thing which the germinating spore does is to form a root-hair. An outgrowth containing little or no chlorophyll arises from the spore, becomes cut off by a cell-wall, and grows down into the soil ; the remaining larger part of the spore grows out towards the light, and divides at first transversely. A few more transverse walls are formed, the end cell being always the one to divide, so that the prothallus soon takes the form of a short green filament (see Fig. 29), each cell of which may produce a root-hair. In most Ferns the rhizoids of the prothallus remain unicellular ; in a few they become multicellular.

Soon the transverse divisions of the filament cease, an oblique wall appears in the terminal cell, followed by another at right angles to it, and thus a wedge-shaped apical cell is marked out. This goes on cutting off segments to the right and left, the segments divide up further, and soon the young prothallus becomes converted into a flat cellular plate, which for a time remains only one cell thick (see Fig. 30). As growth goes on, the prothallus tends to become heart-shaped, the growing-point lying at the base of a depression between two lobes. This is due to the fact that the apical meristem does not grow so fast as the older tissue which has been produced from it on either side.

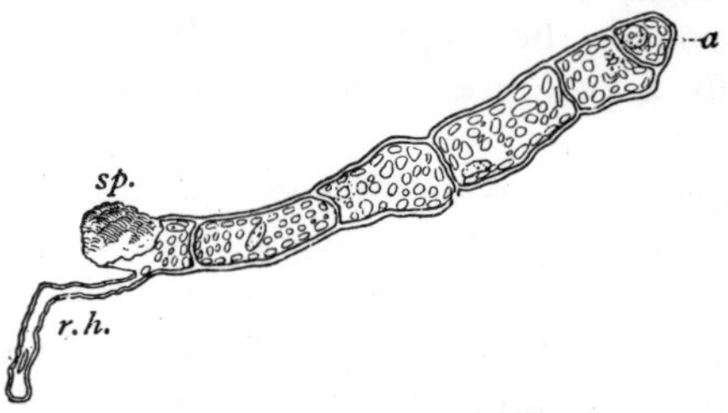

FIG. 29.—Very young prothallus of an *Aspidium*. *sp.* membrane of spore; *r.h*, first rhizoid; *a*, apical cell. Magnified 210. (R. S.)

The single apical cell does not long maintain its independence. It soon divides up into a row of equivalent initial cells, which all take a like share in the subsequent development. The prothallus, which for a short time grows vertically, soon assumes a horizontal position, and henceforth there is a marked difference between the lower side, which is in contact with the soil, and the free upper surface. It is from the under-side that the new rhizoids arise, and to this side also the sexual organs are limited.

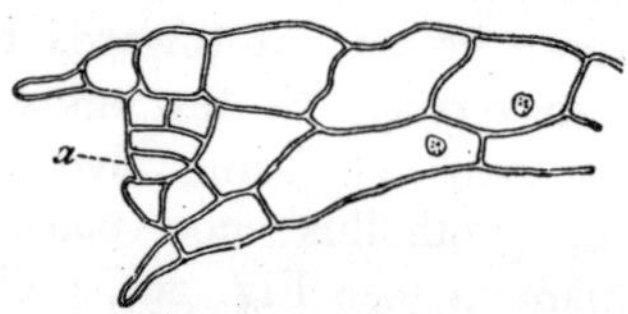

FIG. 30.—Apex of young prothallus of *Aspidium*, older than in Fig. 29, seen in surface view. *a*, apical cell. Magnified 210. (R. S.)

A normal full-grown prothallus seen from below is shown in Fig. 31. The middle part, lying just behind the growing-point, forms a pad or cushion several cells in thickness, while the lateral portions or wings remain one cell thick.

The antheridia or male organs arise chiefly on the

older basal part of the prothallus and sometimes also on the wings. The female organs or archegonia are limited to the cushion. In ordinary cases the prothallus is monœcious, bearing both kinds of sexual organs ; but this is not always so. Male prothalli are not uncommon,

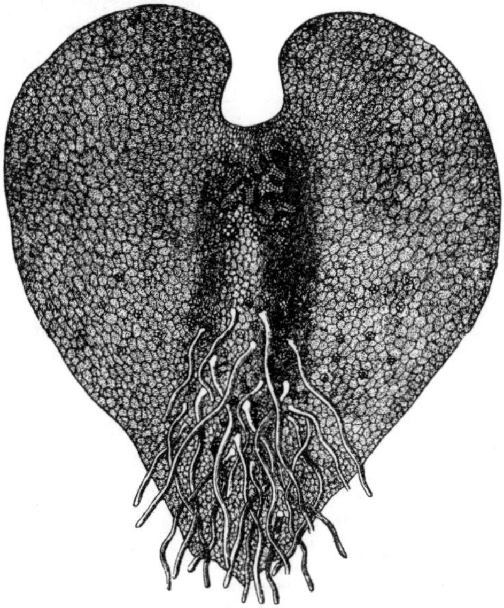

FIG. 31.—Full-grown prothallus seen from below, showing archegonia towards the apex, antheridia and rhizoids towards the base. Magnified about 25 diameters. (After Luerssen.)

and are generally of small size. Sometimes a prothallus at the earliest stage of its development, while still in the form of a short filament, begins to form antheridia. A filamentous prothallus, bearing male organs only, is shown in Fig. 32, but still smaller ones occur.

Specimens with archegonia only are rarer, and are of

the ordinary form. These variations are interesting, because they show how the distinction of sex among the individual prothalli, which has become fixed in *Selaginella* and other heterosporous forms, appears occasionally as a more or less casual phenomenon even in the homosporous Ferns. Small and ill-nourished prothalli suffice for the production of antheridia, which quickly fulfil their function and make no great demands on the food supply. On the other hand, archegonia are useless unless provision be made for the nutrition of the embryo after fertilisation ; and so we find female organs on full-grown and well-nourished prothalli only. In the heterosporous Cryptogams provision is made beforehand, in the spore, for the more abundant nutrition of the female prothallus.

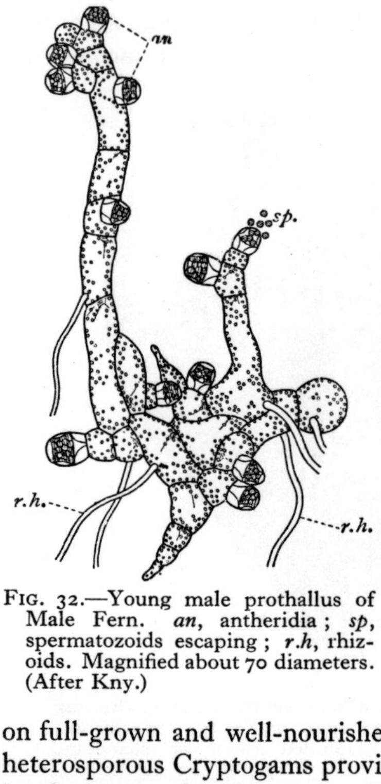

FIG. 32.—Young male prothallus of Male Fern. *an*, antheridia ; *sp*, spermatozoids escaping ; *r.h*, rhizoids. Magnified about 70 diameters. (After Kny.)

B. DEVELOPMENT AND STRUCTURE OF THE SEXUAL ORGANS

1. The Antheridia

Each antheridium arises from a single cell, the upper part of which grows slightly beyond the general surface of the prothallus, and is cut off by a transverse wall. It then undergoes a few divisions, so that the antheridium comes to consist of a central cell, surrounded by two ring-shaped cells, one above the other, and covered in on the top by a cap-cell. Some of the stages of development are shown in Fig. 33, A.

The central cell divides up repeatedly, and gives rise to the spermatozoid mother-cells, the number of which in

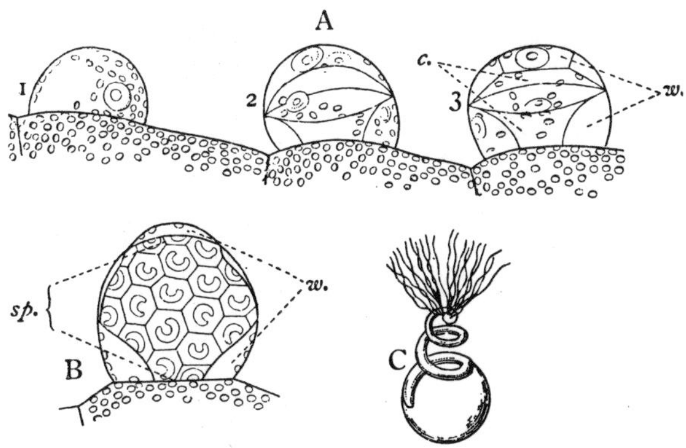

FIG. 33.—Antheridia of Male Fern. A, 1, 2, and 3, antheridia three successive stages seated on prothallus; *c*, central cell; *w*, wall. B, older antheridium; *sp*, mass of spermatozoid mother-cells; *w*, wall. C, a single spermatozoid. Magnified, A and B about 300, C about 700 diameters. (After Kny.)

each antheridium averages about twenty (see Fig. 33, B). In each mother-cell one spermatozoid is formed. The mature spermatozoid consists of a spirally coiled body like a corkscrew, but thicker at one end than the other. Near the thin end a number of excessively fine cilia (contractile protoplasmic threads) are attached (see Fig. 33, C).

The development has been very exactly followed; it is known that the greater part of the body of the spermatozoid is formed from the *nucleus* of the mother-cell ; the cilia, however, and the part of the body to which they are attached, are derived from the cytoplasm. In Fig. 33, B the young spermatozoids are shown enclosed in their mother-cells.

In the ripe antheridium every mother-cell contains its spermatozoid curled up inside it ; as soon as a drop of water comes into contact with the antheridia, they open, by the bursting of their cap-cells (see Fig. 34).

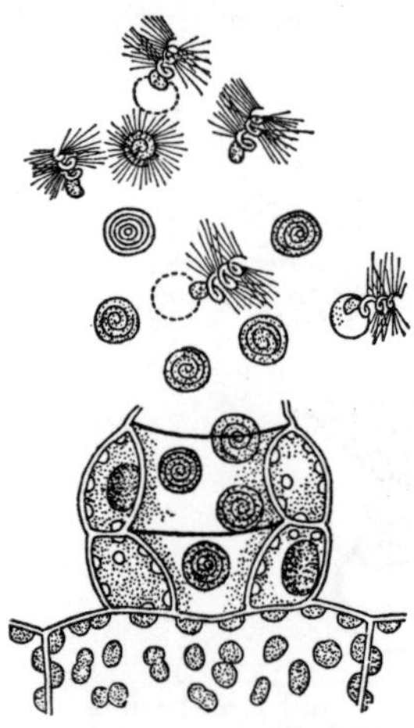

FIG. 34.—Ripe antheridium, showing spermatozoids escaping. Magnified 350 diameters. (After Luerssen.)

The pressure which brings this about is due partly to the swelling of the mother-cells themselves, and partly to that of the ring-cells, which absorb water and press upon the mass of mother-cells, squeezing them out from the antheridium. The whole mass of mother-cells is now set free, but each spermatozoid is still imprisoned within its own mother-cell. The membranes of the latter, however, are soon dissolved, and now the spermatozoids are able to escape, and begin their active career. Each spermatozoid drags with it, attached to the hinder end, a bladder-like sac, which is derived from the inner part of the cytoplasm of the mother-cell (see Figs. 33, C, and 34). The locomotion is very active ; the little spermatozoids go wriggling through the water in all directions, always keeping their thin ciliated ends foremost ; they revolve on their axes, and advance at the same time, not in straight lines, but in varying curved paths. Sometimes the little bladders are left behind, sometimes they hang on all the time, until an archegonium is reached. Before describing the ultimate fate of the spermatozoids, we must now turn our attention to the archegonia.

2. The Archegonia

As we have already mentioned, the archegonia do not rise so indiscriminately on different parts of the prothallus as the antheridia do, but are limited to the sides of the thickened cushion. An archegonium, like an antheridium, arises from a single cell, which at first projects only slightly above the level of the neighbouring tissue. It divides by two transverse walls into three cells : the lowest or basal cell undergoes a few divisions, but takes no important part in the further development ; the middle cell ultimately forms the ovum and the two canal-cells ; while the uppermost of the three grows and divides to form the neck (see Fig. 35). The neck is the only part which projects beyond the surface of the cushion.

The neck-cell first divides, by two longitudinal walls at right angles to each other, into four cells placed cross-wise, as seen in surface view (see Fig. 35, D). Each of these four cells then divides up repeatedly by approximately transverse walls, so that the neck is finally made up of four rows of cells. While these divisions are going on, the neck is increasing in length, and at the same time the central cell grows up between the four rows of neck-cells (see Fig. 35, A and B), which separate a little to make way for the outgrowth. The projecting part of the central cell is presently cut off by a wall, and forms the *canal* of the neck. This canal-cell may itself undergo one or two further divisions, but they are usually incomplete, no cell-walls being formed. A second canal-cell is now cut off below the first ; the remaining part of the central cell rounds off its protoplasm, and now constitutes the *ovum*. The archegonium has by this time reached its complete development. The neck is not straight, but is sharply curved backwards, *i.e.* towards the basal end of the prothallus (see Figs. 35 and 36).

We see that the archegonia are really quite similar to

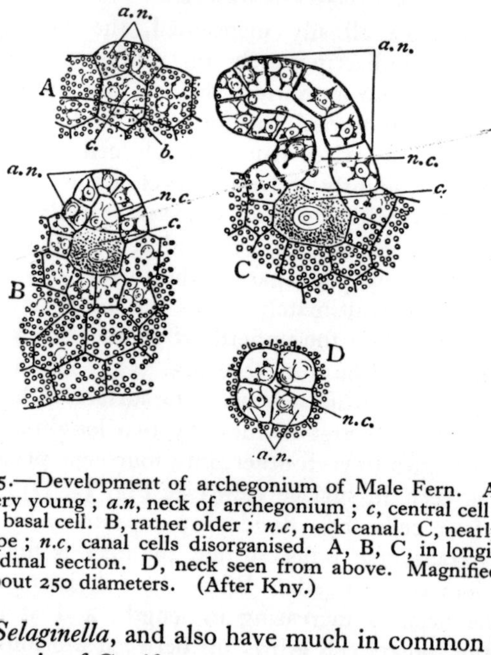

FIG. 35.—Development of archegonium of Male Fern. A, very young ; *a.n*, neck of archegonium ; *c*, central cell ; *b*, basal cell. B, rather older ; *n.c*, neck canal. C, nearly ripe ; *n.c*, canal cells disorganised. A, B, C, in longitudinal section. D, neck seen from above. Magnified about 250 diameters. (After Kny.)

those of *Selaginella*, and also have much in common with the archegonia of Conifers.

C. FERTILISATION

In Ferns, as in Cryptogams generally, fertilisation can only take place under water. In nature this happens after rain or heavy dew, when the under-sides of the prothalli are thoroughly wetted. When we are cultivating prothalli it is necessary to sprinkle them with water from above, when the sexual organs are ripe, if we wish to obtain embryos. We have already seen how the antheridia open under water, and how the active spermatozoids are set free. In like manner the archegonia, when moistened, open to receive them. This happens because the protoplasm of the canal-cells swells up and becomes converted into mucilage, which exercises a pressure on the neck, and causes it to open at the top, the four rows of cells being forced apart. The mucilage now more than fills the canal, and forms a viscid drop at the mouth of the archegonium (see Fig. 36).

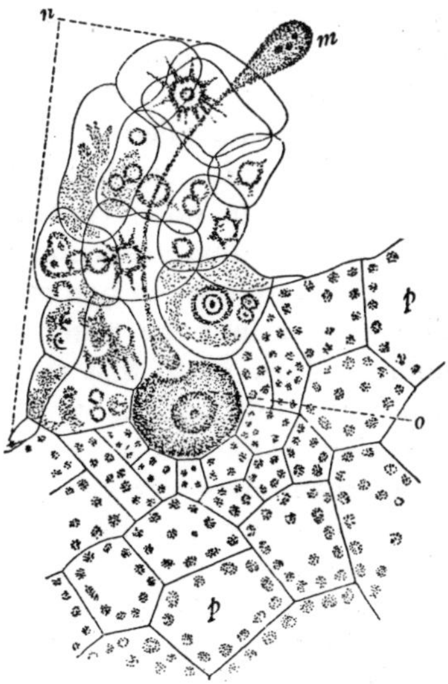

FIG. 36.—Archegonium ready for fertilisation. *o*, ovum; *n*. neck; *m*, mucilage extruded from canal; *p*, cells of prothallus. Magnified 350. (After Strasburger.)

The spermatozoids swimming through the water are attracted by the archegonia. This remarkable fact, which long remained an absolute mystery, is now so far explained that we have good evidence as to the nature of the substance which attracts them. When a spermatozoid, as it makes its devious way through the water, comes within a short distance of the neck of an open archegonium, it turns aside from its course, and makes for the opening. Here it finds the mucilaginous drop, and promptly plunges into it. Its movements do not cease, though in the denser fluid they go on more slowly;

the spermatozoid wriggles its way down the neck, through the mucilage which fills it, and so at last reaches the ovum below. Quite a number of spermatozoids may be seen swarming around the opening of a ripe archegonium, and several may penetrate down the canal, but only one succeeds in uniting with the ovum.

Now it has been shown by experiment that the spermatozoids of Ferns are attracted by certain chemical substances, and especially by malic acid. If artificial archegonia are prepared (consisting of tiny capillary glass tubes) and filled with a mucilage to which a small quantity of this acid has been added, they are found, when placed in water containing fern-spermatozoids, to exercise the same attraction upon them which the real archegonia exercise in nature. The malic acid gradually diffuses out into the water, and the spermatozoids are influenced by it, so that they move in the direction in which the substance is more concentrated, *i.e.* towards the tube. Although it cannot be proved that the archegonia themselves contain malic acid, as they are too small for a recognisable quantity to be obtained from them, yet this substance is known to be present in the prothallus as a whole ; so there can be little doubt that the natural archegonia owe their attractive influence to the same chemical agent which has proved efficacious in experiment.

We see, then, that these minute protoplasmic bodies, the spermatozoids, are not only capable of active movement, but also possess a certain power of perception, by which their movements are guided. This is a remarkable illustration of the great fact that the protoplasm of plants and animals is essentially the same, and that the living matter of a plant may show properties usually regarded as belonging especially to animals, whenever such properties are needed.

Now that we have learnt how fertilisation is brought about, we will go on to consider its results.

D. Embryology

The first change after fertilisation is the formation of a cell-wall around the protoplasm of the fertilised ovum. It now at once begins to grow and divide, becoming the embryo, or young plant, of the sporophyte generation.

The embryo of a Fern differs from that of the plants hitherto described, in having no suspensor; the whole of the fertilised ovum goes to form the embryo. Throughout the Fern-group there is considerable uniformity in the manner of development of the embryo from the ovum. The first wall (called the basal wall) runs nearly parallel to the axis of the archegonium, and at right angles to the axis of the whole prothallus. This divides the young embryo into an *epibasal* and a *hypobasal* half; the former faces the apex and the latter the base of the whole prothallus. Two more cell-walls then appear, all three being at right angles to each other, so that the embryo is now cut up into eight parts or octants. From the epibasal half the apex of the stem and the first leaf arise, while the hypobasal part produces the apex of the root, and an organ called the *foot*, which is of a temporary character and serves to attach the young plant to the prothallus and to take up food from it until the embryonic stage is past (see Fig. 37, *f*).

Growth and accompanying cell-division go on rapidly; the parts which develop quickest are the root and first leaf; for a long time the stem remains very rudimentary. The ventral part of the archegonium becomes much enlarged, to make room for the developing embryo. The root is the first part to break through, whereupon it makes its way down into the soil. It is soon followed by the first leaf, which turns upwards between the lobes of the prothallus, and spreads out its blade to the light. Meanwhile the foot is absorbing the food produced by the prothallus, but this is soon exhausted, and then the embryo becomes an independent plant, which continues its growth, producing fresh leaves and roots. The leaves which are first formed are always of a comparatively simple shape, and it is only gradually that the successive leaves assume the form characteristic of the species. At the same time the stem increases in bulk, and its anatomical structure becomes more complex.

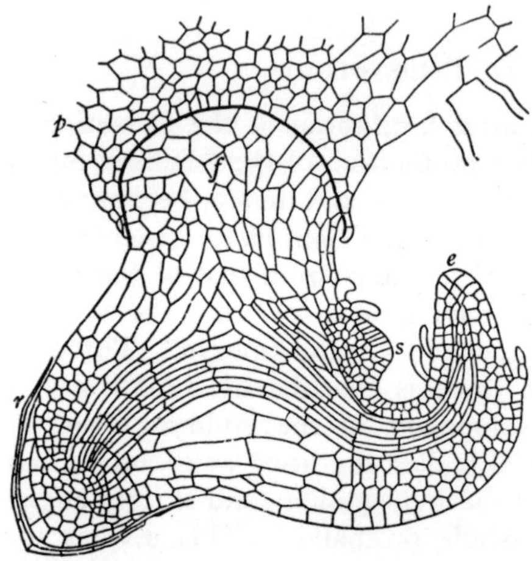

FIG. 37.—Embryo of a Fern (*Pteris*) in median section. *s*, apex of stem; *e*, first leaf; *r*, root; *f*, foot by which embryo is attached to prothallus; *p*, prothallus. Magnified 150. (After Hofmeister.)

We have now traced the normal life-cycle through its complete course, and have got back to the asexual generation, or sporophyte, from which we started.

E. COMPARISON BETWEEN THE LIFE-HISTORY OF FERNS AND THAT OF THE HIGHER PLANTS

In the Ferns, for the first time, the occurrence of a distinct alternation of generations becomes manifest. In these plants the prothallus, though small, is just as distinct an individual, and leads just as independent a life, as does the asexual Fern-plant itself. In fact, we may even say that the prothallus is the more independent of the two, for while the young Fern-plant is for a time dependent for its nutrition on the prothallus, the latter is never dependent in any way on the Fern-plant. At any rate we have in normal Ferns two perfectly definite generations, as distinct as possible from each other; one bearing exclusively the sexual, and the other exclusively the asexual organs of reproduction, and in the ordinary course of life these two generations succeed each other in regular alternation. It was in fact from the Ferns that the idea of alternation of generations among plants first arose, though it had been recognised in the animal kingdom long before.

Of course the same phenomenon really occurs in *Selaginella* and even in Flowering Plants, but in all these it is much less conspicuous, because, as we ascend the scale, the sexual generation becomes more and more dependent on the asexual, so that at last the former is reduced to a mere insignificant appendage of the latter, and can scarcely be distinguished from it.

We have just seen that, even in such Ferns as our type, when the prothalli happen to be diœcious, the male specimens often remain rudimentary. In *Selaginella*, where the difference of sex is fixed, this has gone much further; the male prothallus is reduced to one little cell, and is so insignificant as to be scarcely recognisable. The female prothallus, which has much more work to do, is much less reduced, but remains almost shut up in the coats of the megaspore, and so does not obviously suggest an independent individual. When we come to the more ancient Flowering Plants—the Gymnosperms —we find the male prothallus at an equally low level with that of *Selaginella*, but much modified, in accordance with changes in the method of fertilisation. The fact that the pollen-grains are set free and germinate, clearly indicates that they mark the beginning of a new generation, while the production of active spermatozoids in some Gymnosperms links them directly with the Cryptogams. The female prothallus, though not less bulky than that of *Selaginella*, remains for ever shut up in the megaspore, which itself never gets free from the sporangium ; so on the female side all trace of an independent existence of the prothallus has been lost, and, except for minute developmental research, we should never have suspected the presence of a gametophyte at all.

In the Angiosperms matters are still worse for the sexual generation. On the male side, indeed, there is no great change, except that the homologue of the antheridium is harder to recognise, but in the embryo-sac the prothallus is scarcely to be traced ; the endosperm, which has sometimes been considered to represent the prothallus, is an after-product of fertilisation and differs from a prothallus in origin. In fact, the sexual generation in Angiosperms has become so thoroughly incorporated with the asexual, that it seems almost an affectation here to talk of alternating generations at all, and certainly the existence of such an alternation would never have been discovered except by the comparison with Cryptogams. The clue afforded by the life-history of the Ferns has thus enabled botanists to follow accurately the true course of development in the higher plants, which otherwise we should never have understood.

The regular alternation of sexual and asexual individuals is often modified in special cases among Ferns. The modification may either result in a lengthening or a shortening of the ordinary life-cycle. The life-cycle is lengthened when we get vegetative propagation of the Fern-plant, so that the number of asexual generations interposed between two sexual ones is increased. This happens in those Ferns which form buds on their leaves ; the buds become detached and give rise to new plants, as may easily be seen in *Asplenium bulbiferum* and *viviparum*, so commonly grown in greenhouses. Everybody must have noticed the minute Fern-plants which are dotted about on the fronds of these Ferns, and which in the form of their little leaves are so different from full-grown specimens.

Another way in which the life-cycle may be extended is by vegetative reproduction of the prothallus—just the converse of the process already described. In this case a number of additional *sexual* generations may be introduced into the life-history. This is pretty common among Filmy Ferns, and in some tropical species, in which the prothallus produces little buds from which new prothalli arise, so that the number of sexual individuals may increase indefinitely without the intervention of the sporophyte generation.

So much for the *lengthening* of the life-history. In other cases, it is cut short—that is to say, the one generation passes over into the other, without the aid of the regular sexual or asexual reproductive organs. There are two possible cases of this kind ; either the sexual generation may give rise directly to the asexual (*apogamy*), or conversely the asexual generation may give rise directly to the sexual (*apospory*). We have no space to go into the details of these exceptional modes of development, but it is necessary to mention them, because it is very important to learn at starting that the distinction between the two generations is not absolute, but that the one may sometimes pass directly into the other.

In apogamy, which has been sometimes observed in our type, the Male Fern, and in many other species, the vegetative tissue of the prothallus grows out into the various organs (leaf, stem, and root) of the new Fern-plant, the origin of which cannot often be traced to any single cell, or even necessarily to any definite initial group. At the same time vascular bundles appear in the tissue of the prothallus. Archegonia may be absent altogether, or, if present, may have nothing to do with the production of the new plant, which may arise entirely as a vegetative outgrowth on the prothallus. Every stage of transition between prothallus and plant may be found. In a Fern, nearly related to our type, and in at least one other species, sporangia are sometimes produced on the prothallus itself, among the archegonia and antheridia. Recent investigations have shown that a fusion of nuclei — a kind of false fertilisation — may take place in the prothallus-cells concerned in the apogamous development.

In the converse case, that of apospory, which has been observed in several native Ferns, especially garden varieties, either an abortive sporangium grows out into a prothallus, without first forming spores ; or else the sporangia are altogether undeveloped, and the prothallus arises simply as a vegetative growth from the tissues of the leaf itself. In both these cases the sexual generation is formed from the asexual directly, without the intervention of spores. Frequently apospory and apogamy occur together in the same plant, in which case there is no nuclear fusion.

We thus see that we must regard the regular alternation of sexual and asexual reproduction as the normal course of life-history in Ferns and their allies, but not as a cast-iron scheme which can never be departed from.

TYPE VI

THE FIELD HORSETAIL (*Equisetum arvense*, L.)

The Vascular Cryptogams at present existing in the world belong to three great stocks or Classes. We have already examined representatives of two of them—namely, a Club Moss and a Fern. It still remains for us to make the acquaintance of the third Class, that of the Horsetails. The latter are not now a very important group, for there is only one living genus, containing about twenty-five species. But small as the family is in these days, it is a very ancient one, and seems to stand quite apart from its neighbours among the Ferns and Club Mosses. In early geological times, especially in the far-off period when the coal-beds were being formed, the Horsetail family was in the height of its glory, and was represented by a number of very diverse forms, many of which grew into trees. Hence this good old stock, though now so reduced, is quite as worthy of our study as its more prosperous fellows.

Several species of Horsetail are natives of England, and some are very common. In general habit they all bear a strong family likeness to each other, all having stiff, upright, jointed stems, with whorls of little-developed leaves, those of each whorl being united to form a sheath around the stem. If the stem is branched, its branches are also in whorls, the whole plant having a very formal and regular appearance (see Fig. 38). The fructification is in the form of cones, each of which is borne at the end of an upright stem, or of a branch. In some species (as in our type, shown in Fig. 38) there are special fertile stems which only bear the cones, but do not branch, and are not green. In others, the cones are borne on the ordinary green vegetative stems. Underground, the

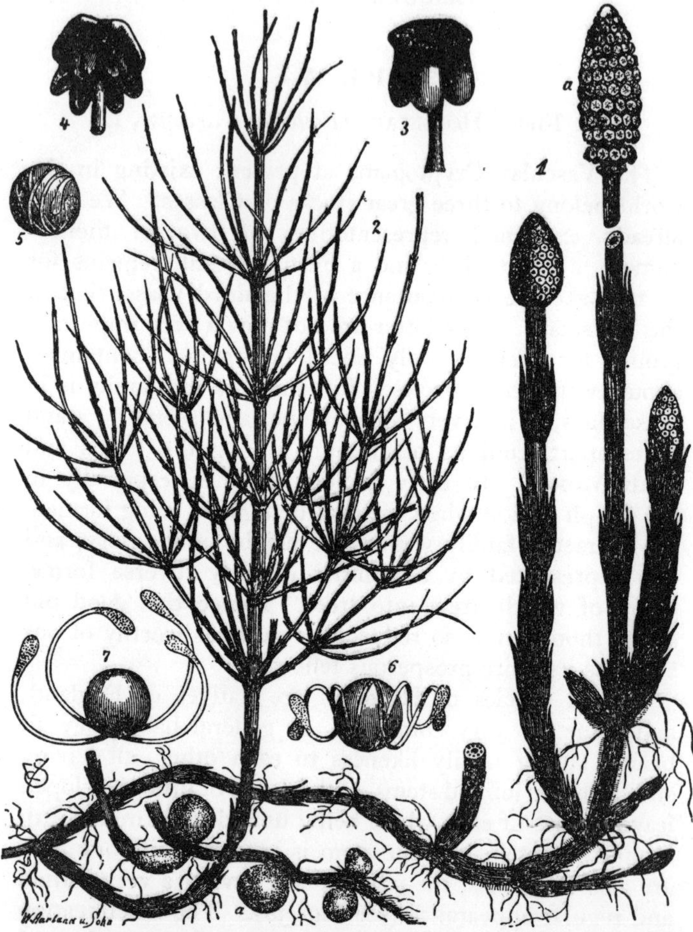

FIG. 38.—*Equisetum arvense*. 1 and 2, general view of plant, showing underground rhizome, bearing roots, with fertile and sterile aërial stems. 1, Fertile stems ; *a*, ripe cone. 2, much branched sterile stem ; *a* (on rhizome), tubers. 3, single peltate scale from cone, showing sporangia. 4, similar scale below ; sporangia dehiscing. 5, young spore, with elaters not yet expanded. 6, mature spore in damp condition ; elaters curled up. 7, the same in dry condition ; elaters expanded. Figs 1 and 2 reduced ; Figs. 3 and 4 magnified slightly ; Figs. 5, 6, and 7 very highly magnified. (After Wossidlo, from Strasburger.)

plant has a much-branched rhizome, which penetrates to a great depth in the soil, and makes these plants most obstinate weeds. If such a species as *E. arvense* or *E. maximum* has once established itself in garden ground, it is almost impossible to get it out again, for its rhizome goes too deep to be easily dug up, and is perpetually giving rise to new shoots.

Equisetum, as we shall find, resembles other Vascular Cryptogams in having a sharply marked alternation of generations. The plant, as we see it, is the asexual sporophyte, and with this we will begin.

I. *EXTERNAL CHARACTERS OF THE SPOROPHYTE*

A. Vegetative Organs

The general habit of the commonest British species, *E. arvense*, is well shown in Fig. 38, but only some of the upper branches of the rhizome are represented. We must picture to ourselves the main part of the rhizome deep down in the soil, perhaps three feet below the surface, sending up branches which alone are visible in the figure. The characteristic leaf-sheaths are obvious on all the stems whether above or below the ground; on the older parts of the rhizome, however, they often wither away. Each sheath consists of a whorl of coherent leaves, the free parts of which are only represented by the teeth at the top of the sheath.

The rhizome bears numerous slender adventitious roots, arising at the nodes, and in this species also produces round tubers, each of which represents a short branch consisting of a single swollen internode. These tubers are capable of giving rise to new plants, and thus form a means of vegetative propagation (see Fig. 38, 2, *a*).

The characters of the stem are best studied in detail on the shoots which rise above the ground. The surface is ribbed lengthwise, each rib lying in the same straight line as one of the leaves of the node next above. Both ribs and leaves alternate regularly in successive internodes. The stems above ground are in this species (*E. arvense*) of two kinds. First, we have the fertile shoots, which show themselves in spring (March) and have no other function than to bear the cones (Fig. 38, 1). These fertile shoots are unbranched, and are of a pale colour, containing little or no chlorophyll. They die down as soon as the spores are shed. The other shoots are sterile, and their branches constitute the assimilating apparatus of the plant, for the leaves are of little importance in this respect (Fig. 38, 2). They are of a deep green colour, and are repeatedly branched, the branches breaking out from the stem through the lower part of the leaf-sheaths. In each whorl the branches are equal in number to the leaves, and alternate with them. The ultimate ramifications are very slender, and only have from three to five ribs, while the main stem may have as many as twenty. The surface of the aërial shoots is very hard and somewhat rough, especially at the ridges.

We see then that our plant has a very characteristic habit, marked partly by the small development of the leaves, and partly by the great regularity of the whorled branches. Other species differ considerably from this type; many have only one kind of stem, the cones being borne on ordinary vegetative shoots, while in others the aërial shoots branch little, or not at all. In *E. maximum*, the largest British species, the barren stems sometimes attain a height of six feet, but some of the tropical kinds, such as *E. giganteum*, a native of tropical South America, are much taller, even, it is said, reaching forty feet.

B. Reproductive Organs

The cone of an *Equisetum* is unlike the fructification of any other living plant, and cannot be mistaken when once seen, though the male flowers of some Coniferæ, such as the Yew, are found to bear a certain resemblance to it when closely examined. The cone is terminal either on the main fertile shoot (as in *E. arvense*) or on a branch (as in *E. limosum*). It consists of a fairly stout axis, giving rise to densely crowded alternating whorls of peltate scales (*sporangiophores*) on which the sporangia are borne (see Fig. 38, 1, *a*). The scales of the cone are usually called *sporophylls*, and their mode of development agrees well with their leaf-nature, but some of the fossil forms throw a certain amount of doubt on this interpretation, so we prefer to call them simply sporangium-bearers. In each whorl there are a considerable number of sporangiophores—about twenty in many cases. Each sporangiophore has a short cylindrical stalk, and expands at the end into a flat disc, to the under-side of which the sporangia are attached—five to ten on each scale. The peltate heads of the sporangiophores are in such close contact that they usually become hexagonal from mutual pressure. The sporangia extend inwards as far as the axis, so as to fill up all the room that is left between the peltate scales. They contain very numerous spores, which are all of one kind. At the bottom of the whole cone is a ring of abortive leaves, called the annulus (see Fig. 38, 1, *a*) ; sometimes there are two such rings. These rudimentary structures are of some interest, because in many of the fossil forms there are whorls of barren leaves or bracts between the whorls of sporangiophores. It is possible that we find the last remnants of these bracts in the annulus of living Horsetails.

II. *INTERNAL STRUCTURE AND DEVELOPMENT OF THE SPOROPHYTE*

1. Vegetative Organs

a. The Stem

The general structure of the stem in the genus *Equisetum* is at once simple and characteristic. Among all the Cryptogams now living, these plants approach most nearly, as regards their anatomy, and especially that of the stem, to the simpler Gymnosperms and Dicotyledons, though in other respects they differ widely from them. The stem of *Equisetum* is invariably traversed by a number of collateral leaf-trace bundles, arranged in a single circle. The course of these bundles is excessively simple ; a single one enters the stem from each leaf, *i.e.* from each tooth of the coherent sheath. It passes straight down the whole length of the internode, without joining on to any other bundle until it reaches the node below. Here it forks into two, and the forks attach themselves to the two adjoining bundles coming up from below, just where they are beginning to bend out into the leaves ; consequently every internode contains just as many bundles as there are leaves at the node above, and as the leaves alternate with each other at successive nodes, so also do the bundles in the corresponding internodes. As all the bundles enter the stem to the same depth, and then turn vertically downwards, it follows that, as seen in transverse section, they always form a single ring. It will be seen that the bundle-system is just of the kind typical for Conifers and Dicotyledons, but it is one of the very simplest examples of this type. The stem is always ridged on the surface, as mentioned above. Each of the ridges corresponds in position to one of the vascular bundles (see Fig. 39).

The Equiseta are characterised by a great development of intercellular spaces, which have a very definite arrangement. There is usually a ring of large spaces in the cortex, and these cortical cavities are alternate in

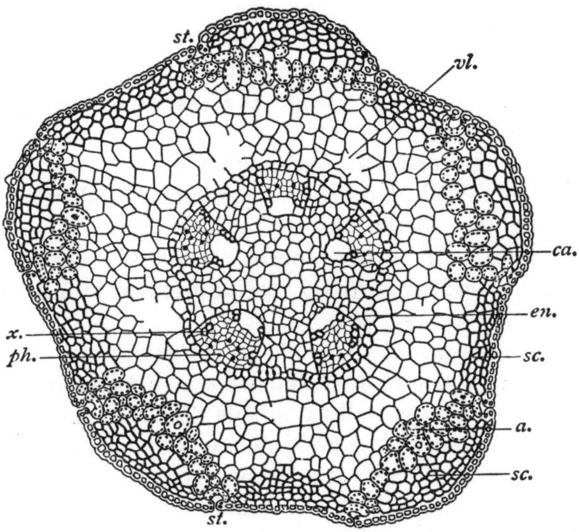

FIG. 39.—*Equisetum arvense*; transverse section of a branch of sterile stem. *x*, xylem; *ph*, phloëm; *ca*, carinal cavity; *en*, endodermis; *sc*, sclerenchyma; *a*, assimilating tissue; *vl*, vallecular cavities (imperfectly formed); *st*, stomata. Magnified 45 diameters. (R. S.)

position with the bundles, and thus lie opposite the depressions or furrows of the external surface. For this reason they bear the name of *vallecular cavities*. Another ring of intercellular canals accompany the bundles, one on the inner side of each; these lie opposite the ridges of the stem, and are consequently called the *carinal cavities*. We shall see presently how they arise. These canals are interrupted at the nodes. Lastly, the whole interior of the pith of the internodes often becomes hollow, leaving only a persistent diaphragm at each node. This almost always happens in the main aërial stems; but in the finer aërial branches (see Fig. 39) and in the rhizome the pith often remains solid, as is the case in *E. arvense*. The intercellular spaces do not all fulfil the same function; the carinal cavities and the central cavity, if present, usually contain water, while the vallecular cavities are always full of air.

In *E. arvense* and some other species the central cylinder is well defined, a common endodermis surrounding the whole ring of vascular bundles on their outer side. In other species, however, there is a separate endodermis round each individual bundle, as in *E. limosum* (see Fig. 40).

In others again there is an intermediate state of things, for a common endodermis is present *inside* the ring of bundles, as well as outside them (*E. variegatum*). These differences, however, do not otherwise affect the anatomy. Although the same general structure is maintained throughout the shoot, yet in the minute ultimate branches the number of bundles becomes much reduced, often down to three, and in these cases the appearance of the transverse section may be very different from that of a main stem or larger branch (see Fig. 40).

We will now consider the tissues rather more in detail, and will begin with the vascular bundles. Each bundle is normally collateral, *i.e.* with xylem on its inner and phloëm on its outer side. The carinal cavity marks the position of the *protoxylem* or first-formed tracheids of the bundle (see Fig. 39). Here a few tracheids have become thickened (in an annular or spiral manner) at a very early stage of growth; consequently they cannot follow the expansion of the surrounding tissues, and a rupture takes place, forming the cavity. Projecting from the walls of this cavity we see the rings or spirals of the disorganised tracheæ (see Fig. 39). The later-formed part of the xylem, consisting of a few scalariform tracheids (not vessels), is usually separated from the protoxylem by a little parenchyma, and forms two groups to the right and left of the bundle. The whole wood, therefore, if continuous and not disturbed by the carinal cavity, would form, as seen in transverse section, a V with the point inwards and the limbs outwards.

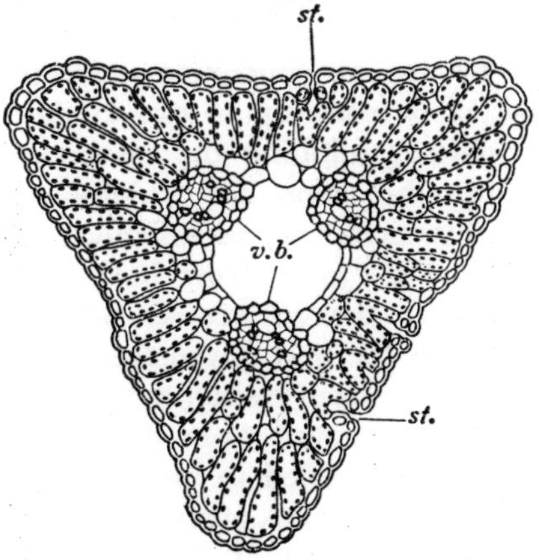

FIG. 40.—*Equisetum limosum*; transverse section of an ultimate branch of the aërial stem. *v.b*, the three vascular bundles, each with its own endodermis; *st*, the depressed stomata. The pith is hollow, but there are no vallecular cavities. Almost the whole cortex is assimilating palisade-tissue. Magnified 100 diameters. (R. S.)

The phloëm lies between the limbs of the V (see Fig. 39, *ph*). It consists of sieve-tubes (with sieve-plates on their oblique transverse walls) and parenchyma. Beyond this, on the outer side, we come to the pericycle and then to the endodermis (with well-marked cuticularised bands on its radial walls) which marks the beginning of the cortex (Fig. 39). The xylem is often very little developed, especially in the rhizomes and the stems of aquatic species. The pith, or what remains of it, when the stem is fistular, consists of ordinary parenchyma, and presents no features of interest.

The cortex, however, at least in the aërial stems and branches, is highly differentiated, as indeed we might expect, considering that it has here to perform the assimilating function usually assigned to the leaves. The inner cortical layers consist of large-celled parenchyma traversed by the air-containing vallecular spaces. The outer cortex is made up of two kinds of tissue, namely, sclerenchyma, fulfilling the mechanical function of strengthening the stem, and chlorophyll-tissue, to which the functions of assimilation and transpiration belong (see Fig. 39). Now both these tissues need to be as near the surface as possible, in order to do their work to the best advantage. The mechanical tissue offers the greater resistance to bending strain, the further it is removed from the centre-line, or "neutral axis," as it is called in mechanics, of the column, here represented by the stem. For this reason we know that iron columns are always made hollow, for the same amount of material can be used to better advantage if brought as near the exterior as possible, than if distributed all over the transverse section. This mechanical principle is constantly illustrated in the construction of plants. Again, the assimilating tissue obviously requires to be as near the surface as possible, so as to be fully exposed to light, without which its work cannot go on.

Now we will see how in the stem or in a branch of *Equisetum* a compromise is made between these two competing interests. Each prominent ridge of the stem is occupied by a strand of sclerenchyma, and there are an equal number of additional strands placed at the bottom of the furrows (see Fig. 39, *sc*). The assimilating tissue occurs in curved bands, each of which lies behind one of the sclerenchymatous ridges, and reaches the surface on either side of it, between the mechanical tissue of the ridge and that of the furrow (see Fig. 39, *a*). The epidermis has stomata at those places only where the chlorophyll-tissue reaches the surface, so they are placed where they are most needed for transpiration and the passage of gases. We notice also that the bands of chlorophyll-tissue lie directly opposite the vascular bundles, so that they are well situated both for receiving the water and mineral substances from the latter, and also for transferring to them in return the products of assimilation. In the very minute ultimate branches, such as that of which a transverse section is shown in Fig. 40, things are simplified. Here there is little need for mechanical strength, as the weight of the branch is trifling, and so we find the whole cortex utilised for assimilation; the vallecular spaces also are absent. Functionally these little twigs do duty as leaves.

The epidermis is chiefly remarkable for its strongly silicified outer cell-walls, which make the surface extremely hard. If all the organic matter be completely burnt away, a perfect skeleton of silex, still showing every marking on the cell-walls, is left behind. The stomata are peculiar, because the guard-cells are completely covered in on the top by the subsidiary cells, so that a double pair of guard-cells, one above the other, seems to be present.

The description of the structure of the stem, which we have just given, refers more especially to the sterile shoots growing above ground. Both the underground rhizomes and the fertile shoots are somewhat modified in structure. In the former the epidermis is destitute of stomata, and the cortex of chlorophyll-tissue, while mechanical tissues are little needed and little developed;

thus the whole differentiation of the outer tissues is much reduced.

In *E. arvense* the pith of the rhizome is solid, and this is often the case in the smaller aërial branches also, as shown in our Fig. 39. The tubers consist simply of parenchyma crowded with starch, and traversed by a few reduced vascular bundles ; each tuber corresponds to a single internode.

The fertile stem being a transitory organ, with no other function than to bear the cone, has a simplified structure, and is destitute at once of stomata, chlorophyll, and sclerenchyma. Throughout all parts of the shoot, however, the vascular system maintains the same structure, and this tissue-system is the most constant and characteristic feature in the anatomy.

b. The Leaves

The leaves of *Equisetum* are of little importance as organs for gaseous interchange, and probably serve chiefly as a protection to the lateral buds which arise beneath them. However, they no doubt take a certain part in assimilation and transpiration, as is shown by their structure. These functions are of course limited to the leaves of aërial shoots, and in the case of species like *E. arvense*, to those of the sterile stems.

The vascular bundles of the leaf-sheaths are of simple collateral structure, and do not have carinal canals. Each bundle is surrounded by its own endodermis, whether this is the case in the stem or not. As in the stem, the bundles correspond in position to the ridges of the sheath ; outside each bundle lies a strand of sclerenchyma. A narrow band of chlorophyll-containing tissue lies between the sclerenchyma and the vascular bundle, and approaches the surface on either side of the ridge. The stomata are placed where the assimilating cells reach the epidermis, so that there are two longitudinal series of stomata corresponding to each vascular bundle. The rest of the leaf-sheath consists of ordinary parenchyma, which thins out between the ridges. The teeth, which alone represent the free part of the leaves, are still further simplified ; a vascular bundle enters each tooth, but gradually dies out.

c. The Roots

The roots of *Equisetum* are always very slender, and must not be confused with the underground parts of the stem, which are much larger (see Fig. 38). All the roots seen on a mature plant are adventitious ; the main root of the embryo only lasts a short time ; its structure is like that of the adventitious roots, and our Fig. 41, which was drawn from the main root, will serve to represent either.

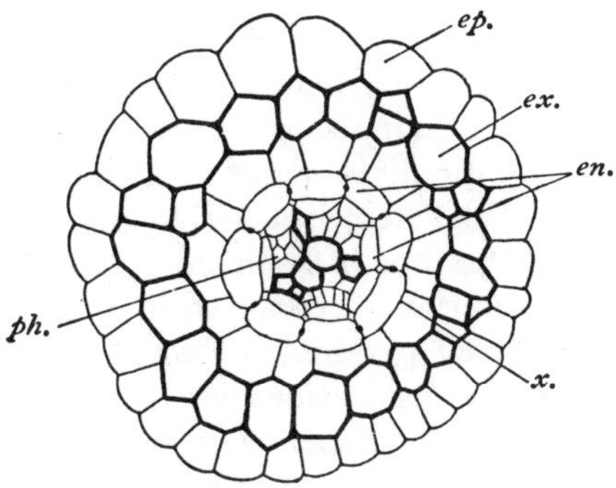

FIG. 41.—*Equisetum* ; transverse section of main root. *x*, triarch xylem ; *ph*, phloëm (three groups) ; *en*, double endodermis ; *ex*, thick-walled exodermis ; *ep*, epidermis. Magnified about 100 diameters. (After Buchtien.)

The young parts of the root bear numerous root-hairs. They have a wide cortex, enclosing a small and simple central cylinder, the structure of which is usually either triarch or tetrarch.

The arrangement of the xylem- and phloëm-groups is that usual in roots ; the centre is occupied by a large tracheid. The chief peculiarity of the root is its double endodermis ; the inner layer taking the place of a pericycle, which is quite absent. That this layer is really part of the endodermis is proved by the development, and by the fact that its cells fit on accurately to those of the outer sheath, which alone has the usual endodermal structure (see Fig. 41, *en*). This double endodermis is a character quite peculiar to the roots of *Equisetum*. The origin and mode of growth of the root will be considered in the next section. Apart from the peculiarity in the endodermis, the structure quite agrees with that of a simple root in the higher plants.

d. Growing-Points and Branching

The growing-points of *Equisetum* afford perhaps the very best examples of growth by means of a single apical cell, by the divisions of which all the tissues arise. The apex of the stem is acutely conical (see Fig. 42), and the top of the cone is occupied by the large apical cell, which has the form, so common in apical cells, of an inverted three-sided pyramid, of which the curved base is free, while the three sides are adjacent to the surrounding meristematic tissue. Divisions take place in the apical cell by walls formed in succession parallel to each of its three sides ; each segment cut off is then divided into two by a wall parallel to the first.

The cells thus formed are again divided by approximately radial walls, and then for the first time division takes place in a plane parallel to the external surface of the growing-point. We now have an outer and an inner set of cells. The former, by their further growth and subdivision, give rise to the whole of the vascular tissue, cortex and epidermis ; the inner cells only form the pith, which in the main stem soon becomes hollow. There is here no trace of the distinct layers giving rise to epidermis, cortex, and stele, such as are sometimes to be recognised in Flowering Plants. The ring of vascular bundles is only marked out at a long distance below the growing-point. About the fifth internode from the apex we find a small-celled zone of tissue, derived from the inner part of the outer layer. This zone gives rise to the vascular bundles, and to the medullary rays between

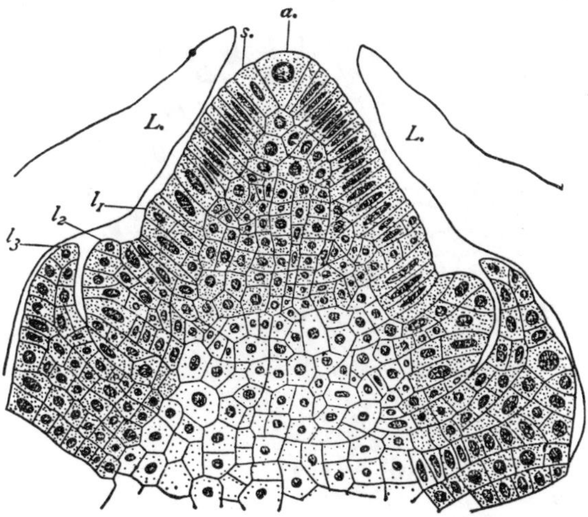

FIG. 42.—*Equisetum arvense* ; longitudinal median section of the apex of the stem. *a*, apical cell; *s*, segment cut off from it; l_1, l_2, l_3, youngest leaves, in order of age. *L*, outline of older leaves. Magnified 180 diameters. (R. S.)

them. The epidermis is also differentiated late, for there is no distinct dermatogen near the apex.

The whorls of leaves are at first crowded closely together; the internodes between them only begin to lengthen some way down the stem. Each whorl arises from the outgrowth of a ring of tissue which extends all round the stem. The circular ridge thus produced, which is at first of equal height all the way round (see Fig. 42, l_1 and l_2), is the young sheath, and soon grows out at certain places to form the leaf-teeth. We see then that the sheath is formed first, and the free part of the leaves later.

The development of the branches in *Equisetum* is peculiar. They are apparently of endogenous origin, and for a long time were thought really to arise below the surface, though this is not the case. The branches are

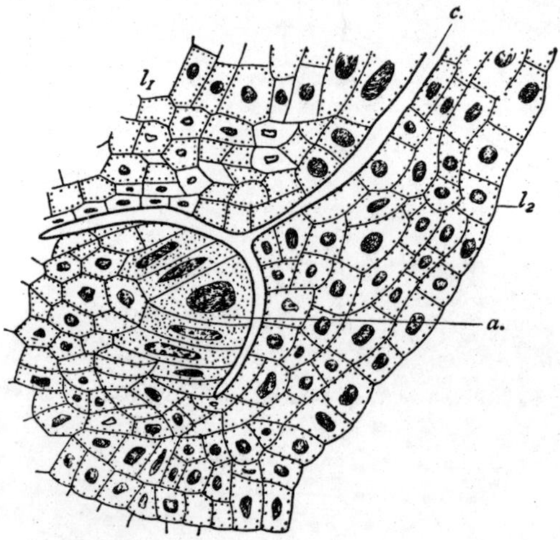

FIG. 43.—*Equisetum arvense*; part of a radial section of stem, just below the apex, to show exogenous origin of branch. *a*, apical cell of branch; l_1, cortex of stem; l_2, base of leaf below branch; *c*, crevice between them, about to close up. Magnified 360 diameters. (R. S.)

arranged in whorls in the axil of each sheath, but alternating with the leaf-teeth. The buds arise near the growing-point, each from a single superficial cell, lying immediately above the junction between leaf-sheath and stem (see Fig. 43).

This cell divides up so as to carve out a pyramidal apical cell like that of the main stem, and the growth of the branch now goes on in the usual way. But while it still consists of a very few cells only, the leaf-sheath grows out above it, and joins on to the tissue of the stem on the upper side of the bud, so as completely to shut it in. Our figure shows the bud just before it is quite enclosed, while there is still a crevice left above it, between the stem and the leaf-sheath. When this passage is once shut, it never opens again ; the bud goes on developing within a closed chamber. It lives to some extent at the expense of the surrounding tissue, and eventually breaks through the base of the leaf-sheath, and at last reaches the light of day. Seen from outside, these branches appear to arise below the node, which of course is not the case really. Endogenous buds are very rare, and we see that those of *Equisetum* are not among them, but only become enclosed after they have started in the usual way, as superficial outgrowths.

Another peculiarity in *Equisetum* is the arrangement of the adventitious roots, which do not grow on the main stems, but are always in connection with lateral buds. As a rule, one root (occasionally more) is formed at the base of each branch, arising on its lower side, just below its first leaf-sheath. On the aërial branches these roots generally remain undeveloped, while the branch goes on growing. On the rhizome the reverse is the case, for, as a rule, the buds themselves are abortive, while the roots which they bear grow vigorously. The root grows at the apex by means of a single apical cell of the same pyramidal shape as that of the stem, from which it differs, however, in forming walls parallel to the free base, in addition to those parallel to the sides. The cells thus cut off at the end increase and multiply very rapidly, and form the root-cap. All the rest of the root is formed from the segments cut off from the three sides of the apical cell. The mode of growth is much the same as in the Fern-root, shown in Fig. 27 (p. 52).

The roots of *Equisetum* branch freely ; the origin of

the branches, as in other roots, is deep-seated or endogenous. In this case it is from the inner layer of the double endodermis that the rootlets are formed, each of them arising from a single cell which lies just on one side of a protoxylem-group. This cell divides up so as to form an apical cell of the usual pyramidal form. The rootlet has to make its way through the whole thickness of the cortex, and in doing so is helped by the presence of a digestive sac (see Part I. p. 171), formed from the outer endodermal layer, which thus constitutes a temporary covering to the young root.

2. Reproductive Organs of the Sporophyte

We have already learnt the main points in the structure of a cone of *Equisetum* so far as they can be seen with the naked eye or a pocket lens (see p. 79). It remains for us to make ourselves acquainted with the more minute characters. The anatomy of the axis of the cone is in all essentials just the same as that of a vegetative stem, and the development takes place in the same manner, though the growth of the cone is limited. The whorls of sporangiophores are in origin somewhat similar to the whorls of vegetative leaves, but in the fertile cone scarcely any sheath is developed, so that sporangiophores are separate outgrowths almost from the first. The upper part of the sporangiophore soon begins to grow in diameter more rapidly than its base, which thus becomes constricted, so that the mature peltate form is already indicated. At about the same time the sporangia begin to show themselves as slight outgrowths projecting from the under-side of the expanded portion. Each of the sporangia, of which there are many, arises from the growth of a little group of cells. The essential part, however, can all be traced to a single superficial cell, which by its repeated divisions gives rise both to the archesporium and to that part of the wall lying over it.

At the stage shown in Fig. 44 the archesporium has already grown and divided up, so as to form a good-sized mass of spore-producing tissue. The wall is several cells in thickness. The layer of cells which immediately

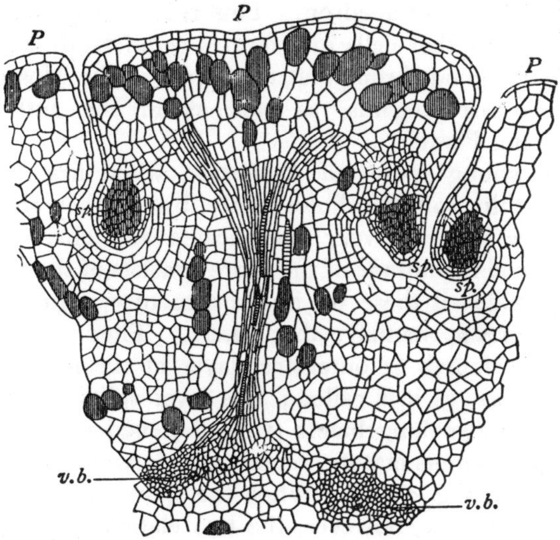

FIG. 44.—*Equisetum maximum* ; part of transverse section of young cone, showing one complete peltate scale, and parts of two others (*P*). *sp*, sporangia ; the shaded part is the archesporium ; *v.b*, vascular bundles ; the scattered shaded cells are tannin sacs. Magnified 50 diameters. (W. C. W.)

surrounds the spore-producing tissue on all sides, and may be called the tapetum, is soon used up for nutritive purposes, and the intermediate layers also disappear, so that the wall of the ripe sporangium is only one cell thick. In the meantime the sporogenous tissue goes on increasing, but all its cells do not become mother-cells ; a good many break down and give up their contents, which serve, together with the tapetum, to feed the survivors.

The remaining mother-cells, of which there are a large number in the sporangium, then divide each into four, the nucleus dividing twice before the partition-walls are formed. Finally, the four daughter-cells are arranged in a tetrahedron. This rule of the division of the spore mother-cells into four is wonderfully constant, and holds good throughout the whole of the Mosses and Vascular Cryptogams, as well as in the anthers of Flowering Plants.

The young spores of *Equisetum*, when first formed, have a thin wall of cellulose only, but as they ripen the structure becomes very complicated and characteristic. The actual membrane of the spore consists of three layers, but outside all these we find a structure quite peculiar to *Equisetum*, namely, the *elaters*. They are formed from the fourth or outermost layer of the membrane—the *epispore*, as it is called ; this layer splits along spiral lines into two long bands (with flattened ends), which, until the spore is mature, remain closely wrapped round it (see p. 76, Fig. 38, 5). When the spores are quite ripe and getting dry, however, the two elaters stretch themselves out, remaining attached only in the middle of their length, and at one point, so that they appear as four distinct appendages (Fig. 38, 7). If it is damp they coil themselves up again (Fig. 38, 6). These extraordinary hygroscopic movements may be repeated an indefinite number of times, as we can see by mounting some spores on a dry slide under the microscope, and then breathing on them. The moist air makes the elaters coil up, and as they dry they stretch out again, setting the spores in motion by their contraction and expansion.

There is some doubt whether these movements tend to disperse the spores. Possibly the chief function of the elaters is to help in the dehiscence of the sporangium. As this loses moisture the spores inside begin to stretch out their elaters; this causes the whole mass of spores to take up more room, and so to press on the wall of the sporangium, which they thus tend to burst. It has also been suggested that they cause the spores to keep entangled together, so that they are obliged to germinate in company. This might be of importance, as the prothalli are usually diœcious.

The outermost layer of the sporangial wall, which alone persists till maturity, consists of spirally thickened cells. Dehiscence takes place by a longitudinal slit (Fig. 38, 4).

The development of the cones, at least in some species of *Equisetum*, is remarkably slow. Thus the cone from which the section shown in Fig. 44 was made, would not have matured for two years. In this species (*E. maximum*, the largest in the British Flora) the cones of three successive years are present on the plant at the same time. In March, when the spores are shed, we have not only the ripe cones rising into the air on the fertile stems, but underground, still enclosed in buds, we find the cones of the next year, and the year after that too. In some species, however, as in *E. limosum*, the development is a great deal quicker.

The spores of *Equisetum* contain not only a nucleus, but chlorophyll-granules. As is usually the case with green spores, they must be sown within a few days after ripening, or they will not germinate at all.

III. *DEVELOPMENT AND STRUCTURE OF THE SEXUAL GENERATION* (*GAMETOPHYTE*)

1. The Prothallus

On the whole, the sexual generation of the Horsetails is much like that of the Ferns, though there are many differences in detail, and the mode of growth and ultimate form of the prothallus are less regular in *Equisetum* than in such Ferns as *Aspidium*. The spore begins by dividing into two unequal cells, and usually the smaller of these grows out into the first rhizoid, while the larger gives rise to the green part of the prothallus. The latter usually divides up so as to form at once a flat plate of

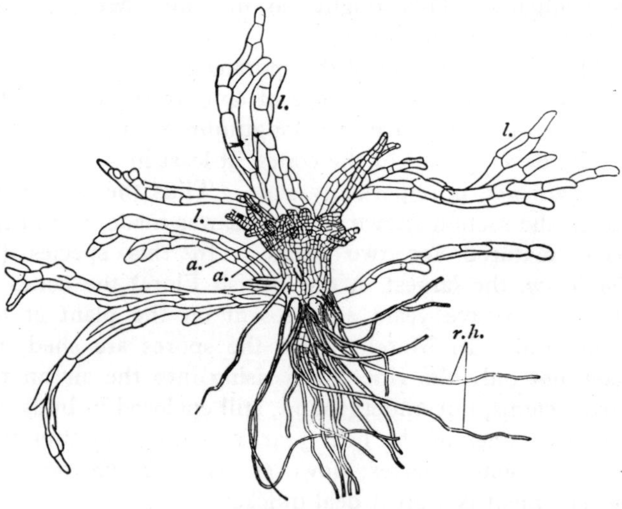

FIG. 45.—*Equisetum maximum*; large female prothallus, seen from below. *l, l,* lobes; *a, a,* archegonia; *r.h,* rhizoids. Magnified about 30 diameters. (After Buchtien.)

tissue; a distinct apical cell is not always to be found. The development is very variable, but generally the prothallus puts out filamentous branches, and forms a kind of cushion with a midrib in the middle, while the sides remain one cell thick. The male prothalli often have no definite growing-point. They remain, on the whole, decidedly smaller than the females, and begin to form antheridia very early. Sometimes antheridia are formed at the ends of branches, while in other cases they arise from the thickened cushion. Fig. 45 represents a very large and complicated female prothallus; the male individuals are much smaller and less branched. The large female prothalli possess a distinct growing-point, which produces a series of lobes on the lower side of the prothallus. Between these lobes the archegonia are placed.

2. The Sexual Organs

a. The Antheridia

The antheridium of an *Equisetum* is a very simple structure (see Fig. 46). It arises from a single cell, which divides into two by a wall parallel to the free external surface. The outer of these two cells simply forms the cover; the inner, after very numerous divisions, gives rise to all the spermatozoid mother-cells. The cover-cell divides up two or three times, by walls at right angles to the surface, usually forming a triangular cell in the middle of the cover, through which dehiscence takes place.

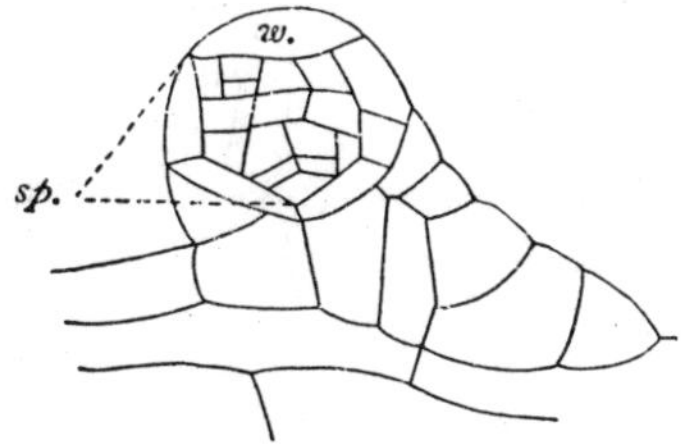

FIG. 46.—*Equisetum maximum*; young antheridium. *w*, wall; *sp*, tissue from which spermatozoids will be formed. Magnified about 200 diameters. (After Buchtien.)

In each of the very numerous cells in the interior of the antheridium a single spermatozoid is produced. Its spirally coiled body is derived almost entirely from the nucleus of the mother-cell, while the cilia themselves, and just that part of the body to which they are attached, are formed from the cytoplasm. The spermatozoids are almost exactly like those of Ferns, and go spinning through the water in the same way. In fact there are some Ferns in which the antheridia as well as the spermatozoids agree in every respect with those of the Horsetails.

b. The Archegonia

The first archegonia arise on the cushion-like part of the prothallus ; as fresh lobes go on forming, additional archegonia are developed at the base of each lobe (Fig. 45, *a*). The archegonia at first point downwards as in the Ferns, but subsequently get turned upwards by the growth of the lobes. The development of the individual archegonium is almost exactly like that in some Ferns. Usually no basal-cell is formed ; the mother-cell of the archegonium divides at once into central cell and neck. The central cell cuts off two canal-cells at the top, and in the meantime the neck goes on growing. Here also there are four rows of neck-cells, each row consisting of three or four cells. The four cells at the top are very long, and bend far back, leaving a wide opening between them when the organ is ripe (see Fig. 47).

It is remarkable that while the Horsetail plant bears no resemblance whatever to a Fern plant, the prothallus and sexual organs are so much alike in both. It is true that they differ a good deal from those of the Male Fern, but other Ferns come very near the Horsetails as regards their sexual generation.

The conditions of life of the sexual prothallus are, as a rule, simple and uniform compared with those to which the asexual plant is exposed, so that the former has less need for varied structural adaptations. Consequently we often find that at this stage there is much in common between families which, so far as their sporophytes are concerned, have lost all traces of relationship.

The most striking point about the prothallus of the Horsetails is its being usually (though not without exception) diœcious. We found that in the Ferns very small prothalli often form male organs only, while the better-grown individuals produce archegonia as well. In Horsetails this difference has gone further and become more constant. Even in Horsetails, however, it is not fixed, but depends a great deal upon nutrition. Prothalli grown on a poor soil (*e.g.* damp sand) will only produce male organs, while those which are better treated and provided with plenty of food (say in the shape of a food-solution such as that described in Part I. p. 202) will generally become females. This is interesting, for here we see the beginning of sexual differentiation of the prothallus, which has become quite a fixed thing in other families of Cryptogams. It is now a well-established fact that some of the relations of the Horsetails, which lived in the very ancient coal period, were heterosporous. Evidently our living Equiseta come from some of the less specialised members of the stock.

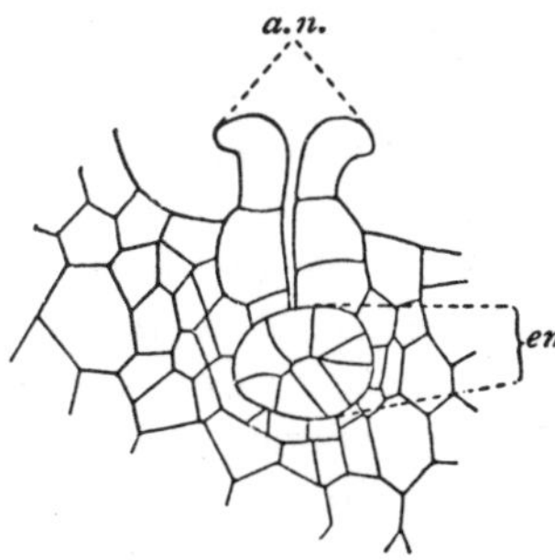

FIG. 47.—*Equisetum maximum*; fertilised archegonium. *a.n*, neck of archegonium; *em*, young embryo, showing first divisions. Magnified about 150 diameters. (After Buchtien.)

3. THE EMBRYO

Fertilisation, so far as is known, takes place in the Horsetails in the same way as in the Ferns. The spermatozoids, like those of Ferns, are sensitive to the presence of malic acid, and are no doubt thus enticed to seek the other sex. The prothalli are by no means easy to cultivate, and only a few observers have succeeded in tracing the whole history. The prothalli generally grow healthily enough up to the time when the first antheridia are formed, but then they often begin to " damp off." However, the development has been followed throughout by a few botanists, so that we know how the embryo arises from the fertilised ovum. The latter first divides into two by a horizontal wall. The first division in the upper half (that towards the neck of the archegonium) is by a somewhat inclined wall, which separates the first leaf from the unicellular rudiment of the young stem. The latter at once cuts off two segments, which give rise to the second and third leaves. These three leaves form the first whorl of the young Horsetail. Though coherent at the base, they are more distinct from each other than the leaves of later-formed whorls. After these first divisions the apical cell of the stem has already assumed the pyramidal form which it keeps all through life.

In the meantime similar divisions have taken place in the lower half of the young embryo. Here an inclined wall separates the cell destined to give rise to the main root from one which merely forms the *foot*, a comparatively unimportant structure in *Equisetum*. The root-cell, which lies exactly opposite that from which the stem is derived, divides up so as to form the usual pyramidal apical cell, from which, by a wall parallel to the free surface, the root-cap is marked off. Thus the young embryo of *Equisetum* is started, and even at this early stage shows something of the characters of the mature plant, such as the whorled leaves and pyramidal apical cells. Fig. 47 shows a very young embryo enclosed in the venter of the archegonium, when only a few divisions have taken place. In Fig. 48 we see the section of an embryo at a much more advanced stage, when two whorls of leaves are already formed.

Up to about this age the embryo remains within the cavity of the enlarged archegonium. So far, the root has not developed much, but now it grows rapidly and breaks through the tissue of the prothallus below it. It is followed by the stem, which bursts the neck of the archegonium. The young plant is now becoming independent, though for a time it remains connected, by means of the foot, with the prothallus. The main axis formed directly from the embryo has only a limited growth. It remains very slender, and stops growing after it has formed from ten to fifteen whorls of three leaves each. It is interesting to note that in the smallest species of *Equisetum*, *E. scirpoides*, three-leaved whorls are formed throughout life. At the base of the main stem a stouter lateral branch arises, and this again produces another still more vigorous shoot, and so on. Thus the mature form of the species is gradually built up by the production of successive branches, each more highly developed than the last. After a time one of the lateral shoots turns downwards and penetrates the ground, thus forming the first rhizome of the young plant. The main root is fairly well developed, though only a temporary organ, and shows the same structure as the subsequent adventitious roots.

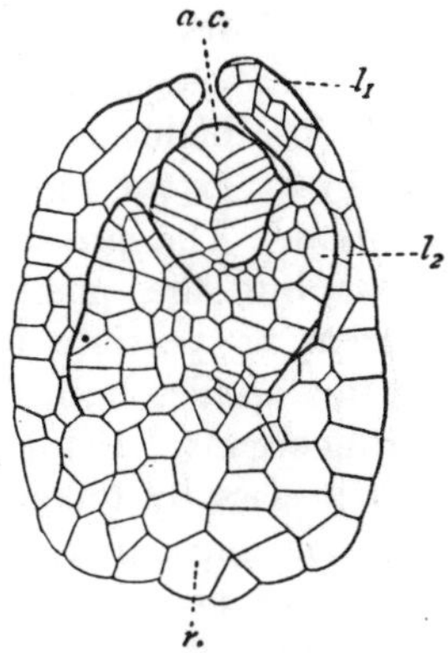

FIG. 48.—*Equisetum maximum*; embryo in median longitudinal section. *a.c*, apical cell of stem; l_1, leaf of first whorl; l_2, leaf of second whorl; *r*, position of root. Magnified about 200 diameters. (After Buchtien.)

Summary

We have now traced a Horsetail through the complete cycle of its life. So far as the general course of development is concerned, we have found an essential agreement with that in the Ferns—namely, a sharply marked alternation of generations, spores of one kind, and a well-developed prothallus. Although the latter is usually diœcious, the distinction between male and female prothallus is not a fixed one, but is dependent to a great extent on external circumstances, especially nutrition. The prothallus is not unlike that of some Ferns, but the plant—the asexual generation—is as different as possible both from Ferns and Lycopods, and this applies both to vegetative structure and spore-bearing organs. Evidently the Horsetails form a perfectly distinct class by themselves. As mentioned above, this class was once, in remote geological ages, an extensive and varied one. Many of its members not only grew into trees, but had the same mode of secondary growth by means of cambium, which is now almost entirely limited to Dicotyledons and Gymnosperms. Their fructifications showed a great variety, some few resembling those of *Equisetum*, while most of them were more complicated, and several produced spores of two kinds. In fact we can form a much better idea of the class Equisetineæ from the study of its extinct members, than from that of the small remnant which has survived to our own times.

We have now come to the end of our types of Vascular Cryptogams, and may very briefly sum up the characters of this great and ancient sub-kingdom of plants. They are quite easily characterised as plants with a clear alternation of sexual and asexual generations, each of which leads a more or less independent life, the asexual stage always being much the more highly developed of the two. The fertilisation by means of spermatozoids, which sometimes occurs even among the Gymnosperms, is here a constant character. The heterosporous Vascular Cryptogams come nearest to the Flowering Plants, as was fully explained in our chapter on *Selaginella*, which is the only heterosporous type which we have had space to describe. Heterospory, however, is by no means limited to the Lycopod series; it occurs also among Ferns (in the widest sense), and, as we have already pointed out, among the fossil Equisetineæ. We cannot say for certain at present which of the heterosporous groups really comes nearest to the Phanerogams; probably none of those now living bear much resemblance to the real transitional forms (still altogether unknown), which existed at an enormously remote period, represented by some of the oldest fossiliferous strata. *Selaginella* serves as well as any other type to enable us to form an idea how Cryptogamic and Phanerogamic modes of reproduction are related.

In finishing our account of the Vascular Cryptogams, we have also come to the last of our series of vascular plants. So far, the same general system of anatomical structure has prevailed all through; henceforth we shall leave it behind altogether, and find ourselves among plants with a much simpler, or at least a totally different, internal organisation.

CHAPTER II

THE BRYOPHYTA

THE step which we are about to take, in passing on to our next type, carries us across one of the widest gaps in the Vegetable Kingdom. So far, the *plant*, in the ordinary sense of the word, has in all cases been represented by the sporophyte generation. We have always found that the stage of the life-cycle, lying between fertilisation and spore-production, is that in which the chief vegetative development is attained. The other stage, namely, that which succeeds spore-production and precedes fertilisation, has up to this point appeared as a comparatively insignificant organism, hardly recognisable as a distinct generation in the Phanerogams or *Selaginella*, though maintaining a more independent position in the Ferns and Horsetails. Henceforth we shall find the relative proportions of the two generations reversed, the chief vegetative growth taking place in the sexual stage, corresponding to the prothallus of the higher plants, while the sporophyte develops as a fruit rather than as a plant, and serves for little more than the mere production of spores.

The sub-kingdom, then, with which we have now to deal, is characterised by the occurrence of a sharply defined alternation of generations, in which the sexual generation is the more important as regards vegetative development, the sporophyte being always dependent upon the gametophyte for a great part of its nutrition, and never becoming free. This sub-kingdom is that of the *Bryophyta*, or mosslike plants. It includes two great Classes, the true Mosses and the Liverworts. The Mosses, the general appearance of which is familiar to every one, have a vegetative growth much like that of the higher plants, with well-formed stems and leaves, but all these organs belong to the sexual generation, and so are not directly comparable with the leaves and stems of the higher plants, which belong to the asexual stage. The Liverworts, perhaps less generally known to those who are not botanists, sometimes have distinct leaves and stems not unlike those of the true Mosses, but many of them have a much simpler organisation, the plant showing no distinction of leaf and stem, but consisting of an undifferentiated body performing the functions of both these organs, and called a *thallus*. We will take one of these simpler Liverworts for our first type of the Bryophyta, because its gametophyte generation is so much like the prothallus of a Fern, a fact which helps us at once to grasp the true homologies between plants otherwise so different.

A. THE LIVERWORTS (HEPATICÆ)

TYPE VII

PELLIA EPIPHYLLA

1. THE THALLUS

Pellia epiphylla is one of the commonest Liverworts, growing in very various habitats, sometimes by the sides of brooks or wells, or in damp woods and hedgerows, sometimes actually living under water ; in other cases, however, it grows on comparatively dry sandy ground. The plant in its vegetative condition is a green, flat, lobed thallus, repeatedly branched, the lobes often overlapping each other (see Fig. 49).

The plants grow socially, and may collectively cover a considerable patch of ground. If we cut off a part of the thallus and examine it, we find that it forks repeatedly, all the branches lying nearly in the same plane. The thallus has an upper and under surface, the former darker green than the latter; it is traversed by a midrib, from which it thins off on either side towards the margins (Fig. 50). On the under-side numerous rhizoids are borne, which spring from the midrib and fix the plant to the ground; for *Pellia*, like other Bryophyta, possesses no true roots.

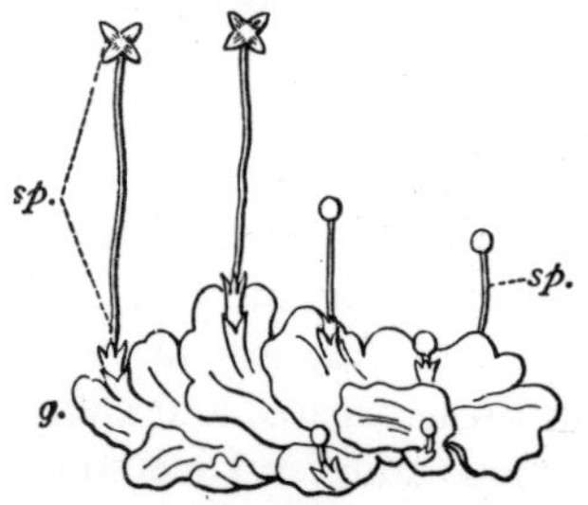

FIG. 49.—General view of a plant of *Pellia epiphylla*. *g*, the lobed thallus, constituting the gametophyte generation; *sp*, the fruit constituting the sporophyte generation. The fruits to the left have opened; those to the right are younger and still closed. Half natural size. (After Cooke.)

The whole character of the plant varies greatly according to the conditions under which it grows; so much so that its different forms would never be supposed to belong to the same species, if the transitional states had not been observed. Under water (where, by the bye, *Pellia* never fruits) the thallus is long, narrow, and strap-shaped, branched at rather distant intervals, with a very distinct midrib, and very thin margins. On damp ground, where *Pellia* attains its greatest luxuriance, the thallus is much broader than in the aquatic form, but still elongated, with the branches spread out nearly flat, and the midrib very strongly marked. On dry sandy soil the plant assumes a very different form; the thallus remains short and stunted, with densely crowded branches overlapping each other. The whole plant is much thicker and tougher, and consequently the midrib becomes indistinct.. In spring, when the plants begin their new growth, they send out a great number of small crowded branches, giving a parsley-like appearance to the growing edge of the thallus.

The anatomical structure of the thallus is excessively simple. It consists entirely of parenchyma, the cells of which are elongated in the midrib, and polyhedral in the rest of the thallus. Chlorophyll-granules occur chiefly in the more superficial cells. They are most abundant in the cells on the upper surface and in all cells of the thinner marginal portions. The whole tissue is rich in starch-grains. The epidermis scarcely differs from the rest of the tissue, but has a thin cuticle, at least on the under surface of the thallus. In the interior of the middle part of the thallus there are sometimes groups of cells with much thicker walls than their neighbours. The walls, however, are of cellulose, and there is no further differentiation. The root-hairs, or rhizoids, are unicellular; the cell-walls of the older rhizoids have a brownish colour, but give cellulose reactions.

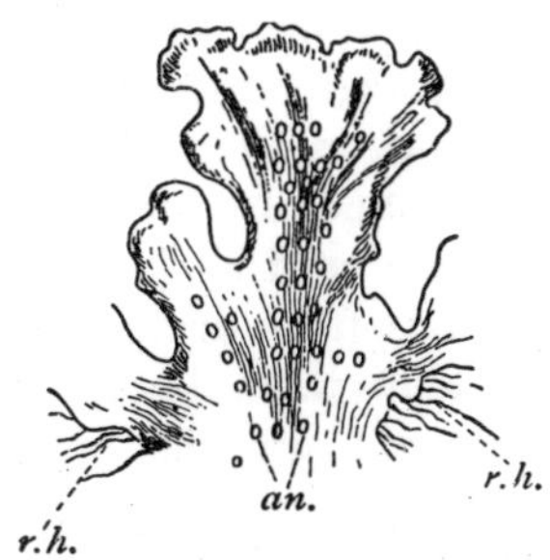

FIG. 50.—Part of the thallus of *Pellia*, seen from above. *an*, the numerous antheridia; *r.h*, *r.h*, the rhizoids. Slightly magnified. (R. S.)

The growing-point of each branch lies at the base of a depression between the lobes, just as in the prothallus of a Fern. The growth here goes on by means of a single large apical cell, which cuts off segments both at its sides and base. The former build up, by their subsequent growth and divisions, the lateral parts of the thallus, while the basal segments are chiefly concerned in forming the midrib. The tissue derived laterally from the apical cell grows more rapidly than the apex itself, which consequently always lies in a recess of the margin.

The branching of the thallus, which as we have seen may take place very freely, is dichotomous, the original growing-point giving rise to two. The way this happens is that a new apical cell is formed from one of the lateral segments, and then both the apical cells go on growing on their own account. The growing-point is surrounded by short glandular hairs, which secrete mucilage and so help to prevent the delicate tissues of this part from drying up.

We see, then, that the thallus of *Pellia* is both in external form and internal structure a very simple organism, bearing no resemblance to any of the plants hitherto considered, so far as their asexual generation is concerned. There is, on the other hand, a very marked agreement with the prothallus of a Fern in form, structure, and general mode of growth. In fact, as we shall find, *Pellia* and a Fern stand on nearly the same level as regards their sexual generations, though the sporophytes of the two are absolutely different.

2. The Sexual Organs

a. The Antheridia

Pellia is usually monœcious, the thallus producing antheridia at first, and then beginning to form the archegonia. Although our plant bears a general resemblance to the prothallus of a Fern, we must not expect to find an exact agreement. In the position of the reproductive organs there is an important difference ; in the Fern-prothallus they are usually limited to the lower surface, while in *Pellia* and the Liverworts generally it is always the upper side which bears them. The antheridia are easily seen with the naked eye, dotted over the upper surface on either side of the midrib (see Fig. 50).

The antheridia when mature are globular bodies, reaching 0·3 mm. in diameter, attached to the thallus below by a very short multicellular stalk. Each antheridium is enclosed singly in a flask-like sheath, leaving only a very narrow opening at the top (see Fig. 52). This sheath is formed by the gradual growing up of the thallus-tissue around the young antheridium. The development takes place in the following way :—

The antheridium arises from a single superficial cell situated on the upper side of the thallus, immediately behind the growing-point. This cell rises above the general level of the thallus, and divides by a transverse wall ; the lower cell thus formed, after undergoing a few further divisions, forms the short stalk. The upper cell divides by a longitudinal wall into two cells, and these rapidly subdivide in such a manner as to form a single superficial layer enclosing a few central cells (see Fig. 51). The former constitute the wall of the antheridium, which remains one cell in thickness ; the central cells undergo a great number of divisions, giving rise to a small-celled tissue, which, when mature, is entirely composed of the mother-cells of the spermatozoids (Fig. 52). During the cell-division rapid growth of the whole organ goes on, and in the meantime a wall of cells grows up around the antheridium, keeping pace with its development, and ultimately closing it in, except for a narrow opening at the top (Figs. 51 and 52).

This is the usual course of antheridial development in the Liverworts ; the sheath, however, is not constantly present. There is also a fairly close agreement with the antheridia of Ferns, though there are some differences in the details of development as well as in size.

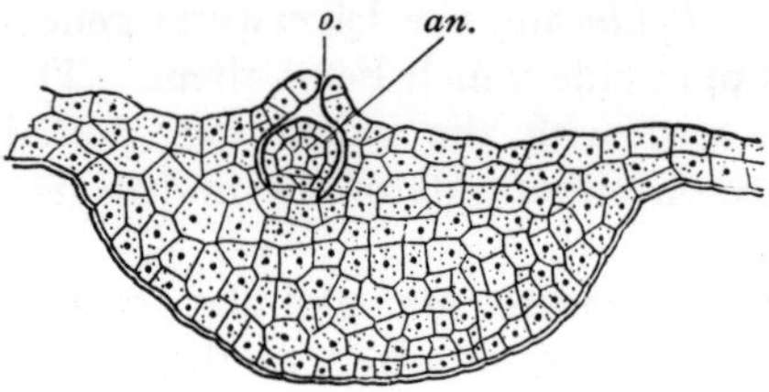

FIG. 51.—Transverse section through the midrib of the thallus of *Pellia*, showing a young antheridium. *an*, antheridium ; *o*, opening of the sheath surrounding the antheridium. Magnified 80. (R. S.)

Each of the numerous cells of the central mass of tissue produces a single spermatozoid, just as in Vascular Cryptogams ; the development is also just the same, for the body of the spermatozoid arises almost entirely from the nucleus, while the cilia, which are here two in number, are derived from the cytoplasm. The presence of two cilia only is constant throughout the Liverworts and Mosses. Among Vascular Cryptogams we find this number in the spermatozoids of most of the Club Mosses, as represented by *Selaginella*, while in the Ferns and

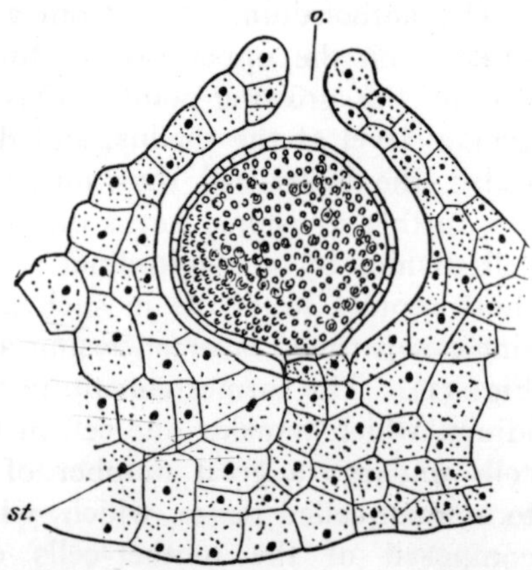

FIG. 52.—Part of a similar section showing a nearly ripe antheridium. *st*, stalk of antheridium ; *o*, opening of sheath. Magnified 80. (R. S.)

Horsetails the cilia are much more numerous. In *Pellia* the body of the spermatozoid is spirally coiled, with the cilia attached just behind the thin end, which keeps in front while the spermatozoid is swimming (Fig. 53). Here also a little bladder, formed from the remains of the cytoplasm, hangs on to the spermatozoid when it is first set free.

b. The Archegonia

The female organs, which here, as in the Vascular Cryptogams, bear the name of archegonia, arise in large numbers just behind the growing-points of the older thalli on the upper side. The thallus always thickens where they are formed. The thickened part comes to a sudden end towards the margin of the thallus, and the archegonia thus come to be seated on a steep slope, facing towards the growing-point. In the meantime the thallus goes on growing below the thickened part, forming a thin membrane, while simultaneously a membranous outgrowth arises above, behind the archegonia, and completely overlaps the whole group, which thus appears to be enclosed in a kind of pocket on the upper surface of the thallus. This pocket is called the *involucre*. The development of the involucre varies much according to the position in which the plant grows; in dry habitats it reaches a great length, while in wet places it remains short.

FIG. 53.—Single spermatozoid of *Pellia*, showing the spirally coiled body and the two long cilia. Magnified 1225. (After Guignard.)

We will now follow the development of the archegonium itself. Like the antheridium, it arises from a single superficial cell. This grows out and cuts off a basal cell by a transverse wall. From the upper cell the archegonium itself is developed. Three vertical walls are first formed, separating a central cell from three peripheral cells. A transverse wall cuts off a cap-cell from the top of the central cell. Then the peripheral cells, or some of them, divide vertically, so that we have a ring of about five cells surrounding the central one. Next, all the cells divide across by a transverse wall, cutting the whole archegonium into two halves, the lower being the venter and the upper the neck (Fig. 54). The principal parts are now marked out. The external cells of the lower ventral part grow and divide, giving rise to a wall two cells thick, while the central cell undergoes a single transverse division into two very unequal portions ; the small upper part is the ventral canal-cell, the lower and larger cell becomes the *ovum* itself (Fig. 54). In the meantime the neck elongates greatly, and all its cells divide repeatedly by transverse walls, so that the ripe archegonium consists of a chimney-like neck, enclosing a row of canal-cells leading down to the ovum at the

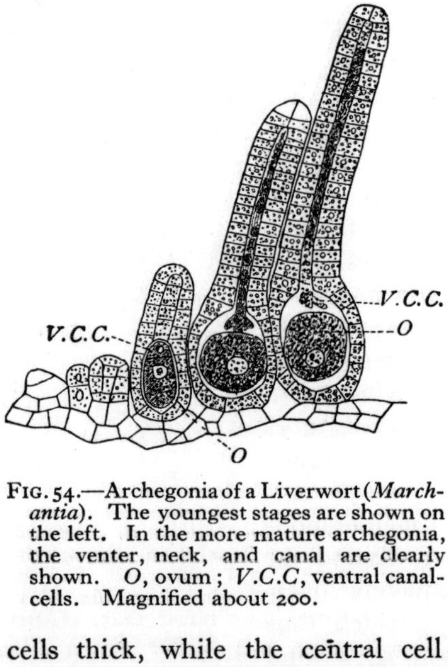

FIG. 54.—Archegonia of a Liverwort (*Marchantia*). The youngest stages are shown on the left. In the more mature archegonia, the venter, neck, and canal are clearly shown. *O*, ovum ; *V.C.C*, ventral canal-cells. Magnified about 200.

bottom (see Fig. 54). The cap-cell at the top of the neck divides into four by vertical walls crossing each other at right angles. We see that the archegonium of a Liverwort differs from the corresponding organ of a Fern or other Vascular Cryptogam, not only in the much greater length of the neck, but also in the origin of the neck-canal. In the Liverworts this is derived from the upper part of the archegonium, while in the Vascular Cryptogams it is formed from an outgrowth of the central cell. The final result, however, is much the same in both cases, and on the whole there is more reason to lay stress on the essential similarity of the sexual organs in plants so remote from each other, than to dwell on their somewhat minute differences.

When ready for fertilisation, the archegonium opens. This is due to the pressure of the mucilaginous substance in the canal, arising from the disorganised neck canal-cells. This substance takes up water, swells, and so forces the four cap-cells apart, causing the neck to open, while at the same time a portion of the mucilage protrudes through the opening.

c. Fertilisation

In *Pellia*, as in the Cryptogams generally, fertilisation must take place under water; after rain or dew the surface of the thallus is wet enough for the spermatozoids to accomplish their journey. The cells of the antheridial wall take up water, swell, and press upon the mass of spermatozoid mother-cells. The antheridium bursts, and its contents are set free. As soon as the spermatozoids are released from their mother-cells, they swim through the water, rotating as they go, in much the same way as those of a Fern. They are also drawn towards the archegonia as soon as they come within their "sphere of influence"; in an allied Liverwort

(*Marchantia*) the spermatozoids are found to be attracted by soluble proteins (see Part I. p. 216), such as are doubtless present in the disorganised canal-cells. They are caught in the mucilage, wriggle down the neck of the archegonium, and one of them effects fertilisation by union with the ovum. In all this process, so far as the details have been worked out, there is exact agreement with the Vascular Cryptogams.

So much the more surprising is the remarkable difference in the ultimate product of fertilisation. The ovum when fertilised surrounds itself as usual with a cell-wall, and begins to divide. The result of this development will be considered in the next section.

3. The Sporogonium or Fruit

a. External Characters

If we examine a fertile specimen of *Pellia* about February, we can easily recognise the young fruits on the upper surface of the thallus. At this stage each fruit appears as a little dark green ball, about one-sixteenth of an inch in diameter, projecting from the involucre (see p. 106, Fig. 49, lower part). It is attached by a short, thick stalk of a lighter green colour, the bottom of which is tightly fixed in the body of the thallus. If we look at the fruit with a lens, we see that the upper spherical part—the capsule—is partly enclosed in a light-coloured membrane, the *calyptra*, which it is just beginning to burst; the capsule, where its surface is exposed, is smooth and glossy. The stalk is called the *seta*.

Later in the season, about April, a great change happens. The seta grows with astonishing rapidity, and in three or four days attains a length of perhaps as much as three inches (see Fig. 49). The seta in its elongated condition is of a pure white colour, rather transparent, and bears the dark green, or now almost black, capsule aloft on its top, the whole looking like a thick pin with a round head (Fig. 49, on right). Shortly afterwards the dehiscence of the capsule takes place, by four longitudinal fissures, splitting the walls into four valves, which straighten themselves out, forming a horizontal cross on the top of the seta (see Fig. 49, on left). The spores are then set free. It is not long before the seta collapses, and the whole structure, when once the spores are shed, soon perishes. We will now go back to the fertilised ovum, and see how the fruit, which here represents the asexual generation, is developed.

b. Development

After fertilisation, the ovum first divides by a transverse wall, but of the two cells thus produced the lower takes no part in the further development. Capsule and seta are both entirely derived from the upper cell, which next divides by a vertical and then by a transverse wall, the latter marking the boundary between capsule and seta. Cell-walls parallel to the external surface now appear in both parts. In the young capsule these divisions separate the external layer from the archesporium which constitutes the whole of the internal group. Innumerable cell-divisions take place, keeping pace with the growth both in seta and capsule, until a stage is reached like that shown in longitudinal section in Fig. 55. The seta at its lower end develops a conical foot, with a flange projecting upwards, firmly fixing the seta in the thallus, with which, however, there is never any real continuity of tissue.

The seta itself consists throughout of very uniform short-celled parenchyma. The capsule has a wall three or more cells in thickness ; from the internal mass of cells the sporogenous tissue is produced, but the whole of this is not used up to form the mother-cells of spores. A certain number of the cells grow in length, while remaining narrow, and ultimately form long tubular structures with a double spiral thickening. These curious elements — very characteristic of the Liverworts— are called the *elaters*. When young, the elaters no doubt serve to convey food-substances from the seta to the developing spores. At a later stage they play a part in the dissemination of the spores, as we shall see further on. In *Pellia* the elaters radiate from the base of the capsule, where they are attached to a mass of shorter cells, likewise spirally thickened. They extend from this part, upwards and outwards, passing between the mother-cells, which chiefly occupy the outer and upper part of the capsule (Fig. 55). Young elaters, with the spiral bands just beginning to form, are shown in Fig. 56, and nearly mature ones are represented in Fig. 57 among the spores.

FIG. 55.—Young fruit of *Pellia*, in longitudinal section. It is enclosed within the calyptra. *n*, neck of archegonium, in which the fruit was formed; *ar*, abortive archegonia; *f*, foot of sporogonium; *sp*, spore mother-cells; *el*, elaters. Magnified 40. (R. S.)

The mother-cells of the spores, which are very numerous, are of a peculiar shape. At an early stage they become very deeply four-lobed, the lobes being tetrahedrally arranged, so that only three are seen in one plane (Fig. 56, *sp*). The lobes are connected in the middle by a quite narrow neck, in which the nucleus, which remains for a long time undivided, is situated. Eventually the nucleus of the mother-cell divides into four, each daughter-nucleus travelling out into one of the lobes, which now become separated from one another by cell-walls, so that the division is complete. We must not, however, suppose that this peculiar form of mother-cell is general in all Liverworts, though it is very common among them.

The venter of the archegonium enlarges very rapidly

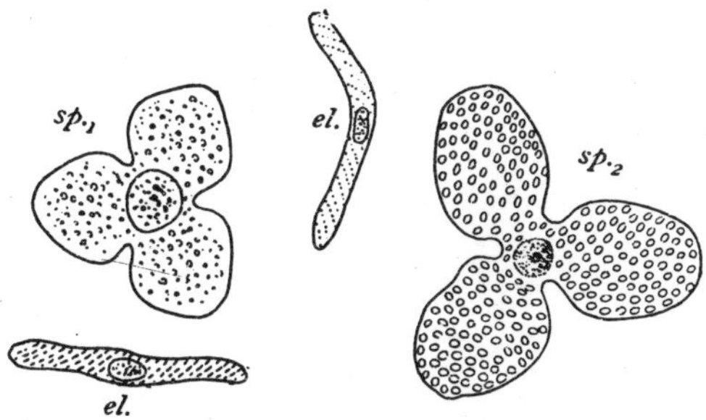

FIG. 56.—Elaters and spore mother-cells of *Pellia*. *sp*.$_1$ and *sp*.$_2$, spore mother-cells about to divide ; *el*, *el*, young elaters with indications of spiral thickenings. Magnified 360. (R. S.)

to keep pace with the growth of the fruit inside, which it completely envelops for a long time ; in *Pellia*, however, this envelope (called the *calyptra*) is not entirely formed from the venter of the archegonium, but the neighbouring thallus-tissue also takes part in its growth, so that we see abortive, unfertilised archegonia carried up on the sides of the calyptra (see Fig. 55, *ar*), while at the top, the neck of the fertilised archegonium itself can still be recognised (Fig. 55, *n*). The cells of the thallus surrounding the foot become especially crowded with starch, and thus provide food for the developing fruit.

The whole structure of the fruit is already complete while the seta still remains quite short. Its elongation, as we have seen, is a comparatively sudden process, and is due to the great stretching of cells which are already formed. The calyptra, which was previously ruptured, is now left behind as a torn membrane at the base of the stalk.

The spores of *Pellia* present a peculiarity which is quite exceptional among Liverworts, for they become multicellular while still enclosed in the capsule. In Fig. 57 the oval spores are shown at various stages of cell-division; when ripe they consist of several tiers of cells. They contain chlorophyll, and continue their germination,

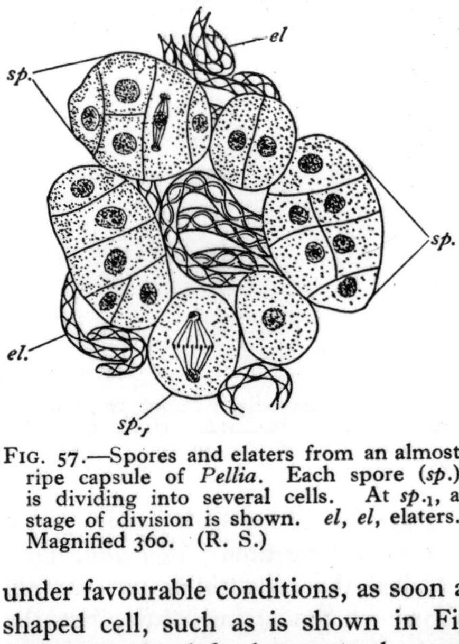

FIG. 57.—Spores and elaters from an almost ripe capsule of *Pellia*. Each spore (*sp.*) is dividing into several cells. At *sp.*$_1$, a stage of division is shown. *el*, *el*, elaters. Magnified 360. (R. S.)

under favourable conditions, as soon as shed. A wedge-shaped cell, such as is shown in Fig. 57 in the lower spore on the left, becomes the growing-point of the new plant, while one or both ends grow out into the first root-hairs. The elaters are of use in loosening the mass of spores, so that they are more easily scattered by the wind. They also perform hygrometric movements, which actively disperse the spores. The development of the fruit of *Pellia* occupies a full year; when the spores are shed new archegonia are already ripe for fertilisation.

Summary

We have now completed the simple life-history of this Liverwort. We must not suppose that all Hepaticæ are equally simple ; the class is a large one, said to include about four thousand species, and embraces a considerable variety of form and structure. In some (*e.g. Marchantia*) the thalloid form is retained, but a great complexity of anatomical structure exists, while at the same time the thallus bears highly-modified branches for the production of the sexual organs. Special organs of vegetative propagation—gemmæ—are also very frequently present, which serve to reproduce the thallus directly. In another very numerous series of Liverworts, we find, instead of a thallus, a delicate leafy stem of great beauty ; in this group (which includes the majority of the species) we have a high external differentiation of the gametophyte, while the anatomical structure remains simple. We will now sum up the essential points in the development of Liverworts, as represented by our type.

The thallus of *Pellia* is obviously comparable to the prothallus of a Fern, while the antheridia and archegonia also are evidently of the same nature as the sexual organs of the Vascular Cryptogams. Fertilisation is accomplished in just the same way, but the product is totally different. In the Ferns the sexually produced embryo grows up into the plant itself, which goes through a long and vigorous course of purely vegetative development, before it proceeds to form the asexual reproductive cells (spores). In *Pellia*, and Liverworts generally, the sexually produced embryo grows, not into a plant at all, but merely into a fruit, which remains all its life attached to and dependent upon the sexual individual. The capsule, it is true, contains chlorophyll in its outer layer, and so can do some assimilation on its own account, but for the bulk of its food it must rely on the store produced by the thallus. Spore-production is the one function of the fruit; all the parts—foot, seta, and capsule—are means subservient to this end; there is no vegetative development worth mentioning. This is the great characteristic, not merely of the Liverworts, but of the Bryophyta generally—the gametophyte is the prominent vegetative generation, while the sporophyte has little more to do than to discharge its purely reproductive functions as a spore-producing organ. The fruit, indeed, is not always so simple as that of *Pellia*, but still the same rule holds good. We see, then, that in this subkingdom we have to do with plants in which the sexual generation is readily comparable to that of the higher Cryptogams, while the product of fertilisation — the sporophyte—is developed on entirely different lines. The Bryophyta are in fact plants with a well-marked alternation of generations, in which the *sexual* generation is the more conspicuous and independent. The distinction between Bryophyta and Vascular Crytogams is so sharp and constant that the gulf between them has been spoken of as the deepest in the vegetable kingdom.

B. THE MOSSES

TYPE VIII

FUNARIA HYGROMETRICA

The true Mosses, the general appearance of which will be familiar to everyone, are more highly organised plants than the Liverworts, but at the same time are more remote from the Vascular Cryptogams ; this is our reason for taking them after *Pellia.* Their greater complexity extends to both generations ; the higher development of the gametophyte removes all obvious resemblance to the prothallus of a Fern, while the sporophyte, though a complicated structure, is still only a fruit, and in no way approaches the asexual generation of the higher Cryptogams. The Mosses, unlike *Pellia*, do not have their vegetative organs in the form of an undifferentiated thallus, but possess a perfectly distinct stem bearing spirally arranged leaves. In fact, the external characters of a Moss plant are essentially similar to those of vascular plants, but in the Mosses leaf and stem belong to the sexual generation, while in the higher plants they form part of the sporophyte.

The special Moss (*Funaria*) on which our description is based, is a very common one, and usually grows on the ground, sometimes occurring on walls also, where, however, it is less abundant than some other kinds. It comes up in great quantities in places where there has been a fire. This Moss grows in close tufts of a bright green colour ; if we separate out a single plant we find that the slender stem is erect, reaching perhaps half an inch in height, and densely clothed with small simple leaves. The lower part, where the plant is kept from the light by the crowding of its fellows, is brown, the leaves here having lost their colour. The upper free part of the stem bears the bright green living leaves, and terminates in a bud. The stem is branched, but not very abundantly, the branches, like the main stem, growing in a vertical direction. At the base of the plant we find a number of absorptive filaments or *rhizoids*, but there is no real root—an organ which does not exist in any of the Bryophyta.

The leaves are arranged in a spiral, with a divergence of $\frac{3}{8}$, that is to say, each leaf is separated from the one next above it by three-eighths of the circumference of the stem, so that in following the spiral three times round the stem we pass eight leaves, and find that the ninth lies vertically above that from which we started (see Part I. p. 14). The leaves themselves are inserted on the stem with a fairly broad base ; they are ovate in form, pointed at the tip, and traversed by a distinct midrib, though not otherwise veined (see Fig. 58). The rhizoids are white when quite young, but soon become brown. The above description applies especially to the barren stems ; we shall refer more particularly later on to those which bear the reproductive organs.

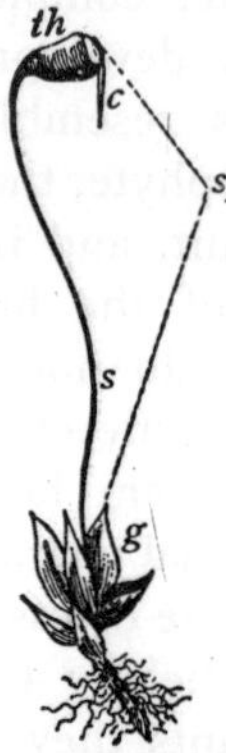

FIG. 58.—*Funaria hygrometrica* ; general view of a fruiting specimen. *g*, the leafy Moss plant or gametophyte ; *sp*. the fruit or sporophyte generation, consisting of the seta, *s*, and the theca, *th* ; *c*, the calyptra. Slightly magnified. (After Sachs.)

1. THE LEAFY STEM

a. Structure

The anatomy of the Moss plant, as represented by *Funaria*, is simple, but yet shows a fairly well-marked differentiation of tissues. In the mature stem three distinct zones can be distinguished,—epidermis, cortex, and central cylinder. The epidermis is one cell thick in most places, though here and there a double row of cells may be found. Its cells are small, and in the older part of the stem become very thick-walled. The cortex is of relatively great thickness, and made up of parenchyma, the outer cells having thicker walls than the inner. When young, the cortical cells contain chlorophyll. The central cylinder consists of a very sharply defined cylindrical strand of long, narrow, thin-walled cells, destitute of chlorophyll. There is evidence that this is a water-conducting tissue. In *Funaria* the central strand is quite uniform throughout ; some of the larger Mosses,

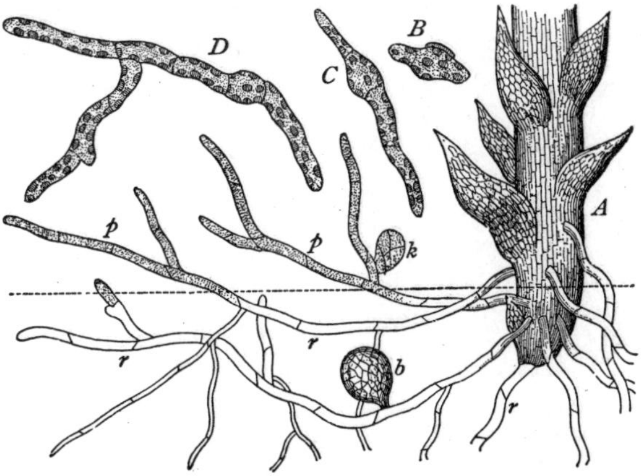

FIG. 59.—*A*, lower part of a Moss plant, bearing leaves and rhizoids (*r*), which grow up above the ground and become secondary protonema (*p*). At *b* is an underground gemma or bulbil. At *k* is a bud from which a new leafy stem will grow. Magnified about 20. *B*, *C*, and *D*, successive stages of germination of a spore of *Funaria*, producing primary protonema. Magnified 200. (After Luerssen.)

however, have a more complicated arrangement, as shown in Fig. 60, which represents the transverse section of the central cylinder of *Catharinea* (*Atrichum*) *undulata*, a large Moss common in woods.

Here there are elements of various kinds. In the middle, the large, central, rather thick-walled cells, which may be analogous to tracheids, and serve to conduct water, are accompanied by parenchyma containing starch ; outside this is a zone of smaller cells which have more abundant protoplasm but no starch, and may fulfil the function of a rudimentary phloëm. In this zone the sections of leaf-trace bundles are seen.

The leaves of *Funaria* are transversed by a conspicuous midrib, while the rest of the leaf is only one cell thick. The cells of the thin part are uniform, except at the somewhat serrated edge, where they are narrower and have rather thicker walls. The midrib is several cells thick, and contains a small strand of narrow cells, like those in the central cylinder of the stem. Probably these cells conduct water and assimilated food, while the function of assimilation belongs to the thin part of the leaf, which is very rich in chlorophyll-granules (Fig. 61, *d*). Moss leaves, by the bye, are very favourable objects for observing the multiplication of the chlorophyll granules by division.

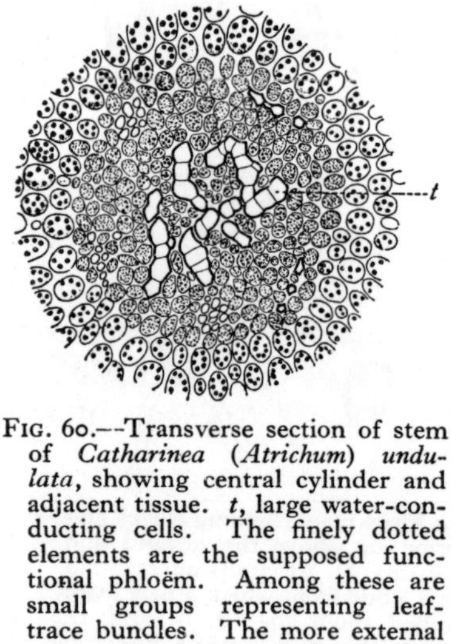

FIG. 60.—Transverse section of stem of *Catharinea* (*Atrichum*) *undulata*, showing central cylinder and adjacent tissue. *t*, large water-conducting cells. The finely dotted elements are the supposed functional phloëm. Among these are small groups representing leaf-trace bundles. The more external cells (containing starch-granules) belong to the cortex. Magnified 190. (After Haberlandt.)

The strand of conducting tissue enters the stem from the leaf, but dies out in the cortex without reaching the central cylinder. Consequently there is no complete leaf-trace system in this Moss, though some of the more complicated Mosses (such as *Catharinea*) have continuous strands connecting the conducting tissue of the stem with that of the leaves. We must remember that the whole mode of life of a Moss plant, especially as regards its water-supply, is very different from that of the higher plants. Many Mosses, and *Funaria* among them, often grow in places such as the tops of walls, or in sandy soil, where they are liable to be completely dried up in hot weather. Yet they are none the worse, and revive as soon as rain comes again. This rapid recovery is due to their power of absorbing water by their leaves—a power which is either absent or which only exists to an insignificant extent in most of the higher plants. Hence less work falls on the conducting tissues of the stem than in the latter, for only a small part of the water-supply is taken up from below, though this part of the supply is important as it carries with it the necessary mineral food-substance.

As we have said, a Moss possesses no true root. The functions of a root are performed by the *rhizoids*, as they are called, multicellular filaments springing from near the base of the stem (see Fig. 59, *r*). These rhizoids are very different from ordinary root-hairs, which are generally unicellular, though multicellular rhizoids occur on the prothallus of some few Ferns. The Moss rhizoids consist of a single chain of very long cells separated from one another by oblique walls. They grow entirely by means of the apical cell at the free end of each filament, and branch repeatedly, the diameter diminishing with the successive orders of branching, so that the final ramifications are very slender indeed in comparison with the main filaments. The whole simulates a regular root system, though totally different in structure.

b. Apical Development

The stem of all Mosses grows by means of a single apical cell, and the plant is built up in the most regular manner from its segments. The cell is of the same inverted pyramidal form which we found in *Equisetum*, and divides at first in the same way by walls parallel (or nearly parallel) to the three sides. Each segment first divides into an inner and an outer cell. From the inner cells thus formed the greater part of the tissues of the stem is derived, while the outer cells give rise to leaves, buds, and the outside part of the stem. Each outer cell divides into an upper and a lower half; from the upper half the leaf is produced, while the lateral buds, where they exist, owe their origin to the lower of the two cells. We see then that every segment produces a leaf, and that each lateral bud stands *below the leaf* to which it belongs, instead of in its axil,—a striking difference from the higher plants, though we find something like it among the Ferns. Each leaf grows in length by means of a two-sided apical cell. The chief points then in the development of a Moss stem are the growth from an apical cell, the origin of a leaf from each segment, and the position of the lateral buds *beneath* the leaves to which they belong.

2. The Sexual Organs

Funaria is monœcious,[1] though this is not the case with all Mosses. The female shoots arise as lateral branches on the male. The male stems are of fair size, reaching a centimetre in height; in the lower part of the stem the leaves are scattered, but at the top they are crowded together to form a conspicuous rosette. This is not unlike a flower, especially as the middle part of the rosette is of a reddish colour. In some of the larger Mosses (such as *Polytrichum*, which includes some very conspicuous Mosses, common on heaths) the resemblance to a flower is still more striking.

[1] As conclusively shown by Mr. L. A. Boodle, " The Monœcism of *Funaria hygrometrica*," *Annals of Botany*, vol. xx. 1906, p. 293.

However, there is of course no direct homology, for these rosettes belong to the gametophyte, not to the sporophyte generation, and the organs which they enclose are antheridia, not stamens. On the growing-point, within the rosette, numerous antheridia arise in long-continued succession without any strict order. Both young and mature antheridia are shown in Fig. 61. As usual, the

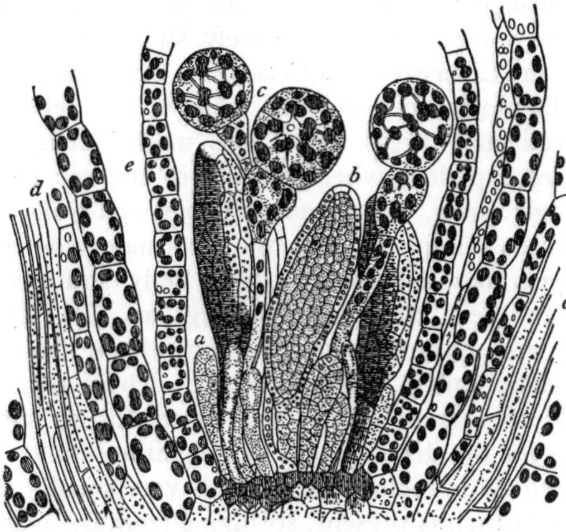

FIG. 61.—Longitudinal section through the apical bud of a male shoot of *Funaria*. *d*, leaves cut through the midrib; *e*, leaves cut through the lamina; *c*, paraphyses; several antheridia are shown; *a*, very young; *b*, nearly ripe antheridium. Magnified 300. (After Sachs.)

antheridium owes its origin to a single cell in which one or two transverse walls are formed, after which the growth goes on entirely by means of the apical cell, which cuts off two rows of segments. It is a good general rule in the Mosses that every organ, of whatever kind, grows by means of an apical cell, whereas this mode of growth is nothing like so general among the Liverworts. By further subdivisions of the segments, and finally of the apical cell itself, the antheridium is differentiated into an external wall one cell in thickness, and an internal mass of small-celled tissue, each cell of which becomes the mother-cell of a spermatozoid (Fig. 61, *b*).

The mature antheridium is club-shaped and reaches 0·3 millimetre in length, containing an enormous number of mother-cells. The development of the spermatozoid within its mother-cell is precisely similar to the same process in Liverworts or Vascular Cryptogams. The antheridia open on access of water; the cells of the wall swell and press upon the mother-cells within, which are expelled, when the top of the sac ruptures, in a single mass. The mucilaginous cell-walls of the mother-cells disappear and the spermatozoids are set free; each has a spirally curved body and two cilia, just as in the Liverworts (see Fig. 62, *C*). The antheridia are accompanied by multicellular hairs with large heads, called the *paraphyses*. Their function is probably to secrete water, and so to ensure sufficient moisture for the development of the antheridia.

The female branches are, at first, very small; usually one such branch is present, springing from the male stem some distance below the rosette; sometimes more branches are produced. The leaves converge together at the top, forming a bud within which the archegonia are contained (Fig. 62, *A*). They arise from the cells of the growing-point, and the apical cell among others is itself used up to form an archegonium, so that no further direct growth of the vegetative axis is possible.

The cell from which an archegonium is to be formed first divides by a transverse wall. The further growth is by means of the apical cell, which in this case gives rise to four rows of segments, three of which are peripheral and form the wall of the archegonium, while the fourth row is central. From the lowest cell of the central row the ovum and ventral canal-cell (see Fig. 62, *B*) are produced; the rest of the series of central cells forms the canal of the neck. The external segments undergo further transverse and vertical divisions; the wall of the neck ultimately consists of six rows of cells surrounding the canal. The ventral part of the wall, enclosing the ovum, becomes two layers in thickness, and the whole archegonium is seated on a multicellular pedicel. Apart from this last point, the final form of the archegonium is similar to that in the Liverworts, the chief difference consisting in the marked apical growth which goes on in the archegonium, as in other organs, of the true Mosses. When ready for fertilisation the terminal cells of the neck separate widely from each other, leaving an open passage into the canal, which now only contains the mucilage derived from the disorganised canal-cells.

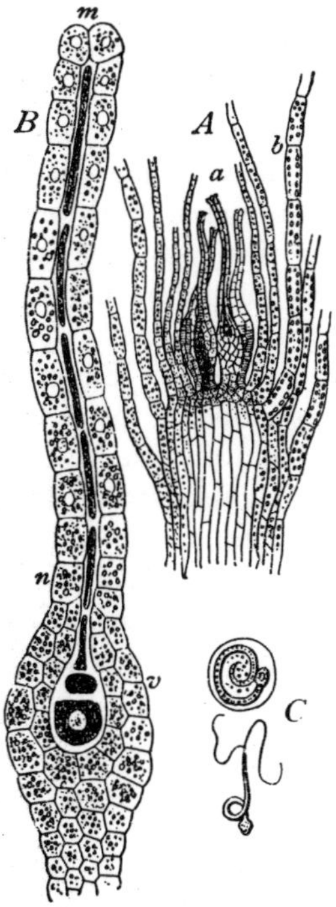

FIG. 62.—*Funaria*. *A*, longitudinal section through the apical bud of a female shoot; *a*, archegonia; *b*, leaves. Magnified 100. *B*, a single archegonium; *v*, the enlarged venter within which the ovum and ventral canal-cell are seen; from *n* to *m* is the neck, enclosing the neck-canal. Magnified 550. *C*, unripe spermatozoid in its mother-cell, and mature spermatozoid with two cilia. Magnified 800. (After Sachs.)

Fertilisation, as in Cryptogams generally, takes place under water. Raindrops, which have fallen on open male "flowers," and become impregnated with the discharged spermatozoids, trickle down on to the lower female plants, and some of the water, carrying the spermatozoids with it, may make its way between the leaves of the archegonial bud, and reach the archegonia themselves. In this way the active male cells are brought into the neighbourhood of the female organs. The rest of the journey they must accomplish by their own movements. Experiments precisely similar to those described in the case of the Ferns have been successfully carried out on Mosses, and here also it appears that the archegonia attract the visits of the swarming sperm-cells by means of a chemical secretion. In Mosses, however, it is not malic acid, but sugar (cane-sugar), which forms the bait. The spermatozoids having been thus lured to the archegonium, penetrate the neck-canal, and one of them ultimately reaches the ovum and effects fertilisation.

3. The Sporogonium or Fruit

Funaria fruits very freely, and if we look at a patch of it, at any time of year, we are sure to find plenty of fructifying plants at one stage or another. In the mature state the fruit consists of a long, thin, red-brown stalk, bearing at its end a nodding pear-shaped capsule (see Fig. 58, *sp*), which at first is green, but finally turns brown. Until almost the last the capsule carries on its top a conical hood (the *calyptra*) split along one side; at an earlier stage this completely envelops the capsule, and is only pushed off as it expands, remaining hanging for a long time. When the calyptra is removed we see the top of the capsule, which is closed by a neat conical lid. The whole of this fruit, including both stalk (*seta*) and capsule, constitutes the asexual, spore-bearing generation, and is derived from the fertilised ovum. The calyptra, however, is formed from the enlarged wall of the archegonium, which is split off at the base, and borne aloft on the fruit as it grows. The calyptra therefore is, by its origin, a portion of the sexual plant. We will now describe the structure of the fruit when fully formed, and then shortly trace its development from the ovum.

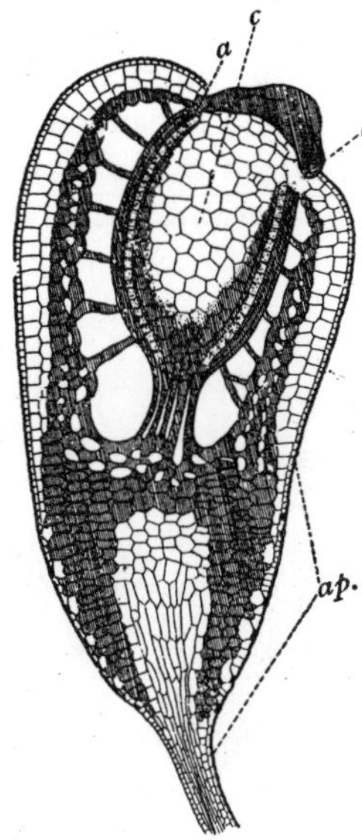

FIG. 63.—Capsule of *Funaria* in longitudinal section. *ap*, the apophysis; *a*, *a*, the archesporium, forming a single layer of cells (lightly shaded); it surrounds the columella, *c*. The green assimilating tissue is darkly shaded. The operculum closes the top of the capsule, above the columella; its individual cells are not shown. The dark spots in the epidermis of the apophysis indicate the stomata. Magnified 14. (After Haberlandt.)

Beginning with the capsule, which is the essential part, containing the spores, we find that its base is solid, while the upper portion contains a large hollow space separating the central mass of tissue from the wall (see the longitudinal section shown in Fig. 63, which is from a young capsule in which all the tissues are already marked out).

It is the upper part of the capsule which is fertile, while the basal solid portion (*apophysis*) performs nutritive functions. The whole capsule is covered by a well-marked epidermis, which, on the apophysis, contains stomata. In the upper part of the capsule the wall is several cells in thickness; the hypodermal layers are colourless, while those towards the interior contain chlorophyll. Connected with these inner cells of the wall are filamentous strands, also containing chlorophyll, which extend across the intercellular space, and form a junction with the internal tissues. The central mass is narrow below and expanded above, assuming a barrel-like shape. It is from this part that the spores are produced. The archesporium forms at first a single cellular zone, which has a hollow cylindrical form, or, more accurately, is shaped just like a barrel with the ends knocked out (see Fig. 63, *a*). The archesporial layer is separated from the intercellular space by a zone of sterile tissue called the outer *spore-sac*. Within the archesporium is the large central mass of sterile colouring tissue (the *columella*), which is connected below by a thinner strand with the tissue of the apophysis. The layer between archesporium and columella is called the inner spore-sac. The lid (*operculum*) is at first continuous with the capsule, but eventually becomes detached by the severance of a ring of cells (the *annulus*) between lid and wall.

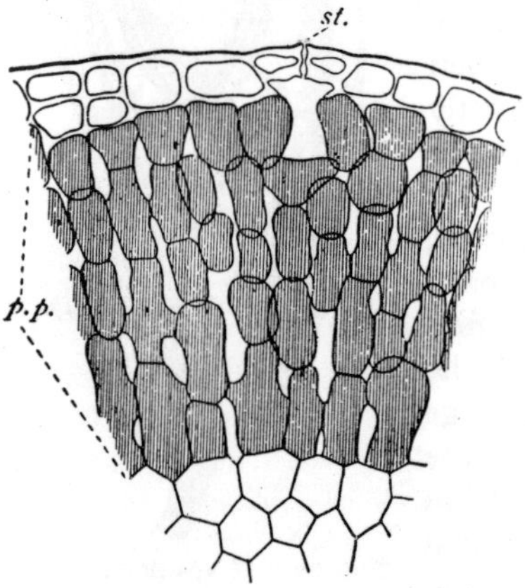

FIG. 64.—Part of the apophysis of a Moss (*Bryum*) in transverse section. *p.p*, the assimilating palisade parenchyma ; *st*, stoma. Magnified 130. (After Haberlandt.)

The apophysis is essentially the assimilating part of the capsule ; beneath the epidermis is a broad zone of chlorophyll-tissue, the cells of which are in many cases of the typical palisade form (see Figs. 63, 64). The epidermis bears well-developed stomata, which are in essentials similar to those of the higher plants. In *Funaria*, however, they generally have the peculiarity that the wall between the two guard-cells breaks down at the two ends, so that the part enclosing the pore is left standing up in the middle of the fused guard-cells, like a chimney-shaft passing through a room (Fig. 65). At an earlier stage, however, the stoma is two-celled, just as in vascular plants, and in many Mosses it remains so all through. In other respects these Moss-stomata are quite typical. The guard-cells differ from the ordinary epidermal cells in containing abundant chlorophyll-granules; the form of the cells, as seen both in surface view and in section, could be exactly matched in the stomata of flowering plants. Beneath each stoma is an intercellular space (Fig. 64). It is remarkable to find these organs so perfectly differentiated in plants like the Mosses, which in all other respects are so remote from the higher groups. There is one Liverwort (*Anthoceros*) which also has well-formed stomata on the fruit. It is worth noting that typical stomata have in no case so far been found in the sexual generation; when the gametophyte bears organs with the same function (as in certain Liverworts), they are constructed on a totally different plan.

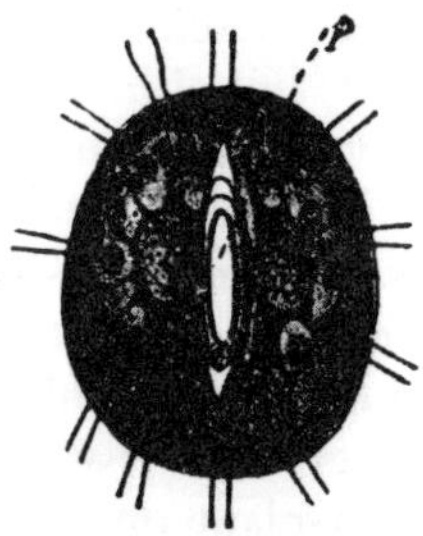

FIG. 65.—Stoma of *Funaria* in surface view. *p*, the pore. Note the nuclei and chlorophyll-granules of the fused guard-cells. Magnified 630. (After Haberlandt.)

It is evident from the anatomical structure that the sporophyte is capable of obtaining a great part of its food for itself, and this has been proved experimentally to be the case. So far as the assimilation of carbon is concerned, a sporogonium such as that of *Funaria* is able to provide for itself, from the time when its assimilating tissue is developed. Water, with the mineral food-substances, is necessarily supplied through the stem of the Moss plant, and passes up to the capsule through the seta, which contains a central conducting cylinder, like that of the stem itself. The cortex of the seta consists of thick-walled tissue, and serves to give the mechanical strength necessary to enable this slender stalk to support the weight of the capsule. The bottom of the seta is fixed in the tissue of the gametophyte by a conical foot, but although the contact is a very close one there is never any organic connection between the two generations.

We see then that the sporophytic generation of *Funaria* is in part parasitic on the sexual plant, in part independent. It resembles in this respect a green parasite such as the mistletoe, which, like the *Funaria* fruit, must obtain all its water and mineral food from the host-plant on which it grows, but can provide its carbonaceous food for itself. In some other Mosses, however, the sporophyte is destitute of chlorophyll, and so has to lead a completely parasitic existence, depending for the whole of its food on the leafy Moss plant.

We will now return to the essential part of the capsule, that, namely, in which the spores are formed. The archesporium is at first only a single layer of cells, and occupies but a small part of the capsule (Fig. 63). Repeated divisions now take place, and the archesporium increases in thickness. Ultimately each of the cells formed by it becomes a spore mother-cell, which, as is so usually the case, divides into four spores, arranged tetrahedrally in each mother-cell. As soon as the spores are ripe the capsule begins to dry up. The columella and all the delicate tissues of the fruit collapse, and when the capsule is fully ripe it consists essentially of the wall only, filled with a mass of dark-green spores. The lid becomes detached, but the capsule after this is not left freely open, for in the meantime a double row of teeth (called the *peristome*) has been formed. These teeth, which project from the edge of the capsule and partly close its mouth, are formed from strips of thickened cell-wall, all other parts of the cells involved having perished. The peristome plays a part, as we shall see later, in the dissemination of the spores.

The whole of the fruit—seta and capsule together—constitutes the sporophyte generation, and is derived from the fertilised ovum. The latter first divides by a transverse wall; further divisions take place, so that a two-sided apical cell arises at both ends of this embryo, and for a time both apical cells are active, each giving rise to two rows of segments. The upper growing-point, however, is the important one, for it produces the capsule and the greater part of the seta, while the lower apical cell only contributes to the foot, which penetrates downwards into the tissues of the gametophyte. The segments derived from the upper apical cell undergo division by walls parallel to the surface. In the part which forms the capsule, the inner cells thus formed constitute what is termed the *endothecium*, and from this central part the columella and archesporium are ultimately derived. Everything outside the archesporium is the product of the peripheral cells or *amphithecium*. This origin of the spore-producing layer from the outer part of the endothecium is characteristic of the great majority of Mosses. In the seta the central group of cells, corresponding to the endothecium of the capsule, simply gives rise to the central strand of conducting cells. The differentiation of the capsule from the seta takes place rather late in *Funaria*, after the whole fruit has grown to a considerable length.

4. Germination of the Spores

As the capsule dries, the walls of the cells of the annulus split across and the lid is detached. The spores are not all scattered at once; the dissemination is regulated by the teeth of the peristome, which, when the air is wet, completely close the mouth of the capsule, only allowing the spores to escape in dry weather. The spores themselves contain abundant chlorophyll, and also have a reserve of oil which serves to provide material for germination. When this takes place, the spore does not at once give rise to a Moss plant, but first of all produces a branched filamentous growth of very simple structure, much resembling some of the simpler plants (*Algæ*), as we shall find later on. This filamentous condition of the young Moss, which thus forms the first stage of the gametophyte generation, is called the *protonema* (see p. 123, Fig. 59). The spore generally sends out filaments in two directions; one remains green and creeps along the surface of the ground, the other loses its chlorophyll and becomes the first rhizoid. The filaments grow in each case by an apical cell; they branch freely but remain one cell only in thickness; often the protonema develops to a great extent, forming a tangled green felt, which may cover several square inches of ground. The young Moss plants arise from the protonema as lateral buds (see Fig. 59, *k*). A cell of the protonema gives rise to a branch; the branch divides by inclined walls so as to form a tetrahedral apical cell, and as soon as this has taken place regular segmentation begins, and the leafy Moss plant is soon built up. The first leaves are simpler than those of the more mature plant, and may be destitute of a midrib.

The protonema which we have just described is

formed directly from the spore, and is therefore called the *primary* protonema. It may also arise in a secondary way from any part of the plant—from rhizoids (see Fig. 59), or stem, or detached leaves, or even from the fruit itself. In the latter case we have production of the sexual direct from the asexual generation, affording, in fact, an instance of *apospory*, such as sometimes occurs in the Ferns (see p. 73). The production of protonemata provides the plant with a most abundant means of vegetative propagation, for every growth of protonema is capable of giving rise to a number of Moss plants. Many Mosses produce special vegetative buds either on their stems or rhizoids (Fig. 59, *A*, *b*), or throw off certain of their leaves as organs of propagation. In most cases, whatever be the nature of the reproductive body, whether spore or bud, it begins by forming protonemata, from which the leafy plants arise at a later stage. This insertion of a filamentous stage of growth in the life-cycle, before the production of the typical form of gametophyte, is very characteristic of the true Mosses; in the Liverworts, the protonema is on the whole much less developed. It may be compared with the early filamentous stage of a Fern-prothallus, with which it is quite homologous.

Summary

If we now briefly sum up the characteristic points in the life-history of the true Mosses, we see that both generations are decidedly more highly organised than in the Liverworts. The gametophyte is here constantly developed as a leafy stem, quite comparable to that of the higher plants, though occupying a different place in the life-history. We find at the same time a considerable degree of anatomical complexity, corresponding to the higher external organisation. In like manner, the fruit, or asexual generation, is far more complex than in the Hepaticæ. Except in the one point of possessing true stomata, its complexity, however, is on quite different lines from that of the corresponding generation in the higher plants. The Mosses, in fact, constitute a remarkable and very isolated group, highly developed in their own way, but with no near affinities to other Classes of plants.

The Bryophyta, as a whole, form a perfectly well-defined sub-kingdom, characterised by the occurrence of a well-marked alternation of distinct generations, of which the sexual is the more highly developed, so far as the vegetative organs are concerned. The sexual organs —both archegonia and antheridia—are constituted on the same general plan as those of the higher Cryptogams, though differing in many details. The Vascular Cryptogams, together with the Bryophyta, are sometimes classed in one sub-kingdom under the name of *Archegoniatæ*, founded on the general similarity of the female organs all through these groups. The mode of development of the spores, by division of a mother-cell into four, is also common to Bryophyta, together with the higher Cryptogams, and indeed Phanerogams also, so far as the microspores (pollen-grains) of the latter are concerned. Although the Bryophyta are, at least in the living Flora, unconnected by any intermediate forms with the vascular plants, yet they have many points in common with them, and the general lines of homology between all the classes hitherto considered are not difficult to trace.

Now we have done with archegoniate plants. The groups which remain to be considered are essentially different, both in the organisation of their reproductive organs and in the whole course of their life-history.

CHAPTER III

THE ALGÆ

THE sub-kingdom with which we have now to make ourselves acquainted differs profoundly from any of those of which representatives have been already considered. In habit all the plants included under the the general heading " Algæ " are totally different from any hitherto described, and at the same time they differ more among themselves than the lowest Liverwort differs from the most complex Dicotyledon. The members of the group most familiar to ordinary readers are the Seaweeds, for with very few exceptions all plants which grow in the sea belong to the Algæ. On the other hand, an immense number of species are inhabitants of fresh water, or can get on, like the American steamboat, " wherever it is a little damp." Generally speaking, the larger and more complex forms are marine ; the fresh-water and terrestrial representatives are both smaller and simpler. Among Seaweeds there are species which rank with the most gigantic members of the Vegetable Kingdom, while there are other Algæ which are entirely invisible as individuals to the naked eye. The higher Algæ often show a complex external form, with organs analogous to the root, stem, and leaf of the higher plants ; at the same time, their tissues are highly differentiated. On the other hand, the simplest Algæ consist of single isolated cells. Amid this vast range of forms it is evident that only a very few types can be dealt with here. As far as possible, our examples are selected with a view to illustrate the most striking variations in the life-history and mode of reproduction of Algæ.

The classification of the Algæ into their principal classes roughly follows the colour—not that colour is in itself of systematic importance ; it happens, however, among these plants that differences in their pigments generally coincide with important morphological distinctions. We will begin with the pure green Algæ, those, namely, in which the chlorophyll, like that of most of the higher plants, is not disguised by the presence of any other colouring matter. This class—the *Chlorophyceæ*—includes the majority of the fresh-water Algæ, as well as many Seaweeds. They are, on the whole, among the simpler Algæ, and many of the unicellular forms belong here ; but simple as they are in structure, some of them in their mode of reproduction may approach nearer to the higher plants than any other Algæ. We will take as our first type a fresh-water Alga which, though anatomically simple, shows a very high form of reproduction.

A. THE CHLOROPHYCEÆ

TYPE IX

Œdogonium

1. Structure

The genus *Œdogonium*, of which there are a great many species, includes some of the commonest fresh-water Algæ, and may be found in almost any pond or tank, though less common in running water. The *Œdogonia* are filamentous, the individual threads being only just distinguishable by the naked eye, and grow attached when young to stones, piles, larger water-plants, or, in fact, to any submerged object, forming a dark-green downy coating upon it. In the mature condition the filaments usually float freely.

Fig. 66 shows the whole of a small plant of *Œdogonium ciliatum*, highly magnified; the specimen is much below the usual size. The main outlines of the structure, however, are always the same, the whole plant consisting of a single row of cylindrical cells, attached at one end, which we may call the radical end. The root-cell contains less chlorophyll than the others. It is expanded into a flattened disc, which forms a holdfast, but probably does not take any special part in the absorption of food. Fresh-water Algæ absorb their food, mineral as well as gaseous, by their whole surface. Both the carbon-dioxide which they require for assimilation, and the oxygen necessary for their breathing, are present in

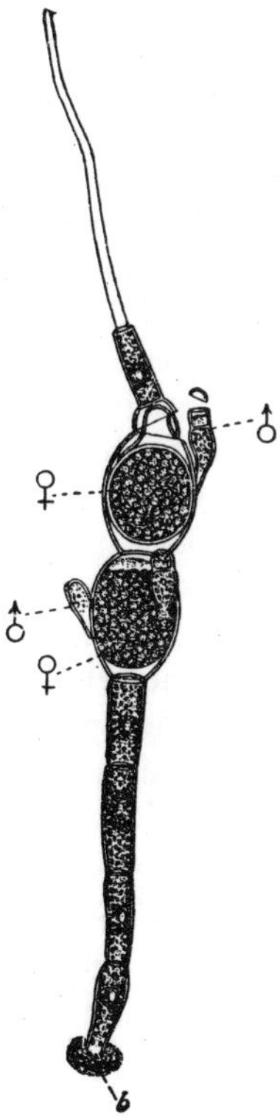

FIG. 66.—General view of a very small female plant of *Œdogonium ciliatum*. *b*, the attaching disc; ♀, ♀, two oögonia, the upper of which has opened by a lid at the top, and contains the fertilised oöspore; the lower is still closed, and the ovum unfertilised. ♂, ♂, dwarf males adhering to the oögonia. The uppermost has opened to discharge a spermatozoid. Magnified 166. (After Pringsheim.)

a dissolved state in the water, which at the same time contains salts in solution quite sufficient to supply the needs of these plants.

The structure of an ordinary vegetative cell of *Œdogonium* is as follows: Within the cellulose wall the cytoplasm forms a hollow sac—the primordial utricle—enclosing a large vacuole. The body containing the chlorophyll is very peculiar. In most plants the chloroplasts are small granules, numerous in each cell (though *Selaginella* forms an exception to this). In *Œdogonium*, however, there is only a single, very large chloroplast in each cell. It lies in the primordial utricle, and extends all round the cell, having the form of a hollow cylindrical network. It is so large as to give a green colour to the whole cell, when seen under low powers of the microscope. Within the chloroplast are several protein granules (the *pyrenoids*), around which starch-grains are deposited as a result of assimilation in sunlight. Each cell contains a single large nucleus embedded in the cytoplasm which lines the wall.

Œdogonium has no apical growing-point. In some species (such as that figured) the end cell grows out into a long hair, and takes no further part in the divisions. Certain cells of the filament, lying between the radical cell and the terminal hair, divide by transverse walls, as long as growth goes on. The formation of overlapping caps on the cell-wall, at the upper end of some of the cells, is due to the fact that after each division the wall of the mother-cell splits near the top, and a new piece of cell-wall is inserted between the broken edges as the daughter-cells grow. As the split takes place repeatedly near the same place, a succession of caps is formed, one corresponding to each cell-division.

We thus see that the vegetative structure of an *Œdogonium* is excessively simple—far simpler than that of any plant which we have hitherto described. We have now to consider the way in which the Alga reproduces itself.

In addition to fragmentation of the filaments there are two distinct methods,—the one asexual, the other sexual. The former serves to propagate the plant rapidly during summer, or as long as the conditions are favourable to its growth; the latter has for its result the production of *resting-spores*, which can survive alike the cold of winter and the periods of drought to which *Œdogonium*, in common with other fresh-water Algæ, is often exposed.

2. Reproduction

a. Asexual

Any vegetative cell may serve as an organ of asexual reproduction, and many individual plants only show this mode of propagation.

The entire contents of a cell are used up to form a single spore. The protoplasm gradually withdraws itself from the cell-wall, the whole mass assuming a rounded form. At the same time a clear, colourless spot appears on one side of the contracted protoplasmic body. From this clear portion of the protoplasm numerous cilia are developed. The cell-wall splits across and the crack opens widely at one side, that, namely, towards which the clear patch of protoplasm is turned. The spore now begins to pass out through the opening, changing its form as it does so, to adapt itself to the width of the passage (see Fig. 67, A). On first becoming free from the mother-cell, the spore is enclosed within a delicate mucilaginous membrane which soon disappears, so that now the reproductive cell is completely at liberty. In shape it resembles a pear, the more pointed end being colourless;

the chloroplast occupies the wider part, in which also the nucleus is contained. There is no cell-wall, and the whole spore is a purely protoplasmic structure. The cilia form a fringe around the narrow end (see Fig. 67, B). Their oscillations set the spore in motion, and now it swims off through the water, rotating on its axis, and advancing with the pointed end foremost.

This is the first instance of an actively-moving *spore* that we have met with; among the higher Cryptogams already described it is only the male cells or spermatozoids which are capable of locomotion; in a large proportion of the Algæ this power extends also to the spores. On account of its active movements such a spore as that of *Œdogonium* is called a *zoospore*, for when first discovered these moving cells were thought to be of animal nature. We now know that spontaneous movement is a power common to all protoplasm, whether belonging to a plant or an animal.

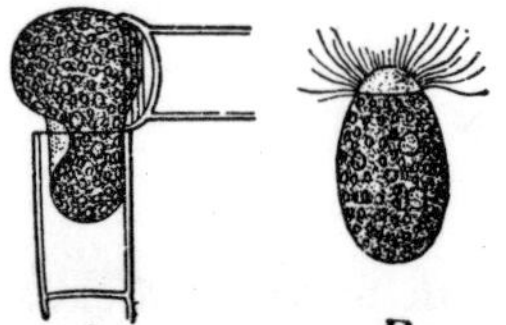

FIG. 67. — Zoospores of *Œdogonium*; A, escaping from the mother-cell; B, free, with the fringe of cilia. Magnified 350. (After Pringsheim.)

The zoospore swims about for some time (24 hours or so); it is sensitive to light, swimming towards light of moderate intensity, and retreating from it when too bright. As the zoospore becomes older it avoids the light more than before, and its movements are then directed towards the bottom of the water or solid objects contained in it; at last it comes to rest, and in doing so attaches itself by its pointed end to some solid body. It loses its cilia, and now for the first time forms a cell-wall of its own. The free end grows out, divides by a transverse wall, and thus starts a new *Œdogonium* filament, like that from which it was produced. This mode of reproduction by actively moving spores, capable of immediate germination, is extremely common among the Algæ. It is characteristic of most of the pure-green group, whether inhabitants of fresh water or of the sea, and extends also to certain other families.

b. Sexual

Œdogonium is propagated very freely by the simple method just described, but it also possesses a mode of sexual reproduction essentially similar to that of the higher Cryptogams, in so far as it consists in the fertilisation of a relatively large and stationary ovum by a small and actively moving spermatozoid. The distribution of the sexes varies much in the different species of the genus. Some are monœcious, others diœcious, while in a third set (the most numerous) a more complex arrangement prevails. In monœcious species, the male organs are formed by successive transverse divisions of one of the thallus-cells, the divisions all taking place near the upper end of the mother-cell, so that a row of rather flat cells is produced. These may divide again further, producing a chain of about a dozen cells in some cases, each of which is an antheridium. In every antheridium the contents divide into two, and each mass becomes a spermatozoid. The spermatozoids resemble the zoospores, and are ciliated like them (see Fig. 69, B). They are, however, much smaller, and relatively poorer in chlorophyll. The spermatozoid contains a single nucleus, which is placed near the end opposite to the cilia. These spermatozoids have much more the character of complete cells than those of the higher Cryptogams. In the latter, as we have already seen (p. 110), almost the whole body is of nuclear origin, only the cilia and that part of the body to which they are attached being cytoplasmic. In *Œdogonium*, however (and in the lower

Cryptogams generally), the greater part of the body is cytoplasmic. The resemblance to the zoospores is a point of the greatest importance.

The female organ, or *oögonium*, like the antheridium, consists of a single cell (see Figs. 66, 68, and 69), and differs herein from the complex *archegonium* of the Ferns and Mosses. The oögonium at its first formation is nearly similar to the other cells of the filament. A transverse wall is formed in the usual way; the upper of the two daughter-cells is the richer in cytoplasm, and has the larger nucleus; this becomes the oögonium; its lower sister-cell, which is poorer in contents and has a relatively small nucleus, is the supporting-cell, which in some species, however, may feed itself up, undergo further divisions, and give rise to another oögonium. The oögonium swells out, assuming a round or oval outline, and further increases the amount of its protoplasm, which thus encroaches considerably upon the central vacuole. The cell-contents meantime withdraw themselves from the wall, and form a free, rounded protoplasmic body—the ovum (see Fig. 69, A)—in the upper part of which the nucleus is placed. The oögonium now opens, either by the formation of a round hole in the membrane, or by the transverse splitting of the cell-wall near the top, in which case the upper part of the membrane acts as a lid (Figs. 66 and

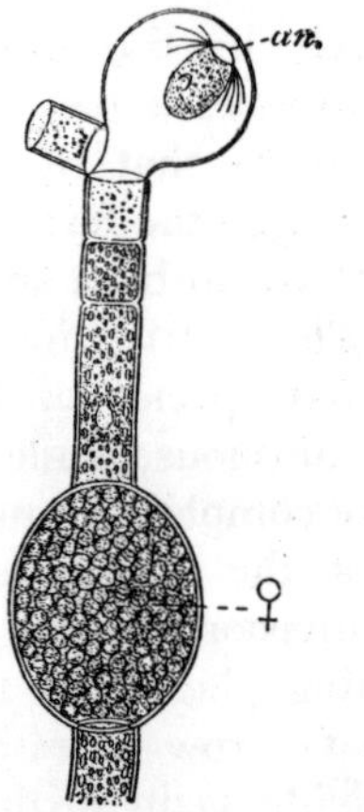

FIG. 68.—Androspore (*an*) of *Œdogonium ciliatum* escaping. At ♀ is an oögonium. Magnified 350. (After Pringsheim.)

69, A). The gap is at first closed by a new membrane formed from the adjacent protoplasm of the oögonium, but this partly disappears again, leaving a free passage to the ovum.

Before describing the mode of fertilisation, we will consider the peculiar distribution of the sexes already mentioned, as differing from the ordinary monœcious and diœcious conditions. It is this form which our figures illustrate. The peculiarity consists in the production of dwarf male plants quite different from the ordinary form of the species.

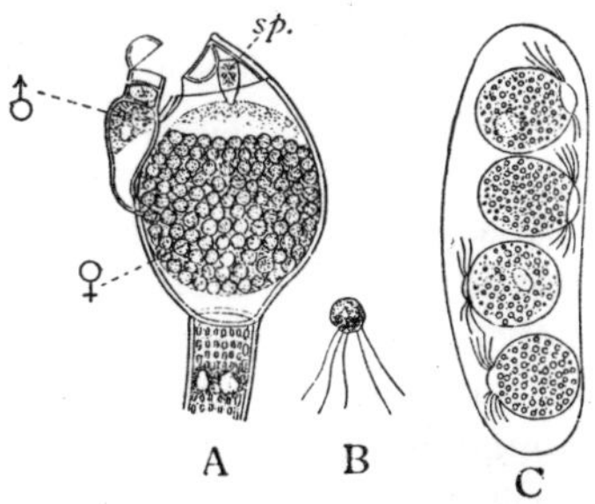

Fig. 69.—A, fertilisation of *Œdogonium ciliatum*; ♂, dwarf male plant from which a spermatozoid has escaped; *sp*, the spermatozoid in contact with the ovum, ♀. Magnified 350. B, a single spermatozoid showing cilia. Magnified 350. C, germination of the oöspore in *Bulbochæte*, showing the contents divided up to form four zoospores. Magnified 250. (After Pringsheim.)

By repeated transverse divisions in parts of the filament a chain of small cells is produced much shorter than the ordinary vegetative cells of the plant. Each of these short cells becomes the mother-cell of a single zoospore of the usual structure, but of a size intermediate between a normal vegetative zoospore and a spermatozoid (Fig. 68, *an*). These small spores (called *androspores*) are most commonly produced from the same filaments which bear the oögonia; more rarely they occur on distinct filaments. Each androspore swims about for a time, and then comes to rest, attaching itself to the female plant either near or actually upon the oögonium (Figs. 66 and 69, A). The androspore surrounds itself with a cell-wall and germinates. The plant which it produces is always of very small size. It may consist of a basal vegetative cell with one or more antheridial cells, or the vegetative part may be absent, and the whole dwarf male be reduced to an antheridium only. In the antheridial cell, or in each of them if more than one be present, two spermatozoids are produced. They escape by a lid-like opening of the antheridium, and make their way to the oögonial aperture (Figs. 66 and 69, A). Like the asexual zoospores, they move by a crown of cilia ; they are also able to help themselves along by the contractions of their whole body. The same power of contraction is of service when the oögonial opening is reached, for the entrance may be much narrower than the body of the spermatozoid, which can only pass through by accommodating its form to the size of the passage.

The act of fertilisation now takes place ; the spermatozoid comes into contact with the cytoplasm of the ovum, and the two cells unite. The details of fertilisation have been exactly followed in *Œdogonium* ; after the cytoplasm of the spermatozoid has united with that of the female cell, the nucleus of the former can still be distinguished. It is much smaller than that of the ovum. The male nucleus passes through the cytoplasm until it reaches the female nucleus, and then the two unite to form a single nucleus. Thus even in so simple an Alga as this, we see that fertilisation is in essentials precisely the same process as in the highest Phanerogams, such as the Lily.

The fertilised ovum (which is now called the *oöspore*) contracts further, surrounds itself with a cell-wall, and gradually passes into a resting state. The contents undergo great changes ; the chlorophyll disappears and is replaced by a brown or red colouring matter, while large quantities of oil appear in the cytoplasm, and at the same time the cell-wall becomes much thickened. The resting stage may only last for a few weeks. It appears that in some at least of the species germination takes place before winter comes on. In the meantime the oöspore has remained enclosed within the oögonial wall.

When germination begins, the inner layers of the cell-wall of the oöspore swell, and burst the hard outer coat; the entire contents surrounded only by a delicate membrane now become free, leaving behind both the outer oöspore-wall and that of the oögonium. In the normal course of development the oöspore does not immediately give rise to a new plant; its contents divide into four cells, each of which rounds itself off and becomes a ciliated zoospore, exactly resembling the zoospores formed in the vegetative cells, except that the contents are at first wholly or partly of a red colour (see Fig. 69, C). These zoospores free themselves from the enclosing membrane, swarm actively for a time and become green, and then come to rest, each giving rise to an ordinary *Œdogonium* plant. It appears that in many cases the plants remain through the winter in the state of young filaments of three or four cells. In a few exceptional cases the oöspore grows out directly into a filament, and the contents of an unfertilised oögonium sometimes behave in the same way. In other cases again the contents divide as usual, but each of the cells germinates directly within the oöspore, without passing through the swarming stage. All these cases, however, are exceptions; the regular process is the production from the oöspore of the four zoospores, each of which gives rise to a new plant.

We have now completed the life-history of this remarkable Alga. In what way can we compare it with that of the higher Cryptogams? One thing is evident; we have here no such clear and regular alternation of generations as we have hitherto found. We cannot draw any sharp distinction between the asexual and sexual individuals, for in most cases the male and female plants themselves give rise also to the asexual zoospores. On the other hand, we may regard the germinating oöspore, producing four asexual reproductive cells, as representing an extremely simple sporophytic generation. The oöspore, however, is in some cases capable of direct germination into a new plant, which is as if the fertilised ovum of a Moss were to grow out directly into a Moss-plant without forming a spore-fruit. It is evident that the well-marked alternation of generations by which all the higher plants are characterised does not occur in many Algæ such as *Œdogonium*, though we can perhaps trace some indication of an analogous process in a rudimentary form.[1]

It has been well pointed out that the great development of the sporophytic generation in the higher Cryptogams may be viewed as an adaptation to life on land. The sexual plant, owing to its mode of fertilisation, is always dependent on the presence of water; the formation of spores suitable for dissemination through the air is advantageously handed over to a distinct generation. The asexual reproduction by locomotive spores is evidently adapted to aquatic life, and is extremely common among Algæ and some of the lower Fungi.

As regards the mode of fertilisation, we still find in *Œdogonium* (which in this respect is one of the highest Algæ) that it takes place within the female organ (*oögonium*), while there is a sharp differentiation of the sexual cells—on the one hand the large stationary ovum on the other the small active spermatozoid. The sper-

[1] In certain Green Algæ, *e.g. Cladophora, Ulva, Enteromorpha*, etc., distinct sexual and asexual plants occur.

matozoid, however, is here much more like an ordinary cell, *i.e.* less specially adapted to its function, than in the higher Cryptogams. At the same time, the unmistakable resemblance between a spermatozoid and an asexual zoospore is very striking.

It is particularly interesting that in *Œdogonium* we have in the androspores of some species a form of cell exactly intermediate between the ordinary zoospores and the spermatozoids. This formation of androspores giving rise to dwarf-males is quite a peculiar case, and only occurs in some species of *Œdogonium* and in the allied genus *Bulbochæte*, which chiefly differs from *Œdogonium* in having branched instead of simple filaments. Its significance is not perfectly understood; possibly by dividing the period of locomotion into two stages (first, the swarming androspore, and secondly, the swarming spermatozoid) the chance of the male cell reaching an oögonium at the right time may be increased.

TYPE X

ULOTHRIX ZONATA

1. STRUCTURE

We will next consider another fresh-water Alga, not at all unlike *Œdogonium* in general habit, but representing a very different and much lower type of reproduction. *Ulothrix zonata* is an extremely common Alga, of the filamentous or confervoid kind,[1] growing by preference

[1] Simple filamentous Algæ are often spoken of as *confervoid*, because all such forms were in the days of Linnæus included in the old genus *Conferva*, which since that time has been subdivided again and again, as the important distinctions among the species originally referred to it came to be better understood.

in running streams, or in water which is constantly renewed, as in brooks, open watercourses, or in open-air tanks with a constant supply.

The thread which constitutes the Alga is very slender, not usually exceeding about 0·04 millimetre (two-thousandths of an inch) in diameter. The filament consists of a single row of cells, often very numerous, amounting to a thousand or more in a single thread. The threads are usually attached at one end to the stones or other things in the water, but they can live equally well detached, floating in tangled masses on the surface.

The cells of *Ulothrix* are somewhat similar in structure to those of our last type. Within the cell-wall is the cytoplasmic layer, or primordial utricle, in which the nucleus is embedded. The chloroplast has the form of a broad transverse band, in which pyrenoids are contained. When, as is often the case, the cells remain short, the chloroplast may occupy almost the whole length of the cell, but when the latter grows longer the green band forms a girdle around its middle part. In the interior of the cell is a large vacuole containing only cell-sap. When the filament is attached by one end, the root-cell is nearly or quite colourless; sometimes this organ of attachment is branched, though the rest of the thread always remains simple. The thallus grows throughout its length, and has no special growing-point. The cells divide by transverse walls to keep pace with the growth.

2. Reproduction

The reproductive cells of *Ulothrix* are active zoospores and gametes. The zoospores are of two kinds, the most constant distinction consisting in the number of the cilia, of which there are four in the one kind and two in the other. Both vary much in size, but on the average the quadriciliate is decidedly larger than the biciliate form. We will first consider the former. The macrozoospores arise from ordinary cells of the filament ; each mother-cell may produce one, two, four, or eight of these bodies from its contents. In the first case the whole protoplasm (with the exception of a thin external layer) is used up to form the single zoospore ; if two are to be produced, a transverse cell-division first takes place, and each half of the contents constitutes a zoospore ; if the number be four, a longitudinal division succeeds the transverse one ; and if eight are to be formed, another division again, at right angles to the other two, completes the partition of the contents.

In any case the resulting zoospores are pear-shaped, contain each one nucleus, and have the chloroplast in the broader part of the cell, while the pointed end is made up of clear cytoplasm. In each zoospore there are two " pulsating vacuoles " which alternately contract and expand about every ten or fifteen seconds ; in this respect these vegetable cells precisely resemble some of the unicellular animals. At the point are the four long cilia by which the locomotion is brought about. The mother cell-wall breaks down at one side, and the zoospores escape. Their movements are very active, and they have formed the subject of some of the best observations on the locomotion of vegetable cells. Relatively to their own length their speed is great. A zoospore will travel through twice or three times its own length in a second, while a fast ship requires from ten to fifteen seconds to traverse its own length ; on the other hand, zoospores are so small that their *actual* speed is slow (about a mètre in an hour).[1] Considering the distance they are required to travel, we may say that their locomotion is very active.

[1] Noll, in Strasburger, Noll, Schenck, and Schimper, *Lehrbuch der Botanik*.

Under the microscope their size and speed are magnified together, so that we are enabled to realise the true relation between their movements and their dimensions.

We saw some time back (pp. 68 and 130) that the spermatozoids of the higher Cryptogams are sensitive to certain chemical substances when dissolved in the water. Zoospores of Algæ are also sensitive, though in different ways; in this case they react especially to the influence of light. In darkness these zoospores wander aimlessly about in all directions, and do not come to rest until they are exhausted and die. In darkness therefore they cannot reproduce the plant; but as they are normally set free in the morning, and come to rest before evening, this difficulty does not arise in Nature. Light, however, shining from one side, at once exercises a directive effect on the movements of the zoospores. If the light be of moderate intensity, they swim towards its source; if the brightness be excessive, they hasten with equal decision in the opposite direction. As the swarming period nears its end, the zoospores tend more and more to avoid the light, and so are brought into contact with dark solid objects in the water, on which they can come to rest and germinate. Hence the reaction to light is of service to the plant, for the zoospores are thus induced to disperse themselves in the open water where their own assimilation can go on, while at the same time seeking shelter from too intense sunlight, which is dangerous to unprotected protoplasm. Finally, their changing sensitiveness ultimately brings them to rest in a position where germination can take place.

The zoospores of *Ulothrix* and of many other Algæ are provided with a red pigment-spot at one side of the clear protoplasm; this has been called the *eye-spot*, a fanciful name, though it is probable that the red pigment may really have something to do with the action of light on the zoospore. These comparatively large, four-ciliated zoospores of *Ulothrix* swim about for some hours, and when they come to rest attach themselves by the colourless end, surround themselves with a wall, and grow out at once into a new *Ulothrix* filament, like the parent.

In other cells, it may be of the same or of different filaments, swarm-cells of a somewhat different kind are formed. These are characterised by having two cilia instead of four, but in other respects are like the former kind, though smaller. These microzoospores are produced to the number of 8, 16, 32, or even more, in a mother-cell, by successive bipartition, just as in the former case. Except in size and for the number of cilia there is no visible difference between the two kinds. On germination the microzoospores grow out into new filaments

In addition to these two kinds of zoospores biciliate sexual cells (gametes) are formed. These look extremely like microzoospores but are somewhat smaller. In some species of *Ulothrix* the gametes are formed under different environmental conditions from those which favour the development of microzoospores. When the gametes escape from their mother-cells they swim off through the water just like the zoospores. If, however, as often happens, one of them meets with another gamete just set free from another cell of the same or of a different thread, an extraordinary process takes place. When two gametes from different mother-cells happen to touch one another they at first become entangled by their cilia, and the pair go on spinning through the water together. They gradually become more closely united ; their bodies come into contact laterally, and soon begin to fuse. The fusion starts at the pointed colourless ends, and after these parts are quite joined up the opposite ends remain for a time separate. Soon however, in fact within a very few minutes, fusion is complete ; the two cells have become one, and now constitute a single four-ciliated spore (see Fig. 70). Its origin from two cells can long be recognised by the two eye-spots and the two chlorophyll bodies.

The movements do not last long after fusion is complete. The spore resulting from this union withdraws its four cilia and comes to rest, attaching itself like a zoospore by the colourless end. In most cases the two cells which unite are of about the same size ; sometimes it happens that a smaller cell unites with a larger (Fig. 70, C), for of course the size of the gametes varies somewhat according to the number produced in a mother-cell, and the dimensions of the latter.

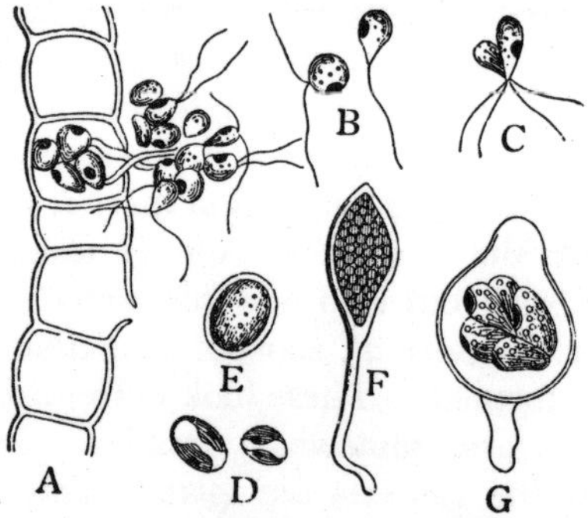

FIG. 70.—*Ulothrix zonata*. A, part of a filament ; most cells are already empty ; from one the biciliate gametes are escaping. B, gametes ; C, two in the act of conjugating ; D, two young zygospores, immediately after conjugation ; E, ripe zygospore ; F, unicellular plant grown from zygospore ; G, similar plant producing cells, which are about to escape. Magnified 482. (After Dodel-Port.)

Apart from such accidental differences, we have here a union of *perfectly similar cells*, whereas in all the plants described above there was a sharp distinction between the male and female cells. We have in fact reached in *Ulothrix* the lowest and simplest stage of sexual reproduction in plants. The process in this rudimentary form requires a special name ; *fertilisation* is the union of *unlike* cells, *e.g.* that of an ovum with a spermatozoid, or with a generative cell from the pollen-tube ; the sexual fusion of *similar* cells is termed *conjugation*, and the result of such a union is called a *zygospore*, to distinguish it from the *oöspore*, which is the product of fertilisation.[1] Every possible intermediate stage between the two processes is, however, known in various Algæ, so we need not hesitate to regard conjugation as really a primitive form of sexual reproduction. One point of importance must be noticed : only gametes from different mother-cells will conjugate, not those from the same. Hence we see that in this very simple plant *self-conjugation* is avoided, just as *self-fertilisation* is so often avoided among the higher plants.

We will now see what becomes of the zygospore. It soon acquires a cell-wall, and may put out a colourless root-hair (of variable length) by which it is attached to the substratum. The germinating zygospore goes on growing slowly for some weeks, but never attains any great length, and always remains unicellular (see Fig. 70, F). Its contents become denser and of a dark-green colour, while the cell-wall is much thickened, and now the little plant enters on a stage of rest.

After some time the plantlets produced from the zygospores wake up, but they never grow into *Ulothrix* filaments. Their contents divide up into a number of cells (from three up to about fourteen), and in this case the division is simultaneous (Fig. 70, G). These cells are usually liberated in a non-motile condition and germinate to form new *Ulothrix* filaments.

Ulothrix, though so simple an organism, has been considered as presenting certain analogies with the life-history of the higher Cryptogams. The resting cells, resulting from the act of conjugation, may be regarded as a dwarf form of plant, limited entirely to reproductive

[1] The sexual union of similar gametes is sometimes termed *isogamy*, and that of unlike gametes *heterogamy*. In the latter the process is spoken of as *oögamy* when the female gamete is non-motile.

functions, and as, in an extremely rudimentary form, so far comparable to the sporophytic generation of the Bryophytes.

Occasionally *Ulothrix* reproduces itself in a manner which is a modification of the process of zoospore formation. Cells of the filaments divide up in all directions into colonies of rounded cells which become surrounded with mucilaginous membranes, the walls of the mother-cells also becoming mucilaginous. This may go on repeatedly, so that large groups of cells embedded in mucilage may be formed. This has been called the *Palmella* condition, because the plant when in this state somewhat resembles another genus, *Palmella.* The *Palmella* state of *Ulothrix* is chiefly found when the plant is left by the receding water on the damp parts of the pool or stream in which it grows. Rarely, too, the cells of the filament may round themselves off, acquire thick walls, and pass into a resting condition. Ultimately these cells reproduce the normal filaments, either by direct germination or by forming zoospores in their interior which escape and ultimately grow into new plants. The thick-walled resting cells are a means of protection against death by drought.

The great scientific interest of *Ulothrix* depends on the variety in its methods of propagation, and especially on the fact that we can here study sexual reproduction in its most primitive form.

TYPE XI

Spirogyra

1. Structure

Our next type is another of the filamentous fresh-water Algæ, but very distinct from those already dealt with.

The genus *Spirogyra* contains about one hundred and seventy species, differing much among themselves both in size and in details of structure. We will give a general account of the genus, noting when necessary the points in which the specific differences are of interest.

Spirogyra occurs chiefly in ponds and lakes, that is to say, in still rather than in running water. It is often present in immense quantities, the filaments forming floating masses, sometimes several acres in extent, buoyed up by the bubbles of oxygen which their assimilation has produced. It is characteristic of *Spirogyra* to float, for the filaments of most species are perfectly free, without attachment to any kind of substratum. There is no distinction of apex and base, and all the cells are alike throughout the thread.

In the larger species of *Spirogyra* the cells are, microscopically speaking, of great diameter (reaching 0·25 mm. = one-hundredth of an inch, in extreme cases). They are therefore very favourable for study. Within the stratified cell-wall is the primordial utricle, in which protoplasmic circulation can be well observed in vigorous plants. Imbedded in this are the very conspicuous chloroplasts, which here take the form of green spiral bands with toothed edges. They form the most striking feature of the plant under the microscope, and from them the genus derives its name. The chloroplasts vary in number from one to ten in each cell (see Figs. 71 and 72) according to the species, but the number is not always constant even in the cells of the same filament. Each chloroplast is studded at intervals with large *pyrenoids*, which can be well seen here, and have a crystalloid form. It is around these protein bodies that much of the starch is formed when assimilation is taking place. Each pyrenoid then becomes surrounded by a coating of minute starch-granules. In every cell there is one large nucleus, usually suspended in the middle of the cell by cytoplasmic fibrils. These fibrils are attached, at the ends remote from the nucleus, to the chloroplast, abutting on it at the points where the pyrenoids are situated. The nucleus contains a large nucleolus.

All the cells are similar, and all take an equal part by growth and transverse division in the development of the plant. *Spirogyra* may therefore be spoken of as *physiologically* a unicellular organism, for all its cells are equivalent, and each appears to be capable of carrying on all the necessary functions for itself. In the ordinary state the plant is *morphologically* multicellular, but occasionally the thread actually breaks up into its separate cells, and we then find that each of these on its own account is capable of independent life, and can start a new plant. In a large number of genera allied to our type, the cells normally lead an isolated existence, separating after each division. This is usually the case in the *Desmids*, a Family of freshwater microscopic plants of great beauty.

In *Spirogyra* the division of the cell coincides with that of the nucleus, which takes place by a complicated process (much like that in the higher plants) resulting in the formation of two exactly similar daughter-nuclei. While this is going on, a transverse septum gradually grows in from the outside of the cell, first appearing as a mere ring, and gradually advancing towards the middle, until a complete disc of cellulose is formed across the cell, dividing its contents into two parts and separating the newly-formed daughter-nuclei (see Fig. 71). The growth of the cell-wall is entirely due to additions from the cytoplasm. This mode of cell-division by ingrowth from the outer wall is common among the lower plants, while in the rest of the Vegetable Kingdom the new wall is as a rule formed simultaneously over its whole area.

2. Reproduction

Except for the occasional breaking up of a filament into short groups of cells, *Spirogyra* possesses no asexual means of propagation. Its normal reproduction is always sexual and is a form of conjugation, consisting, in the union of like cells. As we shall see, however, it is quite different from the mode of conjugation in *Ulothrix*, for in *Spirogyra* and its allies it is the ordinary vegetative cells which unite. Conjugation, in most cases, takes place between two filaments lying side

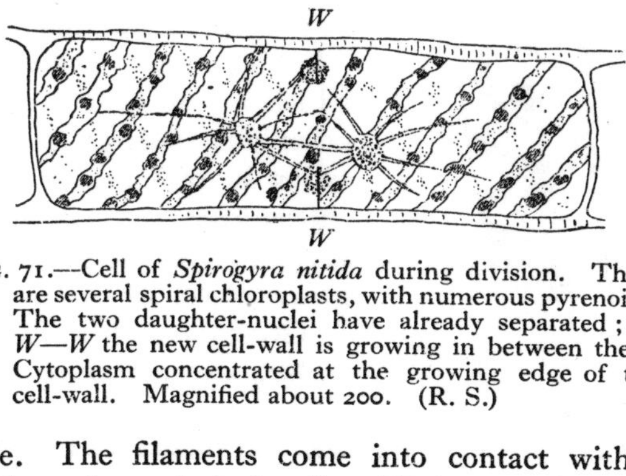

Fig. 71.—Cell of *Spirogyra nitida* during division. There are several spiral chloroplasts, with numerous pyrenoids. The two daughter-nuclei have already separated; at *W—W* the new cell-wall is growing in between them. Cytoplasm concentrated at the growing edge of the cell-wall. Magnified about 200. (R. S.)

by side. The filaments come into contact with each other throughout their length. The cells of each filament then begin to put forth lateral outgrowths which exactly correspond to each other in position. As these outgrowths elongate the filaments are pushed apart. The walls separating the outgrowths from each other become absorbed, thus leaving an open passage between the opposite cells of the two threads (see Fig. 72).

The contents of the cells in one of the two filaments now contract, receding from the cell-wall, and round themselves off; for a time the cells of the other filament remain unchanged. The contracted cell-contents next begin to insinuate themselves into the connecting passage, and gradually work through it, passing into the opposite cell (Fig. 73). They then unite with the contents of the latter, which in the meantime have themselves somewhat contracted.

In this case the fusion of the nuclei of the two conjugating cells has been observed. The united protoplasmic mass assumes a rounded or oval form, and surrounds itself with a cell-wall, which becomes thickened and cuticularised on its external surface. The chloroplasts of the gamete which moves over to join the other one undergo disorganisation so that the number of chloroplasts proper to the species remains constant from generation to generation.

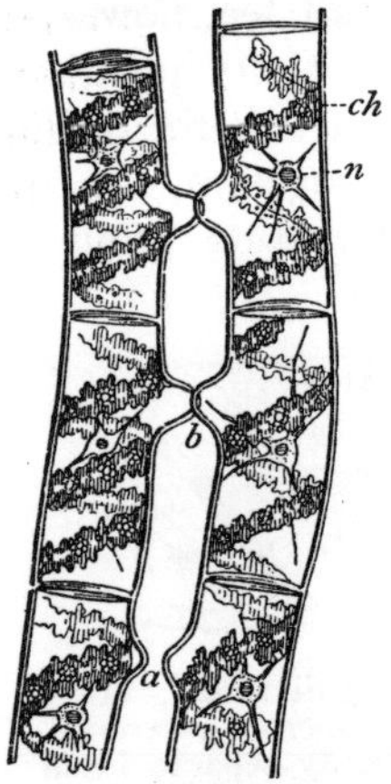

FIG. 72.—*Spirogyra longata* conjugating. The two parallel filaments are putting out the conjugating outgrowths (*a*), which at *b* are in contact; *ch*, spiral chloroplast; *n*, nucleus, with radiating cytoplasmic fibrils. Magnified about 350. (After Sachs.)

That there is actually a certain distinction of sex in *Spirogyra* is evident from the fact that all the cells of each conjugating filament usually behave in the same way, either giving up their own protoplasm or receiving that of the fellow-filament. That the difference, however, is not fixed is shown by the fact that conjugation sometimes takes place *monœciously*, *i.e.* between the cells of the same filament. This mode of union is called "*chain-like*," in opposition to the more usual diœcious or "*ladder-like*" method, and both processes may occur in the same species. In *monœcious* conjugation two adjoining cells put out very short lateral processes, which arise in contact with each other, on either side of the transverse wall separating the two cells. The processes unite, and the contents of the one cell pass through the side passage and fuse with those of its neighbour. In this case there may be a certain sexual difference among the individual cells, but the filament is obviously bisexual.

In most of the allies of *Spirogyra* (*e.g.* in the Desmids) there is no trace of any difference of sex, for the cell-contents meet and fuse midway between the parent cells, each of which thus takes an equal part in the process.

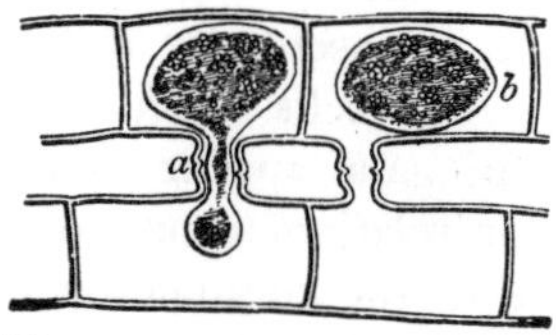

FIG. 73.—Conjugation more advanced. At *a* the protoplasm is in the act of passing over; at *b* the union is complete, one cell having transferred all its contents to the other. Magnified about 350. (After Sachs.)

The contents of the zygospore assume a darker colour, and the starch disappears, giving place to oil, which constitutes the reserve carbonaceous material during the interval of rest. The zygospore can now survive either a period of drought or the cold of winter, as the case may be. On germination the zygospore, unlike that of *Ulothrix* and its allies, gives rise directly to a new plant. The outer wall is burst, the protoplasm grows out, clothed only by the inner cellulose wall, the bright green colour of the chlorophyll reappears, and starch is once more formed. During the germination of the zygospore the nucleus divides twice, but three of the four nuclei thus formed disorganise. On first germinating, the young plant shows a distinction between apex and base, for it remains for a time attached by one end, which is pointed and colourless. This distinction is usually soon lost, and the filament floats freely in the water. In a few cases, where conjugation fails to take place, the single cells have been observed to form resting-spores asexually, and in one or two species this is the rule.

The life-history of *Spirogyra*, which represents a considerable Order of fresh-water Algæ, is, as we see, extremely simple ; there is not the slightest indication of any alternation of generations. All individuals alike are capable of conjugation, and are therefore sexual. The name Conjugatæ is applied to this Order, for it was among them that the union of similar cells to form a zygospore was first observed. In this group it is always the ordinary vegetative cells which unite, so the process is essentially different from the union of actively moving cells such as takes place in *Ulothrix*. The Conjugatæ appear to be an isolated group, showing no near relation to any of the higher forms, whereas from the motile conjugation of *Ulothrix* we can trace a series of steps leading on to the well-marked fertilisation of Algæ like *Œdogonium*. In *Spirogyra*, however, we have really the type of all sexual reproduction, which consists essentially in the fusion of two distinct cells, accompanied by the union of their nuclei. The structural differences between male and female, so generally found in the higher plants, are of only secondary importance, and must be regarded as special adaptations to secure this union with the least expenditure of material and energy. In one word, we may regard conjugation as the primitive form of fertilisation.

TYPE XII

VAUCHERIA

We now come to a group of plants, which in their general structure differ more profoundly from all our previous types than any of the latter differ from one another. So far, all the plants we have considered have been cellular, protoplasmic continuity depending on minute connecting fibrils : in *Vaucheria* the protoplasm is not partitioned by cell-walls at all, but is perfectly uninterrupted throughout the whole plant. A large number of Green Algæ—including the extensive Order Siphoneæ—are distinguished by this non-cellular structure. Peculiar as these plants are in their internal organisation, they are none the less capable of attaining on their own lines a high development. Many of them are of large size, and some possess organs analogous to the stem, leaf, and root of higher plants.

Vaucheria is one of the simplest of the non-cellular Algæ in its vegetative organs, but as regards its reproduction is more highly differentiated than any other genus of this type at present known. Some few species of *Vaucheria* are marine, but most are either fresh-water or terrestrial plants. Some of them are among the very commonest of Algæ, occurring everywhere in ponds and ditches, or on damp earth, as, for instance, on the soil of neglected flowerpots, where they form a tangled green web, and are often troublesome weeds, especially to Fern cultivators. Their threads are rather coarse, darkish-green, and branched. When submerged, they often form dense masses of large size, becoming conspicuous objects in the water.

1. Structure

In the vegetative condition a *Vaucheria* consists of a cylindrical green filament, repeatedly branched at irregular intervals, and often attached by a colourless branched rhizoid (see Fig. 75, *B*). The contents are uniform throughout the green part of the plant. The filament is bounded externally by a cellulose membrane, but internally there is nowhere the least trace of cell-walls, as long as the plant remains in its normal vegetative condition ; the cell-contents are continuous throughout its entire length (which may amount to many inches), and extend without a break into the branches.

The cytoplasm forms a thick layer lining the external wall, and contains an immense number of roundish chlorophyll-granules, the abundance of which accounts for the deep-green colour of the threads. Imbedded in the cytoplasm, just inside the layer of chlorophyll-granules, there are innumerable minute nuclei, which multiply by division as the plant grows. This feature is very characteristic of the non-cellular Algæ and some allied forms, which always have numerous nuclei scattered in their cytoplasm. We see, then, that *Vaucheria* possesses the essentials of cell-structure—cytoplasm and nuclei—though not partitioned into distinct cells. The growth of the stem and its branches is apical ; at the extreme growing end of each filament the protoplasm is colourless and transparent. *Vaucheria* does not form starch ; the product of its assimilation is deposited in the form of a fatty oil.

2. Reproduction

Vaucheria produces reproductive cells of two kinds—asexual and sexual. The asexual reproduction affords a rapid means of propagation, and goes on chiefly when the plant is growing in abundance of water, and generally under conditions that suit it.

The protoplasm accumulates and becomes denser at the end of a branch, which assumes a club-like form. The enlarged end is then separated from the rest of the filament by a transverse septum, for *Vaucheria* forms cell-walls in connection with its reproductive organs, though not elsewhere, except in case of injury. The cell thus formed may be called the *zoosporangium*. The entire contents of the zoosporangium constitute a single zoospore of relatively large size, clothed over its whole surface with numerous short paired cilia (see Fig. 74).

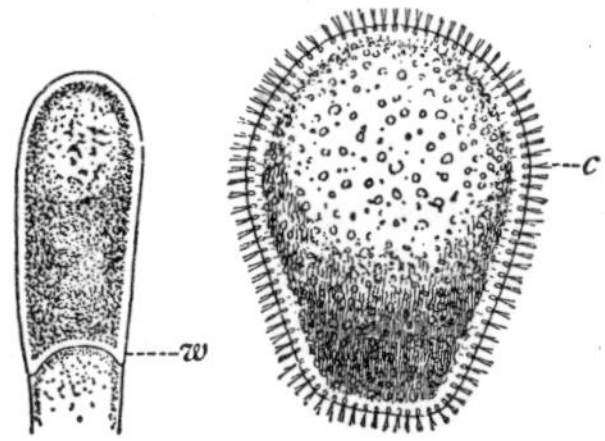

FIG. 74.—Zoospore of *Vaucheria sessilis*. To the left is shown the end of a filament just cut off, by the wall (*w*), to form a zoosporangium. Magnified 25. To the right is a zoospore covered with the numerous cilia (*c*). Magnified 95. (After Strasburger.)

The sporangium opens at the apex; the expulsion of the zoospore is helped by the expansion of mucilage contained in the cell, but depends to a great extent on its own movements. The opening is much narrower than the zoospore, which has to push its way through, and to change its form in the process (see Fig. 75, *A*). Sometimes the front part, which tries to rotate as it becomes free, gets twisted

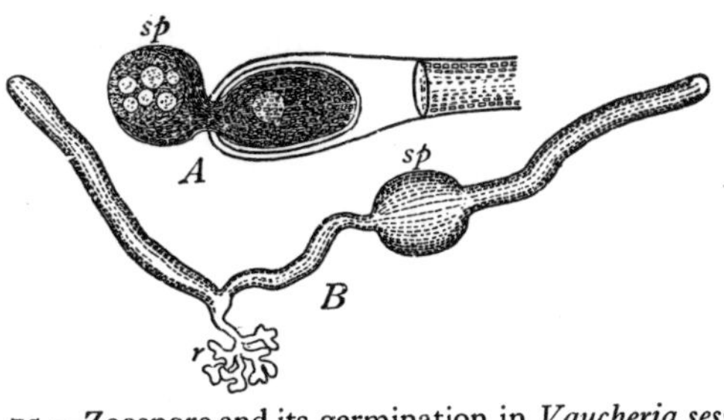

FIG. 75.—Zoospore and its germination in *Vaucheria sessilis*. *A*, zoospore (*sp*) in the act of escaping from the sporangium. Magnified about 30. *B*, germinating zoospore (*sp*), which has formed two green filaments and a rhizoid (*r*). Magnified about 20. (After Sachs.)

off from the rest of the cell, and then two zoospores are formed instead of one.

The zoospore contains a great number of chloroplasts

and of nuclei. In this case the nuclei lie near the outside, in a clear zone of cytoplasm. The numerous cilia are in pairs, each pair corresponding to a nucleus. Evidently the whole of this great zoospore corresponds to a multitude of small zoospores, not separated from each other. In most allied Algæ we find that numerous zoospores, usually with two cilia each, are formed in the sporangium.

The escape of the zoospores of *Vaucheria* generally takes place in the morning. They swim about rather lazily for a quarter of an hour or so, and are so big that their movements can be followed with the naked eye. Then they come to rest, and immediately withdraw their cilia and surround themselves with a membrane. The zoospore germinates at once, sending out a filament which attaches itself to the substratum by a colourless rhizoid and soon begins to branch (see Fig. 75, *B*).

This mode of reproduction may go on indefinitely, from generation to generation, as long as weather and water remain favourable. Sometimes, however, *Vaucheria* proceeds to form reproductive bodies of another kind, especially in bright light and perhaps also when drought threatens. This plant, so simple in its vegetative structure, possesses more sharply differentiated antheridia and oögonia than any of the Algæ already described. Both organs arise as lateral outgrowths. In some species they are seated directly on the main filament; in others a short special branch forms a pedicel on which the sexual organs are borne, as shown in Fig. 76, where the oögonium is terminal on the pedicel, while the antheridium is seated laterally below it. Diœcious forms also occur rarely.

The antheridium at its first origin resembles a young vegetative branch. It contains abundant cytoplasm with chloroplasts and very numerous nuclei. As it approaches maturity, it usually becomes curved like a horn (Fig. 76). The terminal part is cut off from the rest by a transverse wall, and forms the true antheridium. The nuclei assemble in the interior of the cell, leaving

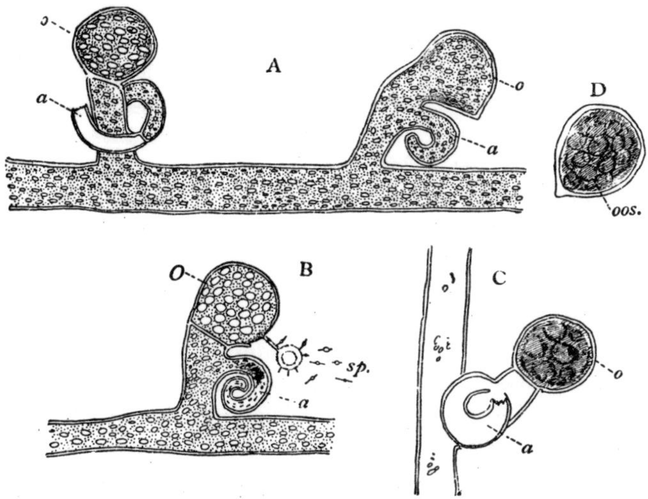

FIG. 76.—Fertilisation of *Vaucheria terrestris*. A, part of filament bearing two reproductive branches; *a*, antheridium; *o*, oögonium. On the right an early stage is shown. On the left the antheridium is already empty, and the oögonium fertilised. B, right-hand group of Fig. A later on, showing fertilisation. The minute spermatozoids (*sp*) are caught in the drop of mucilage expelled from the oögonium. Antheridium of same branch not yet open. C, some time after fertilisation. The antheridium and filament are empty, and the oöspore is ripening. D, ripe oöspore, containing oil-drops. Magnified 110. (R. S.)

the chloroplasts and part of the cytoplasm in contact with the wall. The more internal portion of the contents now breaks up into the excessively minute spermatozoids, each of which consists of one of the nuclei, together with an extremely small quantity of cytoplasm, and is provided with two cilia. They begin to swarm within the antheridium, which now opens at the apex, and expels the spermatozoids, together with a portion of the unused cytoplasm, some of which, however, remains behind.

In the meantime the development of the oögonium has been going on. Arising, like the antheridium, as a lateral outgrowth (or as in the species figured, at the end of the pedicel), it soon assumes a globular form. The protoplasm of the oögonium contains a large quantity of oil, as well as chloroplasts, and at first a very large number of nuclei. After a time a sort of papilla or beak is formed (Fig. 76, A) on one side. A remarkable change now takes place in the contents. A portion of the cytoplasm wanders back into the filament, carrying with it all the nuclei except one, so that the oögonium becomes a uninucleate cell. Alternatively, all the nuclei except one become disorganised in the oögonium. It is now cut off by a transverse wall from the filament. The contents rearrange themselves so that the portion towards the papilla becomes clear and free from chloroplasts, forming the *receptive spot*. The now solitary nucleus (which has grown rather larger than before) with associated cytoplasm lies towards the middle of the cell, connected by a strand of cytoplasm with the clear receptive spot. The wall of the oögonium opens at the papilla, and at the same moment a portion of the colourless cytoplasm is expelled. It is probably due to the pressure of the cytoplasmic mass that the wall is opened; sometimes the extruded part of the contents remains in connection with the cytoplasm within the oögonium. From this point onwards we may speak of the contents of the oögonium as the *ovum*.

Fertilisation now takes place. The minute spermatozoids swarm in at the oögonial aperture, but do not necessarily select the particular oögonium near which they were formed. In Fig. 76, B, for example, an antheridium is shown which has not yet discharged its spermatozoids, although the adjoining oögonium is actually being fertilised. Numerous spermatozoids may enter an oögonium, but only one fuses with the ovum, which it penetrates at the receptive spot where the cytoplasm is clear. The nucleus of the spermatozoid has been traced through the cytoplasm of the ovum, until it reaches the sole remaining nucleus, with which it unites. It appears then that in *Vaucheria*, in spite of the remarkable peculiarities of its organisation, fertilisation is precisely the same process as in the higher plants.

The fertilised ovum surrounds itself with a cell-wall of some thickness, and passes into a resting-stage (see Fig. 76, D) during which a period of drought can be endured with safety. Then germination takes place; the oöspore gives rise directly to a new *Vaucheria* plant, which soon begins forming the asexual zoospores.

We see that there is in this plant no indication of a regular alternation of generations. Sexual and asexual reproduction may both occur in the same individual, and it depends on the external conditions whether the one or the other takes place.

The sexually produced resting-spore is itself a provision against drought, but the plant can also protect itself against this danger in a more rough-and-ready fashion. If the water sinks and the filaments are left stranded on the mud, it sometimes happens that they divide up into a number of portions, each of which surrounds itself with a thick wall. This is so different from the normal form of the plant that specimens of *Vaucheria* in this condition used to be placed in a different genus. We see then that Siphonaceous filaments *can* separate into distinct cells when necessary, though this does not happen in the normal course of their vegetative life.

The Vaucheriaceæ stand quite alone among the non-cellular Algæ in the perfection of their reproductive arrangements. In other cases, where any sexual process has been observed, it takes the form of the conjugation of motile cells. The *Vaucherias* therefore, though among the simplest of these Algæ in vegetative respects, appear to rank highest as regards sexual reproduction.

TYPE XIII

PLEUROCOCCUS VULGARIS [1]

Before leaving the Green Algæ we will take two more examples, the very simplest we can find, as illustrations of unicellular structure. *Pleurococcus vulgaris* is in our climate perhaps the most abundant of all plants. Everyone must have noticed how commonly the trunks of trees, palings, and wet walls are covered by a bright green powdery layer, especially on the side away from the sun. In damp winter weather the green coating is most developed. This substance, though it may include many different organisms, is chiefly made up of *Pleurococcus vulgaris*.

This Alga, which occurs in prodigious numbers, consists of small rounded cells, sometimes quite separate, sometimes grouped together in little packets of two, four, or eight. When adhering together, the sides in contact are rather flattened.

Each cell has a thin cellulose wall, and is densely filled with protoplasm, which at first sight appears to be coloured uniformly green. This, however, is not the case, for the chlorophyll is really limited to a definite chloroplast, usually large and with several lobes, occupy-

[1] Sometimes called *Protococcus viridis*.

ing the outer part of the cell-contents. The rest of the protoplasm is colourless. About the middle of the cell is a nucleus, containing a nucleolus (Fig. 77). The cells divide freely into two; successive divisions take place in all three directions, and are at right angles to each other. The cells may either round themselves off and separate immediately after each division, or may remain grouped together for a few generations. Sooner or later, however, they fall apart. The plant forms small starch-grains in the chloroplasts when exposed to light.

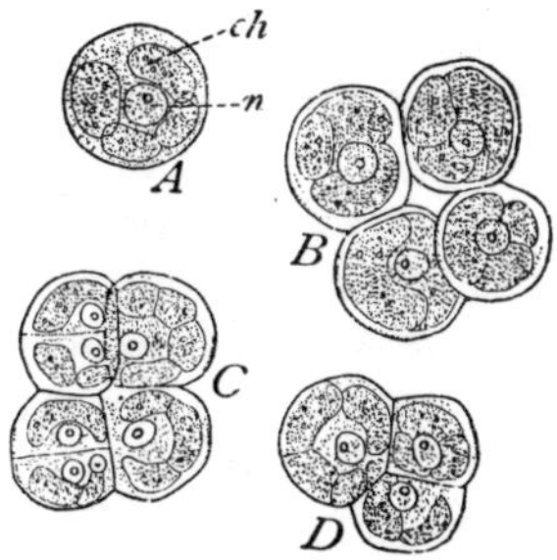

FIG. 77.—*Pleurococcus vulgaris*. *A*, single cell; *n*, nucleus; *ch*, chloroplast. *B*, four cells separating after division. *C*, group of cells remaining in contact. The two to the left have just divided afresh. *D*, tetrahedral group. Magnified 540. (After Strasburger.)

We have in *Pleurococcus* an example of a typically unicellular plant, in which the cells lead a perfectly independent life; each individual cell, however, has the same structure as in higher plants. Evidence has been brought forward to show that *Pleurococcus* may occasionally grow out into short filaments like those of higher green Algæ. The common unicellular condition may be due to reduction from a more advanced type.

TYPE XIV

CHLAMYDOMONAS

This type is introduced as a further example of a unicellular Alga, which is of special interest from two points of view. In the first place, unlike any plant hitherto described, it passes its ordinary vegetative life in a state of active movement. Secondly, simple as it is, *Chlamydomonas* has a perfectly definite sexual reproduction, showing important modifications in different species.

Chlamydomonas, of which numerous species have been described, occurs in ponds and quiet places in rivers, but is especially abundant in rain-puddles, often giving a bright green colour to the water. A species of *Chlamydomonas* is one of the forms of Algæ responsible for "red snow" in Alpine and Arctic regions.

The cells are small; about 20 μ[1] (0·02 mm.) is a common length for the full-grown cell, though some are rather larger. In the ordinary condition the form is usually oval, with one end rather more pointed than the other. To this end the two cilia, the organs of motion, are attached (see Fig. 77**, A; Fig. 77***, G). The cell is surrounded by a delicate cell-wall, often with a papilla at the pointed end. The cilia are continuous with the cytoplasm and pass through minute perforations in the wall. The chlorophyll is contained in a single large chloroplast, occupying most of the cell, and having the form of a cup, open at the top and much thicker at the base than elsewhere. In the lower part of the chloroplast a large pyrenoid (protein body) is imbedded; sometimes there are two or more. Both around the pyrenoid and in other parts of the chloroplast starch-grains are formed.

At the pointed end and in the hollow of the chloroplast there is clear cytoplasm, and imbedded in this is the nucleus. On one side of the cell, near the ciliated end, there is a red eye-spot. Lastly, in the clear protoplasm near the point, there are two contractile vacuoles, which expand and contract, the two pulsating alter-

[1] A μ is $\frac{1}{1000}$ of a millimetre.

nately. This then is the structure of the cell—a highly organised one, as we see.

These cells swim about, through the water, ciliated end foremost, rotating as they go. If we keep them in a saucer near a window, we find that they assemble on the side towards the light. They behave, in fact, just like zoospores, but we cannot call them zoospores, for they are the ordinary vegetative cells of the plant—the first example we have had of a plant which is in active movement through most of its life, like an animal. Yet the structure is altogether that of a plant-cell. The presence of a cell-wall indicates that the active cells are something more permanent than zoospores.

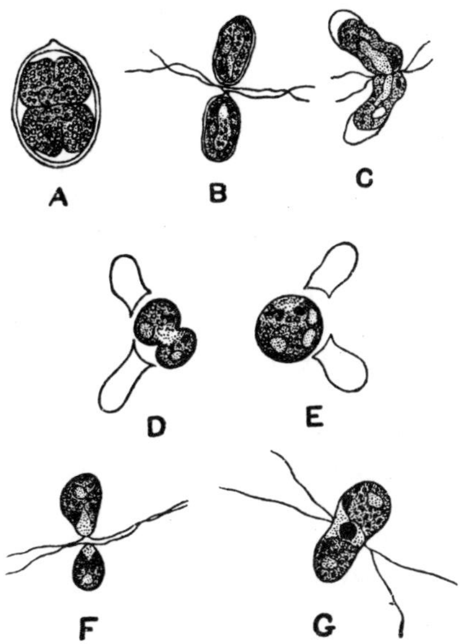

FIG. 77*.—A–E, *Chlamydomonas longistigma*, A, vegetative cell dividing into four. B–E, successive stages of conjugation of the walled gametes, which escape from their walls in the process. F, G, *Chlamydomonas pisiformis*. Conjugation of naked gametes. Magnified about 600. (After Dill.)

During their movements the cells feed and grow and ultimately they divide, division taking place during the evening, night, and early morning. While division is going on the cells lose their cilia and rest for a time. The cell most commonly divides into four (Fig. 77*, A), sometimes into two only, sometimes into eight. The first division is usually transverse, but in some species it is longitudinal, or the dividing plane is at first longitudinal, the cells then shifting so that it becomes transverse. The second division is at right angles to the first, but the daughter-cells arrange themselves so that their long axes are parallel. Each forms its own pair of cilia and cell-wall, and the cells break out through the common wall. This is the ordinary mode of multiplication, and the remarkable thing about it is that a motile cell gives rise directly to new motile cells without anything more than a temporary rest during division.

But there is sometimes another process, in which the division is repeated several times without the daughter-cells escaping or becoming active, the result being a colony of many cells, commonly grouped in packets of four or eight, and held together by a common mucilaginous investment (Fig. 77***, H). This is called the Palmelloid condition, and in some species these masses attain quite a large size. There is a species growing in springs in the Black Forest, in which they are said to be as big as walnuts. Ultimately the cells acquire cilia and escape, again starting the active condition. Thus the stationary phase, in which *Chlamydomonas* lives like an ordinary plant, is as a rule exceptional, though in some species it may be of long duration.

Most interesting of all is the sexual process, which takes place in three different ways, in different species.

The simplest method is like the conjugation of *Ulothrix*. The sexual cells are formed by the division of an ordinary cell into a larger number of parts than usual, so that the gametes are small compared with the vegetative cells; they are also, as a rule, more slender in form, but otherwise there is little difference. The gametes in the first type do not form a cell-wall. They swim about as usual, and when two meet they entangle each other with their cilia, and the pair go waltzing through the water together (Fig. 77*, F). Then they come into contact, and begin to fuse, either end to end, or side to side, according to the species. After union has begun the pair still goes on swimming with its four cilia (Fig. 77*, G). The corresponding parts of the cells unite, and the two nuclei meet and fuse, retaining for a time the two nucleoli. A cell-wall is formed around the whole, and the resultant body is a resting zygospore. There is a British species, *C. Debaryana*, which belongs to this type.

The second type differs from the first in the fact that the gametes, like the vegetative cells, are provided with cell-walls (Fig. 77*, B, C). The meeting takes place as before; for two or three hours the cells keep together, but not in close contact. Then they approach closer, but union is prevented by the cell-walls. Each gamete now contracts a little away from its wall, and then leaves it behind altogether (Fig. 77*, D, E), the contents passing out at the narrow end and rounding themselves off. At this stage the cilia are lost. The two protoplasmic cells, as they are set free, gradually fuse, and as in the previous case, a zygospore is the result (cf. Fig. 77***, E). This form of conjugation occurs in the British species, *C. Ehrenbergii*.

The third type is more remarkable; so far the conjugating sexual cells have been of the same size or with only accidental differences between them. In the third type (represented by *C. Braunii*[1]) there is a marked and constant distinction between the two. The one is smaller than the smallest vegetative cells, the other larger than the largest (see Fig. 77**, A, B). The structure is quite like that of the ordinary cells, and each gamete, as in the second type, is surrounded by a cell-wall. A large and a small sexual cell meet by their ciliated

[1] Sometimes called *C. monadina*.

ends as usual. The walls at the point of contact break down, so that an open canal is formed. The contents of each cell then contract away from the wall, and those of the small cell pass over through the canal, and unite with those of the large cell—reminding one of the conjugation of *Spirogyra*. Fig. 77**, C, shows an early stage of the passage, while D represents a later stage, when union has been accomplished. The two nuclei have now joined into one, but have not yet completely fused — the line of demarcation between the original nuclei is still evident, and the two nucleoli are distinct. The common cell-wall of the united gametes still persists and betrays by its form what has taken place. Then a new and thick cell-wall is developed around the fused contents, and the zygospore is constituted (Fig. 77***, E).

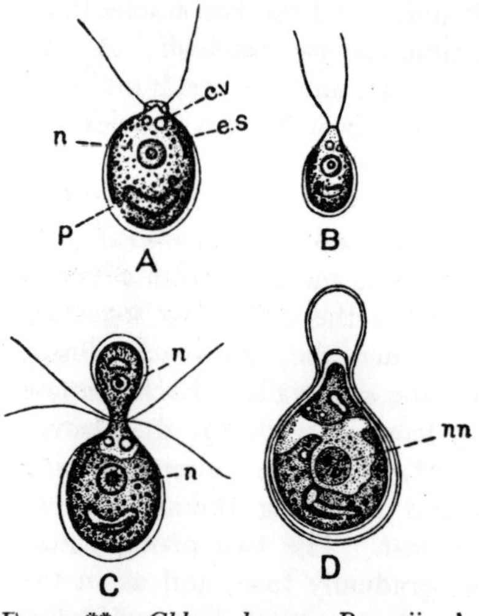

FIG. 77**.—*Chlamydomonas Braunii*. A, female gamete (closely resembling a large vegetative cell); *n*, nucleus; *p*, pyrenoid; *e.s.*, eye-spot; *c.v.*, contractile vacuoles. B, male gamete. C, early stage of conjugation; *n*, *n*, the nuclei of the two gametes. D, advanced stage of conjugation; *nn*, the united nuclei of the two gametes. Magnified about 500. (After Goroschankin.)

This process appears to be much in advance of that in the other species, as there is a marked distinction between the sexes—we might even speak of ovum and spermatozoid, though the former as well as the latter has the power of active locomotion. At the same time we must remember that the female cell is practically identical with a large vegetative cell, and not, as far as we can see, specially differentiated as an ovum. But however we may regard it, the sexual reproduction of *Chlamydomonas Braunii* is certainly on a remarkably high level for a unicellular plant.

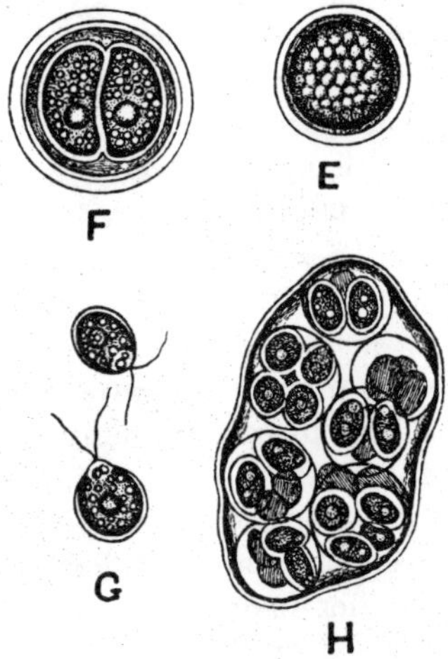

FIG. 77***.—*Chlamydomonas Braunii.* E, zygospore, containing oil-drops. F, germinating zygospore, divided into two cells. G, vegetative cells produced from a zygospore H, palmelloid condition consisting of numerous groups of cells within a common wall. E–G, magnified about 500; H, magnified about 400. (After Goroschankin.)

A few words may be added as to the fate of the zygospore, which is much the same in all the three types. It surrounds itself, as we have seen, with a cell-wall of its own which becomes much thickened in the resting stage (Fig. 77***, E). A red or red-brown oily pigment forms in the contents and hides the chlorophyll. In this condition the zygospore, like other resting spores, can survive a time of drought or cold. When conditions are favourable the contents gradually regain their green colour and begin to divide. The first stage of division is shown in Fig. 77***, F—division may be into two, four, or more. The resulting cells escape as ciliated individuals not essentially different from the ordinary vegetative cells (see Fig. 77***, G) and resume the regular life of the plant.

There is another unicellular Alga (*Sphærella* or *Hæma-*

tococcus lacustris), commonly found in water-butts and puddles, which is nearly related to *Chlamydomonas*, but differs from it, among other characters, in the fact that the inner part of the cell-wall is transformed into a wide mucilaginous layer which is traversed by radiating branched strands of cytoplasm. In this plant a red pigment is often present, in addition to the chlorophyll, even during the actively moving stage. The vegetative cells frequently occur in a resting condition, in which the cell-wall becomes thickened and in which the contents appear to be wholly red. Such resting cells are termed *cysts*. Sexual reproduction is by the fusion of similar gametes, without cell-walls, as in the first type of *Chlamydomonas*. In this genus the resting stage is more lasting than in *Chlamydomonas*.

These plants are important, because, on the one hand, they represent some of the simplest known types of green plants which, with many others, are sometimes classed as unicellular animals under the name of Flagellata, while on the other hand they are nearly related to more highly organised groups of plants (see pp. 300–301). Some of their nearer allies spend their lives associated in definite and complex colonies of regular form, as in the beautiful *Volvox globator*. In another direction it is probable that plants of the *Chlamydomonas* group may have given rise to some of the lines of the filamentous Algæ, such, for example, as *Ulothrix*. The actively moving stage, which in *Chlamydomonas* is the normal vegetative condition, tends in the higher Algæ to be more and more restricted to reproductive cells.

B. THE BROWN ALGÆ (*Phæophyceæ*)

The Brown Algæ, almost all of which are seaweeds, are probably better known to the ordinary observer than even the green group, owing to the large size which many of them attain, and the extraordinary abundance in which they occur on our coasts. In dimensions and structure they present an even wider range than the Chlorophyceæ, for though no Brown Algæ are quite so small or so simple as *Pleurococcus*, yet many of them very much exceed any of the former division both in size and complexity. The peculiar colour of their thallus results from the fact that the chlorophyll has a different composition from that of ordinary green plants ; an additional yellow pigment, *fucoxanthin*, is present besides the various chemical bodies which make up ordinary chlorophyll. The brown colour is the expression of this complex pigmentation. Treatment with hot water (as well as other methods of killing) has the effect of turning brown seaweeds green.

The Phæophyceæ certainly form a natural group, for from the lowest to the highest there are certain points in their organisation which are common to all. The colour in itself is not a character of much importance, but it coincides roughly with structural features, and affords a useful external mark by which the group can in most cases be recognised. This mark must, however, be used with caution, for some Algæ (the Diatoms) resemble the Phæophyceæ in colour but have otherwise nothing in common with them.

Many of the Phæophyceæ are reproduced by motile cells (zoospores and gametes) ; these are called the *Phæosporales*. A second Order only forms sexually produced oöspores ; this is the *Fucales*. We will take one example of each, for space will not allow us to do more, though really a large number of types would be necessary if our object were to gain any adequate idea of the diversity of the brown seaweeds.

TYPE XV

Ectocarpus siliculosus

1. Structure

The genus *Ectocarpus*, various species of which are extremely common on our shores, includes some of the simplest forms of the Phæophyceæ. The thallus is filamentous, and repeatedly branched. It consists of two parts ; a creeping primary portion by which the plant is attached to the substratum (usually one of the larger seaweeds), and a tuft of branched threads, arising from the creeping part and waving freely in the water, often reaching a length of several inches. Throughout the plant the filaments usually remain one cell thick, though occasionally longitudinal divisions occur. Each cell contains a single nucleus and a varying number of plastids, to the pigment contained in which the brown colour of the whole plant is due.

The mode of growth of the free filaments is peculiar. Instead of having an apical growing-point, each branch grows chiefly near its base. In this part short meristematic cells are found, which multiply by transverse division, while the more apical part of the branch consists of long, full-grown cells, which have ceased to divide (see Fig. 78). These *intercalary* growing-points, as they are called, are characteristic of this group of plants, though not by any means constant among them.

2. Reproduction

The reproductive organs are of two kinds, which are distinguished as *unilocular* and *plurilocular* sporangia. They are usually borne simultaneously or consecutively on the same plant, but they sometimes occur on separate plants. Both kinds of sporangia arise as lateral branches, either stalked or sessile.

The unilocular sporangia are simply ovoid or ellipsoid cells, borne at the sides of the branches (Fig. 79). They become densely filled with protoplasm, and their contents divide up into a large number of small gametes. Each gamete contains a nucleus, two brown plastids and an eye-spot, and bears two cilia, which are inserted at the side, not at one end of the cell. The sporangium dehisces at the apex, the contents are shot out in a gelatinous mass, and after a short interval the gametes disentangle their cilia and swim away. During their movements one cilium points forward in the direction of movement, and the other trails behind. In the great majority of the Phæophyceæ the motile cells have two cilia each, which are always inserted laterally, a point which distinguishes them from the corresponding bodies in the Green Algæ. These gametes, which are all of the same size, fuse

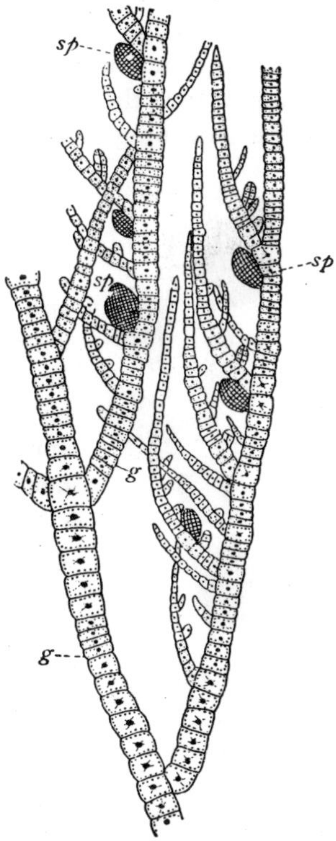

FIG. 78.—General view of part of the thallus of *Ectocarpus*, showing several branches. *sp*, *sp*, plurilocular sporangia, borne laterally on the branches ; *g*, *g*, intercalary growing-points, where cell-division is going on. Magnified 56. (R. S.)

together in pairs. The products of fusion surround themselves with a wall and germinate forthwith, giving rise to new plants exactly like the parent.

In *Ectocarpus siliculosus* there seems occasionally to be a certain functional difference between the sexes, though in appearance the conjugating cells are all alike. Certain of the swarming cells (gametes) come to rest before the others, and withdraw their cilia. Such a cell, which we may regard as female, exercises a remarkable attraction on the others which are still swimming about. They swarm round it in great numbers (Fig. 80, B), and eventually one of the swarmers begins to fuse with the stationary cell. The two are at first connected by means of one of the cilia of the active cell. This cilium gradually contracts, and the protoplasmic bodies are thus brought together, until finally the active and the stationary cell completely fuse into one. We see that this process, by which a zygospore is produced, is a step in advance of the conjugation of *Ulothrix*, for in the *Ectocarpus* there may be so far a difference between the two cells, that at the moment of fusion one is at rest and the other active, though previously they had both behaved quite in the same way. We may look upon this as the first slight approach towards the differentiation of stationary ovum and motile spermatozoid.[1] In spite of this distinction the sexual cells are

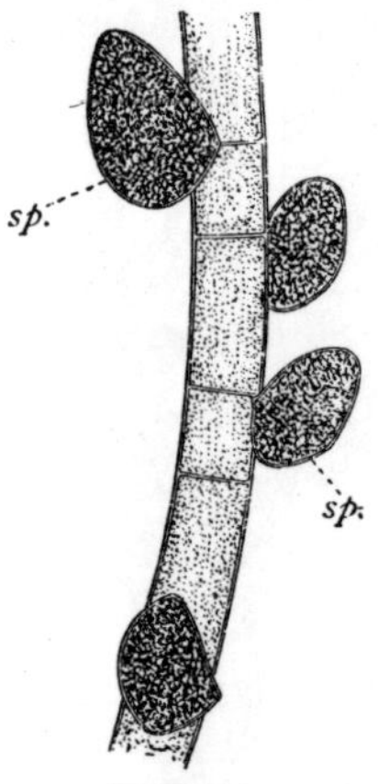

FIG. 79.—Unilocular sporangia (*sp*) borne laterally on a filament of *Ectocarpus ovatus*. Magnified 300. (After Reinke.)

[1] In another species, *Ectocarpus secundus*, there is a marked difference in size between the male and female cells.

capable of independent germination, although when germinating alone, *i.e.* without fusion, they produce small, weakly plants which usually quickly die. In an allied species, *Pylaiella littoralis*, plants arising from unfused gametes may come to maturity and form other gametes.

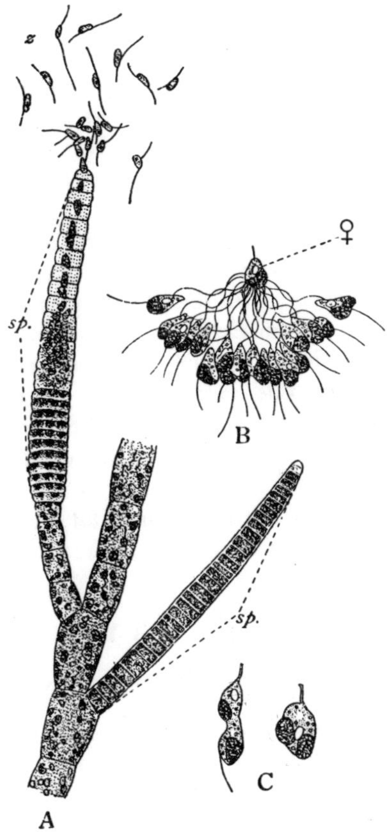

FIG. 80.—*Ectocarpus siliculosus*. A, part of a branch, bearing two plurilocular sporangia (*sp*), from one of which the zoospores (*z*) are escaping. Magnified 330. B, female cell (♀) which has come to rest, with numerous male cells swarming around it. C, two stages of conjugation. B and C magnified 790. (A after Thuret; B and C after Berthold.)

The *plurilocular* sporangia, unlike the *unilocular* sporangia, are multicellular structures. In British waters they are much more abundant than the *unilocular* sporangia. The *plurilocular* sporangium is divided by numerous cell-walls, usually longitudinal as well as transverse, into a multitude of small compartments (Figs. 78 and 80, A). In each of these compartments one or two zoospores are formed, which do not differ obviously from the gametes arising in the unilocular organ except that they are somewhat smaller. The zoospores usually escape from the plurilocular sporangium by a single opening at the end, the walls between the different compartments being absorbed, so that the swarming cells can pass out, one after the other, through the same aperture. The zoospores soon germinate.

Such is the life-history of *Ectocarpus siliculosus* and its near allies in British waters. In the Mediterranean, however, it is the plurilocular sporangia that give rise to gametes in *Ectocarpus siliculosus*, which usually show the functional sexual differentiation referred to above; the motile cells formed in the unilocular sporangia are zoospores. This difference in behaviour of the same plant growing in widely separated localities is of great interest, and is an indication of the plasticity of the developmental phases of some of the lower organisms.

TYPE XVI

Pelvetia canaliculata

Among the commonest and most conspicuous seaweeds on the coasts of cold and temperate countries are the members of the Order *Fucales*. The species chosen for our type is distinguished from all others on our shores by the position in which it grows, which is always close to high-water mark of the spring tides. The plants are thus only under water for a comparatively short time, not more than a quarter of the day during the spring tides. During the neap tides *Pelvetia canaliculata* is often not covered by the sea, and it may be only splashed by spray. This seaweed is able to bear a state of comparative drought without injury.

1. Structure

Pelvetia canaliculata is usually found in abundance on any rocky shore, forming a well-defined band, just below the highest level reached by the spring tides. The plants are only a few inches high (smaller than most of their relatives), and have a forked, flattened thallus

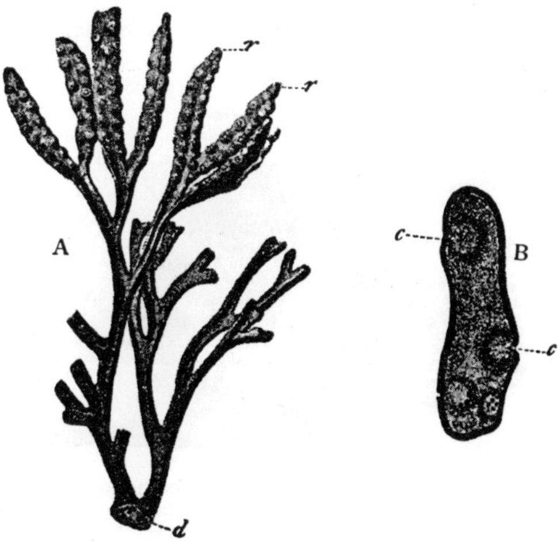

FIG. 81.—*Pelvetia canaliculata*. A, small fertile plant ; *d*, attaching disc ; *r*, *r*, receptacles, each of which bears a number of wart-like conceptacles. $\frac{4}{5}$ of natural size. B, transverse section of a receptacle, passing through several conceptacles (*c*). Magnified about 4. (After Thuret and Bornet.)

attached to the rocks by a rounded disc (see Fig. 81, A). The thallus shows a conspicuous groove or furrow along one side, to which the species owes its name. In addition to the regular dichotomous branches, adventitious shoots may arise on any part of the thallus.

When the plant is in fructification, which happens chiefly in the late summer and autumn, the ends of some of the branches become enlarged and studded with wart-like projections, each of which has a minute pore at the top (see Fig. 81, A, *r*). The swollen ends of the branches are called the *receptacles* ; the wart-like bodies mark the position of the *conceptacles*, which are cavities in the tissue, containing the organs of reproduction (see Figs. 81, B, and 82).

Pelvetia, though one of the simplest of the Fucales, is a very highly organised plant compared with the

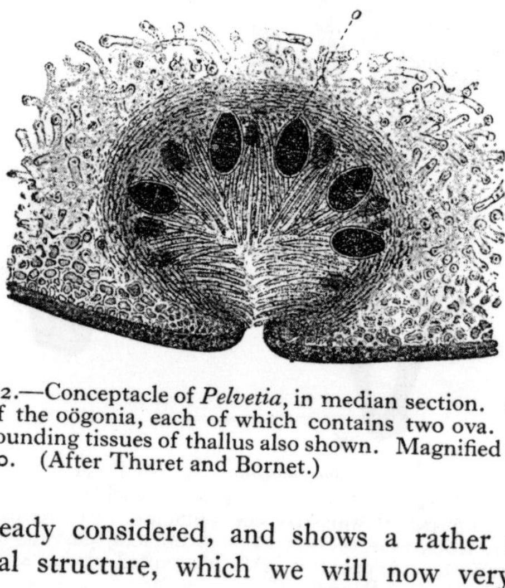

FIG. 82.—Conceptacle of *Pelvetia*, in median section. *o*, one of the oögonia, each of which contains two ova. Surrounding tissues of thallus also shown. Magnified about 10. (After Thuret and Bornet.)

Algæ already considered, and shows a rather complex anatomical structure, which we will now very briefly describe.

The external layer of tissue consists of small cells with abundant plastids, giving their contents a dull brown colour. This superficial layer is no doubt the chief assimilating tissue. Within this we come to large parenchymatous cells, less deeply coloured, and as we approach the middle of the thallus the cells become elongated (cf. Fig. 82). In the lower part of the plant the elongation of the central cells is so extreme that they form a tangled web of branched filaments or *hyphæ*. The elongated cells appear to serve the purpose of conducting food-substances, for they possess regular sieve-plates perforated like those of the vascular plants. Such sieve-plates occur both in the transverse and longitudinal walls of the long cells. The hyphæ of the basal disc and lower parts of the thallus are generally thick-walled. Their function is to strengthen the plant mechanically, and it will be found that the lower portion of the thallus, where these hyphæ are most abundant, offers the greatest resistance to tearing.

Although it is thus possible to distinguish several systems of tissue in mature parts of the thallus, it must not be supposed that the different layers remain permanently distinct one from another. As a matter of fact the cells of one system constantly give rise to those of another. For example, the outermost assimilating cells divide tangentially, and the inner daughter-cells, thus cut off, contribute to the more internal cortical parenchyma, which appears to discharge the function of storing the assimilated food. Again, the cells of the inner cortex grow in length and may give rise to hyphæ, thus adding to the bulk of the central tissue. The elements which correspond to sieve-tubes may subsequently undergo further elongation, thicken their walls, and assume the part of mechanical elements.

In the older parts of the thallus the assimilating layer dies away, and is replaced by a secondary tissue answering the same purpose, produced by the repeated divisions of the underlying cortical cells. Thus we see that the various kinds of tissue, which appear so distinct when fully developed, can be derived the one from the other. Owing to the cell-formation in the superficial and other layers, and to the growth of new hyphæ which insinuate themselves among the old, a constant increase in the thickness of the thallus goes on.

True starch is not formed as the result of assimilation ; it is represented by another carbohydrate occurring in granules, but not capable of being stained by iodine.

The growth in length of the thallus takes place with the aid of a definite apical cell, by the division of which all the tissues originate. Such a cell is situated at the apex of each branch, and lies at the bottom of a slit-like depression, which can be detected on examining the tip of the branch with a lens. When branching is to take place, the apical cell simply divides into two by a wall down the middle ; so here we have a true instance of dichotomy, the two branches being on exactly equal terms from their first origin.

Adventitious shoots arise chiefly as the result of accidental injuries. The internal cells lying beneath the wound are stimulated to renewed growth and division, and give rise to a new thallus which may become an independent plant. Thus the Alga ensures itself against permanent loss in consequence of violence, for the damaged parts are replaced by fresh and vigorous shoots. The formation of these new growths, if they become separated from the old plant, affords a simple but effective means of vegetative propagation. The regular reproduction, however, of *Pelvetia*, in common with all other Fucales, is exclusively by the sexual method.

2. Reproduction

As we have seen, each *receptacle* or enlarged end of a branch contains numerous *conceptacles* or cavities in which the reproductive organs are placed. When ripe, the conceptacle is an approximately spherical hollow, communicating with the exterior by a narrow pore (see

Figs. 81, B, and 82). It arises as a depression on the surface of the thallus, the part which is to form the bottom of the hollow becoming arched over by the greater growth of the surrounding tissue. The formation of the cavity is, however, in some cases at any rate, helped by the breaking down of a central cell, so as to leave a gap in the tissue.

In *Pelvetia* the conceptacle contains organs of three kinds: (1) sterile hairs or *paraphyses* arising all over the wall of the conceptacle, with their free ends converging towards the pore (see Fig. 82); (2) branched filaments, on the lower parts of which the *antheridia* are borne; (3) the sessile *oögonia*, which are placed chiefly in the lower half of the conceptacle (see Fig. 82). This species is therefore hermaphrodite, for the organs of both the sexes occur in the same conceptacle. In most other Fucales the plants are diœcious, all the conceptacles of each individual containing organs of the same sex.

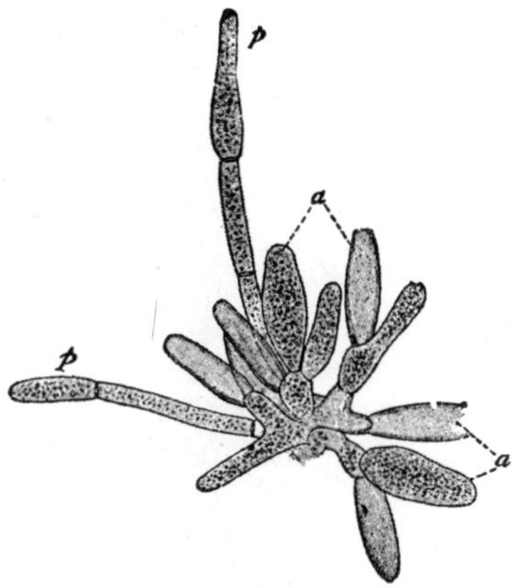

FIG. 83.—Antheridia of *Pelvetia*. *p*, *p*, hairs; *a*, *a*, antheridia, some of which are already emptied. Magnified about 260. (After Thuret and Bornet.)

The antheridia arise as single cells, borne laterally in small numbers near the base of the branched filaments (see Fig. 83, *a*). At first the antheridium, like other cells in Fucales, contains a single nucleus. This undergoes repeated division into two until the total number of sixty-four nuclei is reached. Each of these nuclei now becomes the centre of a distinct cell, the contents of the antheridium dividing up simultaneously into as many cytoplasmic bodies as nuclei are present. These bodies become *spermatozoids*, each of which consists of cytoplasm, a nucleus, and a plastid ; the latter, however, contains but little colouring matter. The spermatozoids are of oval shape, and bear two lateral cilia of unequal

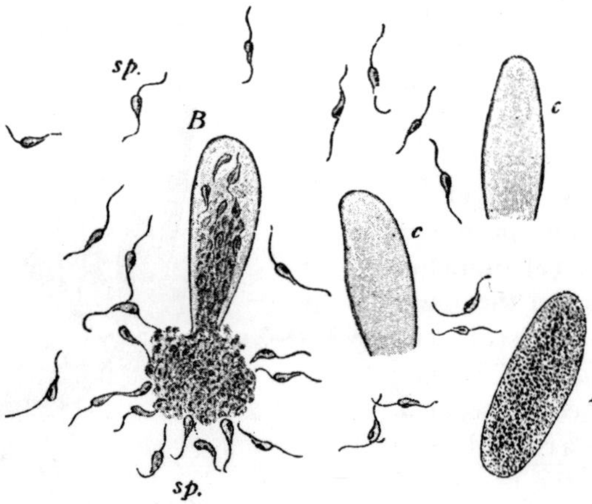

FIG. 84.—Spermatozoids of *Pelvetia*. *A*, unripe antheridium, already freed from outer membrane ; *B*, antheridium opening to emit the biciliate spermatozoids (*sp*) ; *c*, *c*, empty membranes. Magnified about 450. (After Thuret and Bornet.)

length (see Fig. 84, *sp*). The antheridial wall is double, and when the organ is ripe the outer membrane bursts at the top, and the whole contents, which may be already developed into spermatozoids, but are still enclosed within the delicate inner cell-wall, are expelled.

The oögonia are single cells, of large size, seated on the tissue at the base of the conceptacle (Figs. 82 and 85) ; they contain a great many plastids, and turn a very dark colour as they become ripe. In each oögonium there is at first a single nucleus, which divides successively into two, four, and eight. The cell contents, however, divide into two cells only (see Fig. 85) by a transverse wall. Each of the two daughter-cells has one central nucleus; the remaining six nuclei are expelled from the ova, together with a little cytoplasm. The bodies seen close to the transverse septum in Fig. 86, A, are some of these rejected nuclei. In *Fucus* itself all the eight nuclei are utilised, for the oögonium there divides into eight cells; in another genus (*Ascophyllum*) four cells are formed and four nuclei rejected, while in other genera no division of the cell contents takes place, and of the eight nuclei formed in the oögonium only one serves as the functional nucleus of the ovum.[1]

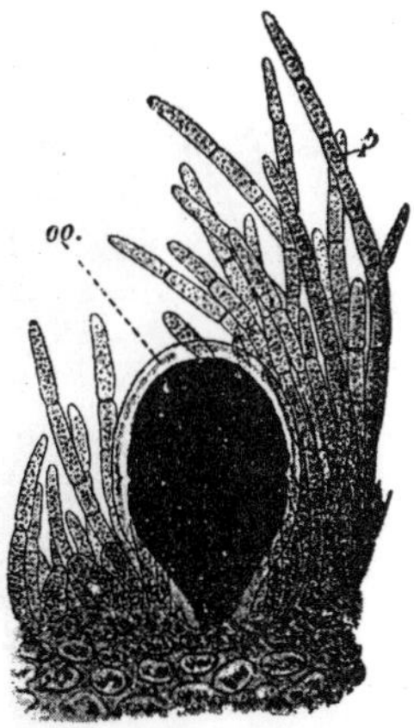

FIG. 85.—Oögonium (*oo*) of *Pelvetia*, already divided to form the two ova. *p*, paraphyses. To the left of the oögonium some antheridia are also shown. Magnified about 110. (After Thuret and Bornet.)

In *Pelvetia*, then, the oögonium forms two ova which are surrounded by a thick and very gelatinous cell-wall showing three distinct layers (see Fig. 86). When ripe, the outer layer of the oögonial wall gives way, and the two ova, surrounded by the thick mucilaginous inner layer of the cell-wall, are set free.

The expulsion of the spermatozoids and ova from the

[1] The fact that a number of nuclei are always formed suggests that the oögonium of the Fucales may have been originally a structure of the nature of a sporangium.

13

conceptacles generally takes place in *Pelvetia* when the tide has gone down and left the plants exposed to the air, though it may also go on under water. The cavity of the conceptacle is full of mucilage secreted by the hairs which line it. The surrounding tissue presses on the

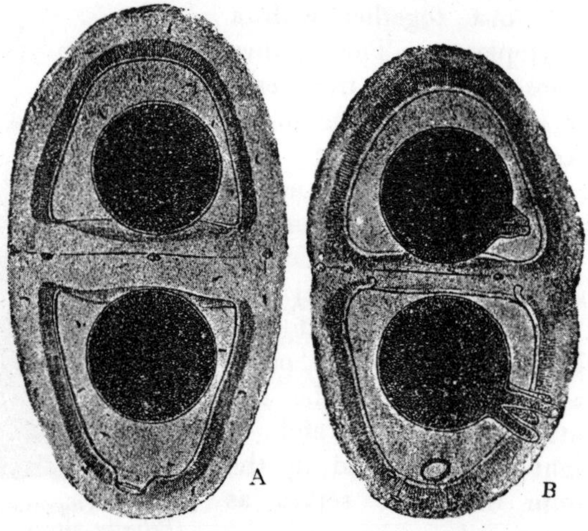

FIGS. 86, A and B.—A, oöspores of *Pelvetia* beginning to divide immediately after fertilisation. They are still enclosed in the gelatinous inner wall of the oögonium. The small bodies in the gelatinous mass are spermatozoids; the larger bodies near the septum are the rejected nuclei of the oögonium. B, later stage. The oöspores are dividing to form many cells, and are sending out rhizoids. Magnified about 120. (After Thuret and Bornet.)

full conceptacle and forces out the mucilaginous contents through the pore; mixed with this extruded mucilage are the spermatozoids and ova. If we hang up *Pelvetia*, or some other Fucaceous seaweed, freshly taken from the water, we can see the little slimy drops appearing at the pores of the conceptacles; and these drops, if examined under the microscope, are found to contain spermatozoids or ova or both, according to the diœcious or hermaphrodite character of the species. In *Pelvetia* we should find both organs in the same drop. The spermatozoids are expelled while still enclosed in the inner antheridial membrane; the ova are in pairs, held together by the inner oögonial wall. There are many packets of spermatozoids and many pairs of ova sent out from each conceptacle. As the spray dashes up over the plants with the returning tide, their reproductive cells are washed down from the receptacles, sometimes on to the rocks, sometimes only on to the lower part of the plant itself, where they often come to rest in the groove of the thallus.

It is a constant rule among the Fucales that fertilisation takes place outside the parent plant, after the sexual cells have been set free. The remaining antheridial membrane bursts after expulsion from the conceptacle, and the spermatozoids are at liberty to swim off by means of their cilia (Fig. 84, *B*). In this species, however, the ova remain enclosed within the soft mucilaginous membrane derived from the oögonial wall. In most Fucales this is not the case; the ova are set free as bare masses of protoplasm; the peculiar state of things in *Pelvetia* probably has to do with the long exposure to the air; the mucilaginous envelope protects the protoplasm within from danger of drought.

The spermatozoids during their movements come across the pairs of ova, and swarm around them in large numbers. Some of them make their way into the mucilage, and penetrate to the cytoplasm, which it seems is generally approached at the side, where the two ova are in contact. It has been shown that ultimately only a single spermatozoid succeeds in entering the cytoplasm, and making its way to the nucleus of the ovum. The details of the process have now been thoroughly worked out, and the fusion of the small male nucleus with the large nucleus of the ovum observed, as shown in Fig. 87, which represents the act of fertilisation in another member of the Fucales. The proof of sexuality has also been afforded by experiment. If the ova are kept apart from the spermatozoids (as can be easily done in the case of the diœcious species) they soon perish, making perhaps some slight attempt at germination, which comes to nothing. If, however, the spermatozoids have access, fertilisation occurs and the result is quite different. The ovum now surrounds itself with a cell-wall of its own, and after a few hours begins to divide. The direction of the first cell-wall formed across the fertilised ovum is said to be determined by light, and to be always at right angles to the incident rays. Other cell-walls follow, and soon the oöspore (which here does not pass through a resting-stage) is converted into a little mass of tissue, but without at first changing its external form. After eight or ten days, several rhizoids begin to grow out at the end away from the light (see Fig. 86, B). They burst through the oögonial wall, which has lasted all this time, and attach the embryo to the rock or whatever else it may be lying

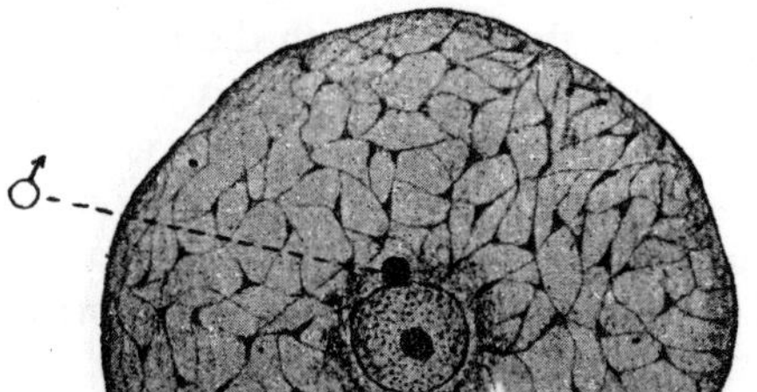

FIG. 87.—Ovum of one of the Fucales (*Ascophyllum nodosum*), seen in section at the moment of fertilisation. ♂, small male nucleus of a spermatozoid, which has traversed the cytoplasm, and is now in contact with the large nucleus of the ovum. The cytoplasm of the ovum shows a distinct foam-like structure. Magnified about 650. (After Farmer.)

upon. The upper part of the embryo now elongates and becomes first cylindrical and then flattened at the free end ; a depression soon arises at its apex, in which a definite apical cell appears, and now we have in all essentials a new *Pelvetia* plant fairly started in the world.

In these Algæ the result of fertilisation is a plant just like the parent. There is no kind of asexual reproduction, and therefore no possibility of any real alternation of generations.

We see, then, that in the Order Fucales we have the simplest possible life-history combined with a very perfect form of sexual reproduction. The plants are altogether very highly organised, as shown not only by the elaboration of their reproductive arrangements, but by their whole structure. Some members of the Order bear perfectly distinct and well-formed leaves, and rival the flowering plants in the perfection of their external morphology. This is the case notably in the genus *Sargassum*, of which everybody has heard, from the fact that the plant forms prodigious floating masses in the mid-Atlantic, giving its name to the well-known Sargasso Sea, which is many thousands of square miles in extent. In anatomical complexity also we have seen that even *Pelvetia* approaches the level of the vascular plants. It is well to realise at once that Algæ may attain a very high organisation. On their own independent lines some of them have reached a degree of differentiation not much inferior to that of the higher land plants, with which, however, they have no direct relationship.

C. THE RED ALGÆ (*Florideæ*)

The great majority of the red group of Algæ are seaweeds, though some genera are limited to fresh-water streams. The marine Florideæ, though they do not reach the great dimensions of some of the brown sea-weeds, are well known to every observer, owing to their beauty of colour and form, and are always especially favourite objects with collectors at the seaside. Most of them flourish rather low down on the shore, especially frequenting the sides of deep rock-pools, while many are only found growing below low-water mark.

In the majority of the Florideæ a red (phycoerythrin) and a blue (phycocyanin) pigment, soluble in fresh water, accompany and usually disguise the green colour of the plastids. The chlorophyll itself is similar to, if not absolutely identical with, that of the higher plants. The shade of colour, produced by the combination of these pigments, varies much in different species, and in different conditions of the same plant. Sometimes a bright rose colour is the result, sometimes a rich purple, sometimes a reddish brown, while in a few cases so little red colouring matter is developed that the green pigment becomes externally visible.

All the plants which have any good claim to rank as Florideæ agree closely in their minute organisation, methods of reproduction, and life-history; so that the whole group is a manifestly natural one, though in the degree of complexity of the thallus, and in the elaboration of the reproductive processes, there is a great range of variation. The Florideæ stand almost completely isolated in the vegetable kingdom as at present known to us. They form a perfectly well-characterised group, which attains a remarkably high development on its own lines, especially as regards the process of sexual reproduction. We shall only be able to describe one representative, and that one of the simpler members of the Group as regards vegetative structure.

TYPE XVII

Callithamnion corymbosum

1. Structure

The form of the thallus among Red Algæ is subject to very great variations; in some the thallus is finely, in others more coarsely filamentous; in some, again, it is of stouter build and cylindrical form, while in others the whole plant assumes a flattened leaf-like shape, or consists of a short axis, bearing leaf-like appendages.

The type which we have chosen is one of the simplest, the whole plant consisting of a repeatedly branched filament, the main axis of which is comparatively thick, the successive branches becoming more and more slender, while the ultimate ramifications terminate at the tips in delicate hairs (see Figs. 88 and 89). The filament is at first only one cell in thickness throughout. In the lower part of the thallus, however, a peculiar kind of secondary cortex is formed, as the plant grows older; the basal cell of a lateral branch gives rise to delicate filaments, which grow in a downward direction, attach themselves closely to the membrane of the main axis, and eventually form a complete coating over it. This mode of forming a cortex by means of adherent branches is by no means uncommon among filamentous Algæ, both of the red and brown divisions.

Each cell of the thallus contains, in addition to the colourless cytoplasm, a number of plastids (the bearers of the combined pigments) and, at least when young, a single nucleus. The cells communicate with each other by pits in their transverse walls; the pit-membrane is covered on either side with a pad of callus like that in sieve-tubes (see Part I. p. 58). Fine strands of cytoplasm extend through the callus and pit-membrane, thus connecting the contents of the adjacent cells. We see, then, that the protoplasm is continuous in these Algæ, as well as in higher plants.

The growth of the thallus goes on entirely at the apex of the various branches ; each branch terminates in an apical cell, which divides by transverse walls to form the successive segments composing the filament. When a fresh branch is to be formed, an oblique wall is produced in a segment which has just been cut off from the apical cell. By the oblique wall two unequal cells are separated, the smaller of which grows out and becomes the apical cell of the new branch. When a branch terminates in a long colourless hair, its growth is at an end ; these branches therefore are of limited length, and in this respect resemble leaves, while other branches retain their apical cell, and are thus capable of indefinite growth. Hence the thallus of a *Callithamnion* comes to have a regular conformation like that of many higher plants, depending on the relative position of its unlimited and limited branches.

2. Reproduction

a. Asexual

The reproduction of *Callithamnion* and of most Florideæ is of two kinds, asexual and sexual. The asexual reproductive cells are called *tetraspores*, because they are almost always produced four together in one sporangium. In this case the tetrasporangia occur on very short lateral branches (see Fig. 88, A), the end cell of which swells up and becomes filled with exceptionally abundant protoplasm and plastids, assuming a very dense red colour. The contents then divide up into four spores, arranged in this particular plant in a tetrahedron (see Fig. 88, A).

The membrane of the sporangium ruptures, and the

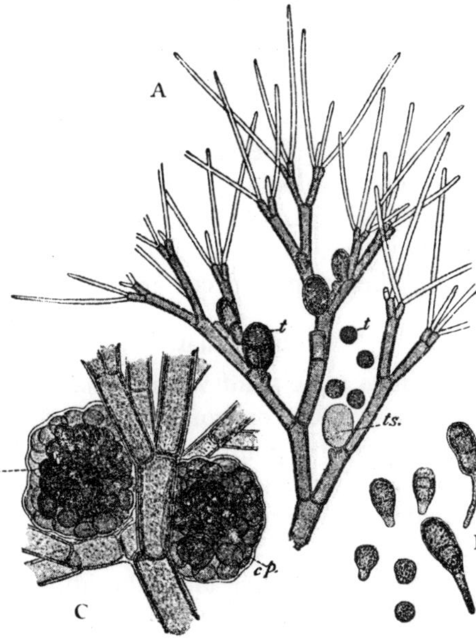

FIG. 88.—*Callithamnion corymbosum*. A, part of an asexual plant, bearing tetrasporangia ; the branches terminate in long colourless hairs ; *t* (on left), tetrasporangium containing the tetrahedrally arranged spores ; *ts*, empty sporangium, from which the tetraspores (*t*) are just discharged. B, germinating tetraspores. C, part of a female plant, bearing two cystocarps (*cp*), both of which are the product of a single procarp. Magnified about 80. (After Thuret and Bornet.)

tetraspores are set free ; when they escape they are without any cell-wall, each spore containing a single nucleus. These spores have no cilia, and usually appear to be quite incapable of any spontaneous movements ;

no doubt they are disseminated by currents in the water. When a tetraspore comes to rest it forms a cell-wall and germinates, sending out a rhizoid at one end and dividing up at the other to form a filament, as shown in Fig. 88, B. This mode of propagation is almost universal among red seaweeds, but the position of the tetrasporangia and the arrangement of the tetraspores in each vary greatly.

Many Florideæ are diœcious, and as the tetraspores also are produced on distinct plants, we commonly have three forms of each species,—asexual, male, and female. Sometimes, however, as has occasionally been observed in our type, all these organs occur on the same individual.

b. Sexual

If we examine a male plant we find that the antheridia occur in dense clusters on some of the thallus-cells—usually just below the point where a branch is given off (see Fig. 89). Each cluster is really a little system of densely crowded and very short branches, all springing from the same point. Each terminal cell of all these crowded branchlets becomes an antheridium, and there are so many of these that they form a continuous mass, quite hiding the short branches on which they are borne. It is often easy to recognise the clusters of antheridia in red seaweeds with the naked eye, for they have no pigment, and so appear as white patches on the red thallus. Every terminal cell of the cluster, then, is an antheridium. Its contents round themselves off, becoming free from the cell-wall, which splits open at the end, often detaching a little lid. Then the cell-contents, which have a single nucleus, escape through the opening; they have only a cytoplasmic membrane at first, but no cell-wall. Thus each antheridium produces a single male cell, which in this case is called a *spermatium*. It has no cilia (in fact cilia are altogether unknown among red seaweeds), but is borne passively along by the movement

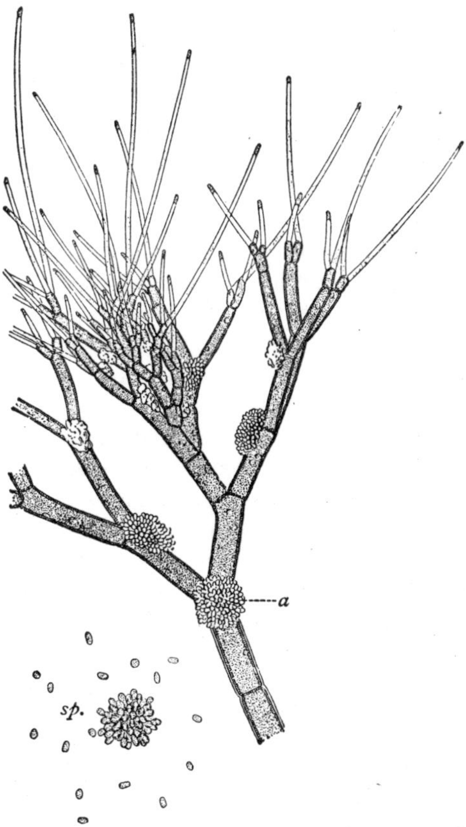

FIG. 89.—*Callithamnion corymbosum*; part of a male plant bearing the clusters of antheridia (*a*). Magnified about 150. *sp*, detached group of antheridia surrounded by free spermatia. Magnified about 240. (After Thuret and Bornet.)

of the sea. Often, after an antheridium has discharged its contents, the cell next below grows up into the empty cavity, and thus forms a new antheridium inside the membrane of the old one. The spermatia are excessively minute, not more than $\frac{1}{200}$ millimetre in diameter.

We see, then, that the production of the male cells is a fairly simple process ; the female structures, however, are much more complicated, and unlike anything we have met with, so far, in any of our types. The whole apparatus destined to form the fruit constitutes a special branch borne laterally on an ordinary cell of the thallus (scc Fig. 90). The fertile branch, which is called the *procarp*, consists in *Callithamnion* of six cells, of which four form a curved central filament, while the other two are situated laterally, one on each side. The four-celled central filament arises from one of the two lateral cells. The uppermost cell of the central filament (see Fig. 90, *a*) is called the *carpogonium*, and is prolonged into a slender hair of relatively great length ; this is the *trichogyne* (Fig. 90, *t*), or receptive organ. The three cells next the carpogonium remain small, and with the carpogonium form the carpogonial branch ; the two lateral cells are the *auxiliary mother-cells*, from which later are produced the *auxiliary cells* which contribute in a very important way to the formation of the fruit. The description of Fig. 90 should be carefully studied, to render the complicated arrangement intelligible. The carpogonium has a single nucleus ; the trichogyne which forms the upward prolongation of the same cell has, in this case, no nucleus of its own, but contains a strand of cytoplasm continuous with that of the lower part of the cell ; the outer layers of its membrane are gelatinous.

In this condition the procarp is ready for fertilisation. The long trichogyne is specially adapted for receiving the spermatia. The young fruit is generally placed in a sheltered part of the thallus (as, for example, in this case, among the densely crowded branches of the bushy stem), where its future development is most secure. In such a position, however, it is not readily accessible to the male cells, and consequently we always find in connection with it the hair-like trichogyne, which projects far out towards the exterior, and thus reaches the exposed part of the thallus to which the spermatia are likely to be conveyed by the chance currents of the surrounding water.

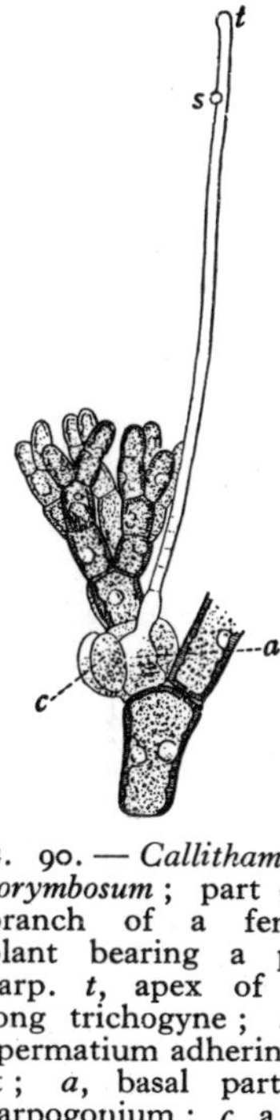

FIG. 90. — *Callithamnion corymbosum*; part of a branch of a female plant bearing a procarp. *t*, apex of the long trichogyne; *s*, a spermatium adhering to it; *a*, basal part of carpogonium; *c*, a cell of the carpogonial branch. The two cells showing to the extreme right and left are the auxiliary mother-cells, which later cut off the auxiliary cells. Magnified 250. (After Thuret and Bornet.)

When a spermatium happens to reach the trichogyne it adheres to its gelatinous cell-wall, and is thus held fast (see Fig. 90, *s*). In the meantime the spermatium has formed a membrane round its protoplasm. At the point of contact between spermatium and trichogyne the cell-walls are absorbed, and so the contents of the male cell are enabled to enter the receptive organ. All the parts concerned are very minute, but the details of fertilisation have now been completely followed in certain instances. This was first accomplished in *Nemalion*, a genus rather simpler in its arrangements than our type. The nucleus of the spermatium travels down the trichogyne and fuses with that of the carpogonium. So far, therefore, as the act of fertilisation is concerned, the Florideæ do not differ from other sexual organisms.

The result of fertilisation, however, is not the formation of a single oöspore, but the development of a whole fruit. The fertilised carpogonium is cut off from the trichogyne by a plug of cell-wall, and each auxiliary mother-cell divides unequally by a transverse wall, the upper larger portion constituting the auxiliary cell. The carpogonium divides into a row of four cells and the two end-cells come into close contact with the auxiliary cells on either side; there is an actual union of the contents of the carpogonial cells with the respective auxiliary cells, but the nuclei do not fuse. Each auxiliary cell now divides by a transverse wall (see Fig. 91, *c*); the upper of the two cells in each case becomes a placenta which gives rise to the spores. *Callithamnion* is different from most of the simpler red seaweeds in so far as it regularly forms *two* fruits from each procarp, whereas its near allies usually form one only.

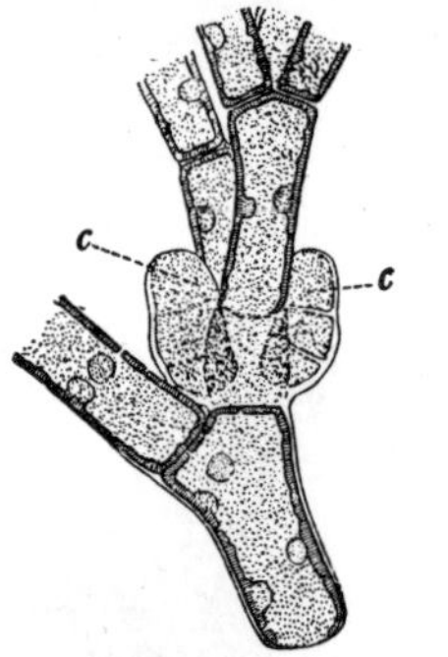

FIG. 91.—*Callithamnion corymbosum*; part of a branch of a female plant, showing the earliest stage of development of the cystocarps (*c*, *c*) after fertilisation. They arise by the growth and division of the two auxiliary cells. The lower cell, formed by the division of the auxiliary mother-cell, can be seen at the base of each group. Magnified 250. (After Thuret and Bornet.)

Each placenta grows out into a dense tuft of threads, groups which divide repeatedly, so that eventually two large groups of cells arise, one on each side of the filament (see Fig. 88, *C*). The groups are really built up of a system of very short and densely crowded branches, those of each cluster all springing ultimately from the placenta belonging to it. The whole fruit is enclosed in a gelatinous envelope, and nearly all the cells of the group give rise each to a *spore*, called for distinction a *carpospore*, as it forms part of the sexually produced fruit, or *cystocarp*. When ripe, the envelope of the fruit bursts and the spores are set free from their membranes. They are large, uninucleate, deeply pigmented cells, destitute at the time of their escape of any cell-wall.

The main points in the development are : (1) fertilisation by means of a special receptive organ, or trichogyne ; (2) union of the cytoplasm of the fertilised cell with that of neighbouring cells ; (3) the formation, as the result of this union, of two complex fruits, producing a great number of spores.

Callithamnion occupies an advanced position among the Florideæ as regards the complexity of its spore-formation. The trichogyne is common to all Florideæ, but some few members of the group (*e.g. Nemalion* and the fresh-water genus *Batrachospermum*) are simpler, in that the carpogonium directly gives rise to the spores, without any preliminary cell-fusion. In many red seaweeds, moreover, the fertilised carpogonium grows out into a number of threads which undergo repeated cell-fusions with auxiliary cells which in such cases occur on more or less distant branches, with the result that a number of fruits may be formed in consequence of a single act of fertilisation ; these fruits often arise at a considerable distance from the directly fertilised cell.

In some respects the process of sexual reproduction in the Florideæ is more complex than in any other plants. It offers the advantage that a single spermatium, if it once reaches a trichogyne, may ensure the production of a very large number of spores. In many cases a further complication is due to the fact that a multicellular envelope grows up around the spores. To the frequent presence of such an envelope, the sexually produced fruit owes its special name of cystocarp.

In so far as the result of fertilisation is the production, not of an oöspore, but of an entire fruit, there is a certain analogy between Florideæ and Bryophyta; but in the former the fruit is always in organic connection with the sexual plant, and can therefore scarcely be regarded as a new generation, though it has sometimes been spoken of as the 'sporophyte,' whereas in the Mosses the sporogonium always remains distinct from the gametophyte, though dependent upon it. It is not probable that the Florideæ have any direct relation with the Mosses, or with any of the higher plants. At the same time it is quite likely that the tetraspores of the Red Seaweeds may really be homologous with the spore-tetrads universally present in the Archegoniate series.

The carpospores germinate in precisely the same way as the asexual tetraspores, but give rise to plants bearing tetraspores. Red seaweeds are difficult to cultivate successfully, and our knowledge of their life-history is still limited; there is undoubtedly a regular alternation of asexual (tetrasporic) and sexual individuals in many species, although this is not invariably the case.

D. THE CYANOPHYCEÆ

TYPE XVIII

Nostoc

There remains to be considered a group of rather obscure plants of simple structure, which resemble the Algæ in their habit and mode of life, and are therefore best described in this place, though their real relationships are open to doubt. Of the Cyanophyceæ some are terrestrial, some aquatic, occurring both in fresh water and in the sea. Many of the Cyanophyceæ are bluish-green in colour, hence these organisms are sometimes called the Blue-green Algæ. The representative of the group which we have chosen—a species of *Nostoc* (see Fig. 92)—is filamentous. The threads are associated in colonies held together by the soft gelatinous outer walls of the cells and by mucilage secreted by the cells. Such colonies often form conspicuous masses of bluish-green jelly on poor ground, especially in wet weather. Within the mass the filaments wind about in every direction. The cells of which they are made up are rounded, so as to give the whole thread a beaded appearance.

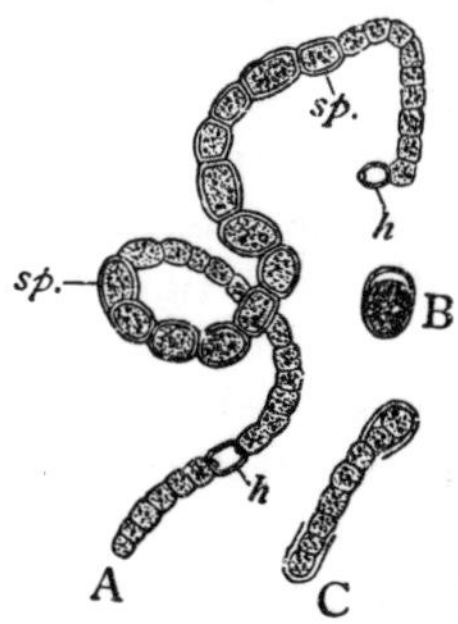

FIG. 92.—*Nostoc Linckii.* A, part of a filament; *h*, *h*, heterocysts; *sp*, *sp*, spores. B, isolated spore beginning to germinate. C, young filament formed from a spore, the burst cell-wall of which is shown at the ends. Magnified 470. (After Bornet.)

The ordinary vegetative cell has a thin inner cell-wall, which is alone visible in Fig. 92, A, the confluent gelatinous layers scarcely showing under the microscope owing to their transparency. The interior of each cell is full of protoplasm, which appears to be coloured throughout its whole mass. No definite plastids have been found to exist in Cyanophyceæ, not has the presence of a nucleus been finally determined, though in certain cases a colourless central body certainly exists, which resembles a nucleus in some of its characters and reactions. In the protoplasm are numerous granules.

Many of these plants float on the surface of water, where they sometimes appear rather suddenly in vast quantities in ponds and lakes, covering many acres with a bluish-green scum. It is said that in these floating species *gas-vacuoles* are present in the cells, that is to say, little cavities in the protoplasm containing a gas, the nature of which has not been determined. These gas-vacuoles appear to be important, as they make the plant lighter and enable it to float.

The apparent simplicity of the histological structure, due to the want of well-defined nuclei and plastids, is the chief reason why the Cyanophyceæ are often separated from the Algæ. They certainly show no near affinity with other Algal groups. The colouring matter appears to be a mixture of substances consisting of blue-green, yellow, red, and pure green constituents. The tint varies greatly in different forms, but we never find the pure green of chlorophyll. In *Nostoc* the filament is interrupted at intervals by larger cells with better defined cellulose walls; they usually appear almost devoid of contents and colourless, but their function is at present entirely obscure. These cells are called the *heterocysts* (Fig. 92, A, *h*,) and are characteristic of *Nostoc* and its nearer relations. Sometimes the filaments break across at the heterocysts, and the short rows of living cells between them become isolated. These detached filaments (called the *hormogonia*) are capable of creeping movements, though *how* they move is quite unknown. They escape from the gelatinous mass, and start new colonies for themselves.

This is one mode of propagation. Another is by means of resting-spores, formed directly from some of the vegetative cells, which grow larger than the rest, accumulate more abundant protoplasm, and surround themselves with a thick cell-wall (Fig. 92, A, *sp*). The spores (Fig. 92, B) can pass through a resting-stage, and endure drought; when water is supplied they germinate, forming new filaments (Fig. 92, C).

Such is the simple history of *Nostoc*. Neither in this genus nor in any of the blue-green Algæ has any kind of sexual reproduction been observed. The plants of this class must rank, according to our present knowledge, as among the lowest members of the vegetable kingdom, the only others which are equally simple being the Bacteria (to be subsequently described).

CHAPTER IV

THE FUNGI

THE Fungi are an immense group—by far the largest of all the cryptogamic Classes. Up to the present time about 50,000 species have been described. The whole of this vast mass of most heterogeneous forms is distinguished by one physiological character—the absence of chlorophyll. Hence all Fungi alike are incapable of assimilating their carbonaceous food from the carbon-dioxide of the atmosphere; they must obtain it ready made, as it were, from other sources. So far as carbon-compounds are concerned, Fungi are entirely dependent on *organic* food. This they obtain either directly from other living creatures, on which they prey, or from dead organic substances produced by living organisms. In the former case we call them ***parasites***, in the latter ***saprophytes***.

Parasitic and saprophytic plants wholly or nearly destitute of chlorophyll occur in other classes of the vegetable kingdom, as members of very diverse families. Thus among flowering plants, for example, we have the Dodder (*Cuscuta*) and the Broomrape (*Orobanche*) as parasites; the Bird's Nest (*Monotropa*) and the Bird's Nest Orchid (*Neottia nidus-avis*) as saprophytes. In all these cases, however, the parasitic or saprophytic forms are near relations of normal chlorophyll-containing plants, and we attach no great systematic importance to the change in their mode of life. Among Fungi, on the other hand, there is a real analogy, if not relationship, throughout the entire class (if we leave a few doubtful cases out of consideration)—so that here the common physiological character coincides with a common type of structure. We must not, however, suppose that all Thallophyta which lead the life of parasites or saprophytes are necessarily Fungi. Towards the close of this book we shall have to consider two such groups which cannot be classed under this head.

Many Fungi are of the greatest practical interest, though chiefly in a disagreeable way. Very many of them are destructive parasites, causing the worst diseases of our field and garden crops and of forest trees. We may mention the rust, smut, and bunt of wheat, potato blight, and apple scab, to which innumerable others might be added. Others, such as the dry-rot fungus, do harm by injuring timber in buildings; others, again, destroy articles of food, which are constantly attacked by "mould." Hence, Fungi have been more studied from a practical point of view than any other Cryptogams, and a vast mass of knowledge has now been accumulated as to their physiology and mode of life. Our own point of view is chiefly a morphological one, and we have chosen the few types which we have space to describe, in order to illustrate some of the most striking facts in the comparative structure and life-history of certain of the more important groups.

It must not be supposed that Fungi are altogether to be regarded as injurious to the higher creatures. Not to mention, what everybody knows, that several of the larger kinds are exceedingly good for food, we may point out that the saprophytes, at any rate, do good service by causing decay, and so ridding the world of the useless remains of dead animals and plants. Masses of dead material would otherwise accumulate to such an extent as to interfere seriously with the life of succeeding generations. Fungi and other saprophytes (notably the Bacteria) bring about the decomposition of dead organic matter, use the products for their own nutrition, and ultimately convert its substance into simple inorganic bodies (such as ammonia and carbon-dioxide), thus rendering it available for the nutrition of green plants, and, through them, for the support of other organisms.

We will begin our illustrations of the Fungi with two examples of a simple type, representing a group which is of special scientific interest, from its evident analogy with certain of the Algæ.

TYPE XIX

Pythium de Baryanum and Peronospora Parasitica

A. PYTHIUM DE BARYANUM

We have chosen as our first type of Fungi a genus which much reminds us of the Algæ, showing an unmistakable analogy with *Vaucheria.* Some species of *Pythium* are parasitic on seedlings, and often do great havoc among them, especially if the seed-beds are kept too damp, and not sufficiently exposed to air and light. One of the commonest species, *P. de Baryanum*, can be obtained almost with certainty by growing Cress seedlings in soil under a bell-glass, and giving them an excessive amount of water ; but, unfortunately, the parasite appears often enough when it is not wanted. The disease caused by it is well known to gardeners as the " damping off " of seedlings. The stem of the seedling when attacked by the Fungus soon tumbles over on to the ground. This is because the outer tissues of the stem, at the part where it gives way, have been so much damaged by the parasite that the stem has not the strength to stand upright. The fallen plants lose their colour and soon completely rot away.

1. Structure

In its vegetative condition, *Pythium* consists of long, fine, irregularly branched filaments or *hyphæ*; the latter name is given to the filaments of Fungi in general. These hyphæ are for a long time without any transverse walls; they are in fact *non-cellular*, just like the filaments of a *Vaucheria*. The inside of the hypha is occupied by vacuolated cytoplasm, in which very numerous nuclei are embedded. Unlike *Vaucheria*, however, *Pythium* has no chlorophyll and no plastids. Neither is starch formed, either in this genus or in any other Fungus. There is the further fundamental difference that in no Fungus does the cell-wall consist of cellulose; its substance contains nitrogen, and resembles the *chitin* of some animals.

It is usual to speak of the whole vegetative body or thallus of a Fungus as the *mycelium*. The mycelium of *Pythium* penetrates the tissues of its victim or "host," and spends most of its vegetative life within them. A hypha can make its way into the stem either by way of a stoma or by boring directly through the outer cell-walls; it goes on growing and branching inside the host plant, where it is not confined to the intercellular spaces, but can enter the cells themselves. Thus the whole plant comes to be infected, and is traversed throughout by the branched mycelium of the parasite, which lives at its expense.

In the First Part of this book (p. 215) we learnt that there are chemical bodies in plants called *enzymes* which have the power of changing the constitution of other organic substances, converting solids into liquids, and indigestible substances into such as are available for nutrition. The example we specially mentioned was *diastase*, which converts starch into sugar, but numerous other *enzymes* also occur in plants. Now parasitic Fungi have the power of secreting *enzymes*, which play a very important part in bringing their victims into subjection. The advancing tip of a hypha secretes an enzyme or a toxin on coming into contact with the host cells, which forthwith kills the protoplasm of the host in the immediate vicinity. The hypha grows into the cells which it has thus killed, more of the poisonous substance is secreted and kills additional host cells, and a considerable mycelium is established by the growth and branching of the infecting hypha. *Pythium* thus feeds on the nutritive substances in the host cells which it has itself killed, enzymes secreted by the mycelium probably playing an important part in rendering the food materials available to the fungus. The mode of life of *Pythium* represents a very simple and destructive type of parasitism in which the fungus kills its host immediately on coming into contact with it. Having killed its host, *Pythium* lives on the dead remains as a saprophyte.

In this way, then, the *Pythium* makes itself thoroughly at home in the body of its victim, infests it in every part, and eventually completely destroys its tissues, converting their materials to its own use. Often the hyphæ leave the host, and grow out upon the soil until they reach other victims, which they then infect. In the meantime the Fungus does not neglect to make provision for future generations. The reproduction takes place in two ways—asexual and sexual. We will first describe the former.

2. Reproduction

a. Asexual

The hyphæ which are to produce the asexual organs of reproduction grow out from the host into the open air. They there form a number of spherical *sporangia* which are terminal, being seated on the ends of short branches or of the main hyphæ (see Fig. 93, A). The sporangia are beaked at the apex, and, when ripe, the entire protoplasm passes out into the beak, which swells up into a bladder-like sac (see Fig. 93, B). The whole process can only go on when there is water enough to immerse the sporangia. The contents of the sac now divide up into a number of membraneless cells which become *zoospores*, each bearing two cilia. The zoospores escape and swim away through the water. After some time they come to rest, surround themselves with a delicate wall, and germinate, producing a hypha, which finds its way into a fresh seedling if opportunity offers.

This mode of reproduction, we see, is altogether that of an Alga. *Pythium*, though a Fungus, is not thoroughly adapted for growth on dry land, for its normal reproductive processes can only go on under water. This is one reason why seedlings attacked by *Pythium* are said to *damp off*, for it is when they are kept too damp that their enemy is best able to attack them ; the moisture enables the *Pythium* to spread. This method of propagation by zoospores allows of an enormously rapid multiplication under favourable conditions ; its success, however, is entirely dependent on the presence of water. It is true that only a little water is necessary, but still *Pythium* is entirely powerless to propagate its kind in this manner, under such conditions as prevail in nature when the weather is at all dry.

The great majority of the Fungi, however, are adapted to the same conditions of life as the ordinary land plants, on which so many of them are parasitic, and this implies that their reproductive bodies are fitted for dissemination through comparatively dry air. In *Pythium* and among its near allies we can trace the steps by which this adaptation to an aërial environment has been attained. In some species of *Pythium*, as, for example, in the species *P. de Baryanum*, which is so common on Cress seedlings, it sometimes happens that the sporangium does not form zoospores at all, but grows out directly into a hypha, thus starting a new plant at once, without the intervention of the active aquatic cells. Evidently this allows of propagation taking place, even though there should not be water enough to float the zoospores. The same thing happens in the closely-allied genus *Phytophthora*, to which the Fungus causing Potato Blight (*P. infestans*) belongs.

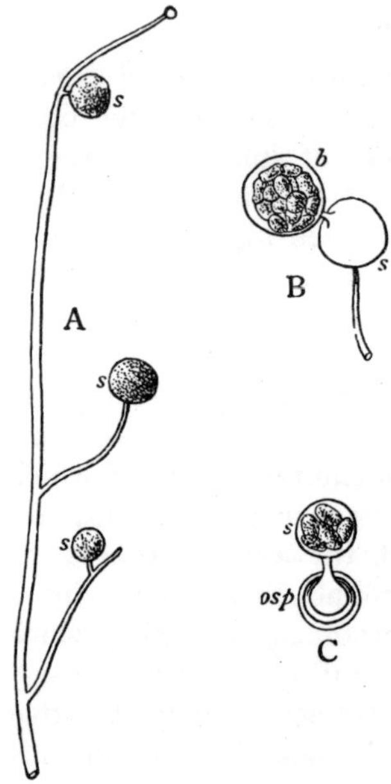

FIG. 93.—*Pythium*. A, branch of the mycelium, bearing three zoosporangia (*s*). Magnified 145. B, zoosporangium (*s*) discharging its contents (*b*), which are still enclosed in the enlarged papilla, but have already divided to form the zoospores. Magnified 145. C, germinating oöspore (*osp*) forming an asexual sporangium (*s*). Magnified 300. (After De Bary.)

Here, and also in the allied genera, *Plasmopara*, *Bremia*, and *Peronospora*, the sporangia always become detached from the hypha which bears them, before germination (whether by the development of zoospores, or by the direct formation of a hypha) takes place. The illustrations in Fig. 94 are taken from different species of these genera.

In *Pl. nivea* (parasitic on Umbelliferæ) the contents of the sporangium divide up at once into a number of

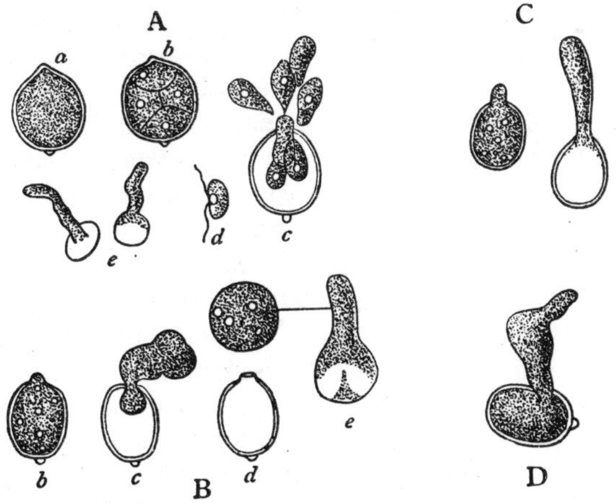

FIG. 94.—Germination of the sporangia in *Plasmopara*, *Bremia* and *Peronospora*. A, *Pl. nivea*; *a*, *b*, *c*, stages in the formation of zoospores; *d*, free biciliate zoospore; *e*, zoospores germinating. B, *Pl. densa*; *b*, commencement of germination; *c*, expulsion of undivided contents; *d*, empty sporangium; *e*, first formation of a hypha from the contents. C, *B. Lactucæ*; direct germination through the apical papilla. D, *Per. Radii*; direct germination, hypha formed laterally. Magnified 400. (After De Bary.)

biciliate zoospores, which escape by the opening of the terminal papilla (Fig. 94, A). In *Pl. densa* (Fig. 94, B), growing on Scrophulariaceæ, the protoplasm of the sporangium is expelled entire through the apical opening without dividing into zoospores; it surrounds itself with a new cell-wall, and germinates directly, to form the mycelium of the next generation. In *B. Lactucæ* (Fig. 94, C) (which infests Lettuces and their allies) a further step is taken ; the contents do not escape at all, but the sporangium simply puts out a hypha which arises at the apex. Lastly, in *Per. Radii* (occurring on flower-heads of certain Compositæ), with which the majority of species of *Peronospora* agree, the apical opening has ceased to be functional, and the hypha grows out laterally (Fig. 94, D). In these latter instances the sporangium has, in fact, become a *spore*. Such asexual spores of Fungi, germinating directly, bear the name of *conidia*.

We will now return to our type *Pythium*. Zoospores and conidia afford a rapid means of propagation so long as a plentiful supply of victims, in the shape of young seedlings, is forthcoming. *Pythium* is not, however, altogether limited to a parasitic mode of life, for if host-plants are wanting, it can live for some time as a *saprophyte* on any decaying organic matter which may happen to be at its disposal. Provision, however, has to be made for bad times when food fails altogether, or when there is not enough moisture for active growth to go on. Such contingencies are provided against by the formation of *resting-spores*, which are the result of a sexual process.

b. Sexual

The organs of sexual reproduction (oögonia and antheridia) may be produced either inside the tissues of the host plant, or on hyphæ which have grown out into the air. An oögonium arises as a spherical swelling on a hypha, and may be either terminal, as shown in our Fig. 95, or *intercalary*, *i.e.* produced at some intermediate point in the course of the filament. The young oögonium is cut off from the rest of the hypha by a transverse cell-wall, or by two, if intercalary.

Its protoplasm now separates into two parts—a

central granular portion which becomes the ovum, and a peripheral layer, lining the cell-wall, called the *periplasm*. The behaviour of the nuclei has now been made out in several Fungi of this group ; in *Pythium*, the oögonium at first contains a large number of nuclei, nearly all of which pass out into the periplasm, leaving behind, in the central mass, a single nucleus, which is the functional female nucleus, and is alone concerned in the act of fertilisation and the subsequent development.

In the meantime the antheridium is formed. It is usually a lateral, club-shaped branch, arising either from the same filament which bears the oögonium (see Fig. 95) or from a different one, and separated from the hypha on which it is borne by a transverse wall. The antheridium directs its growth towards the neighbouring oögonium, to which it closely applies itself.

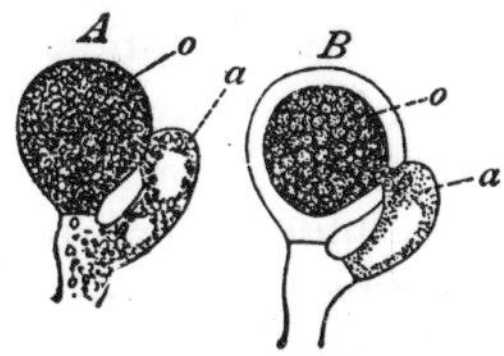

FIG. 95.—Fertilisation of *Pythium*. *A*, early stage; oögonium (*o*) and antheridium (*a*) still immature. *B*, moment of fertilisation. The contents of the antheridium (*a*) are passing through the fertilising tube, to unite with the ovum (*o*). Magnified 800. (After De Bary.)

It may be mentioned here that the mycelium of *Pythium* and its allies, which is non-cellular during its vegetative growth, generally becomes irregularly partitioned up, by a few scattered transverse walls, as the period of reproduction approaches. Previous to this the bulk of the protoplasm has travelled into the more terminal portions of the mycelium, where the reproductive cells are to be produced ; the transverse walls may serve the purpose of keeping it where it is wanted.

The protoplasmic contents of the antheridium, like those of the oögonium, undergo a severance into a central fertile portion and an external layer of periplasm, and here also it is the former alone which plays an active part in the reproductive process. There is no division into spermatozoids, and in fact these bodies are extremely rare among Fungi,—another point in which the adaptation to a terrestrial habit of life has involved the disappearance of motile reproductive cells. The antheridium sends out a short branch, the *fertilising tube*, which penetrates the wall of the oögonium, and reaches the ovum (Fig. 95). The fertilising tube opens at its end, and now the whole contents of the antheridium (with the exception of the periplasm) pass through the tube, and unite with the protoplasm of the ovum (Fig. 95, *B*). The whole process can be directly followed with ease, under a high power of the microscope, and, indeed, *Pythium* is one of the most favourable plants for the immediate observation of the fertilising act. It appears to be now established that only a single nucleus passes over with the male protoplasm, and unites with that of the ovum.

As the result of fertilisation, the ovum surrounds itself with a thick cell-wall, the outer layer of which is derived from the periplasm by which it is surrounded. The ovum has now become an *oöspore*; its contents form a quantity of oil, as a reserve of carbonaceous food, and it next passes into a period of rest.

The germination of the oöspore takes place after a long interval, and only when it is brought into contact with water. The process shows very remarkable variations, both among different species and among individuals of the same species, comparable to the variations in the behaviour of the asexual sporangium described above. In some cases the outer thick layers of the oöspore membrane are burst, and the contents, surrounded by a delicate cell-wall, grow out into a hypha, thus starting a new mycelium directly. In other cases the process begins in the same way, but the hypha at once forms a sporangium, into which the whole contents pass, dividing up into a number of zoospores (see Fig. 93, *C*). In a third mode of germination, the formation of the hypha is suppressed, and the zoospores are produced in the interior of the oöspore itself. The zoospores swim about like those formed on the vegetative plant, and on coming to rest reproduce the ordinary form of the Fungus.

These are differences to which considerable systematic importance would be attached in other groups of plants ; here the different modes of germination are not even constant for the species, but appear to depend partly on the nutrition, direct germination often taking place when food is abundant, while under less favourable conditions zoospores are formed at once, thus allowing additional chances of a suitable habitat being reached.

Pythium and its allies are more like Algæ than most other Fungi ; in fact they were once themselves placed in the former class, and it is evidently with non-cellular Algæ of the type of *Vaucheria* that they have the closest analogies. In histological structure the two genera are almost identical, if we leave out of account the chlorophyll-bodies, which *Pythium* has no need of in its parasitic or saprophytic mode of life. There are great differences, however, between the zoospores of *Pythium* and those of *Vaucheria*. In the reproductive processes there are various deviations from the aquatic type, the most important being the suppression of the spermatozoids, and the gradual replacement (only just indicated in *Pythium*, but more marked among its allies) of zoosporangia by directly germinating conidia. As we advance towards the higher Fungi we shall find the alga-like characters dropping more and more out of sight.

B. PERONOSPORA PARASITICA

1. Structure and Mode of Life

This fungus is closely related to *Pythium de Baryanum*, but its mode of life is entirely different, as it does not destroy its host. The mycelium lives a life together with the host cells, gradually withdrawing food but not killing them. Such a fungus, which can only continue to live in contact with living host cells, is said to be an *obligate parasite*. *Peronospora parasitica* is commonly found on Shepherd's Purse (*Capsella Bursa-pastoris*) and other plants belonging to the family Cruciferæ. It attacks the stem, leaves, and inflorescence. In the reproductive stage it protrudes from the host as a delicate whitish down or fluff, and hence it is sometimes called a " downy mildew." The parts of the host invaded by the mycelium are sometimes deformed. This *hypertrophy* is due to the mycelium exercising a stimulus on the growth of the host cells.

The asexual spores of *Peronospora* are conidia (see below), which are blown about by the wind. If they alight on a moist surface they germinate directly by the formation of a hypha, and if this occurs on a suitable host plant the hypha obtains entry into the tissues by growing through a stoma. Having entered the space below the stoma, the hypha advances until it comes into contact with the wall of a host cell. A portion of the hypha then puts out a protuberance which penetrates the wall of the host cell and expands within it to form a branched, sucker-like process, the *haustorium* (Fig. 95*). Food material passes from the host cell by way of the haustorium back to the hypha in the intercellular space. The hypha continues to grow, it branches, and soon an extensive mycelium is established in the intercellular spaces of the host, new haustoria being formed as growth proceeds.

The mycelium of this fungus, like that of *Pythium*, is devoid of cross-septa. As pointed out above, *Peronospora* lives a kind of common life with its host without gravely damaging it. After the fungus has spread considerably in the tissues it begins to reproduce itself.

2. Reproduction

a. Asexual

Branches arise from the mycelium near the stomata, through which they grow out. These hyphæ divide dichotomously many times on reaching the air and become *conidiophores*. The tip of each branch of a conidiophore swells up to form a single oval or nearly spherical conidium which ultimately falls off (Fig. 95*). It is in consequence of the formation of innumerable conidiophores that the surface of the host appears to be covered with a kind of down when the parasite is in the reproductive stage.

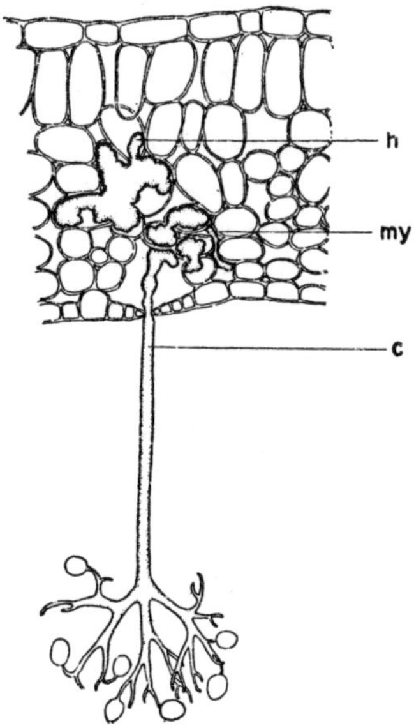

FIG. 95*. — *Peronospora parasitica*. Vertical section of leaf of Shepherd's Purse attacked by fungus. *h*, haustorium; *my*, mycelium; *c*, conidiophore. Magnified 160. (R. W. Marsh.)

b. Sexual

After the development of conidiophores and as the host tissues become senile, sexual organs, similar to those of *Pythium*, may be formed on the mycelium in the host. The process of fertilisation is the same as that in *Pythium*, but the fusion of the male and female nuclei is delayed until the oöspore wall has been formed. The oöspores, which have a brown exterior, remain embedded in the tissues until these disintegrate, when they pass into the soil. The oöspores remain dormant for a considerable time. When temperature and moisture conditions are suitable they germinate by the formation of a germ tube. The oöspores may be splashed by rain on to the surface of plants just above soil level. In *Peronospora parasitica*, sexual organs are not invariably formed in the host tissues, and the precise conditions which lead to their development are not yet clearly understood.

TYPE XX

PILOBOLUS CRYSTALLINUS

Among the Algæ we found in some of our types (as, for example, in *Œdogonium* and *Vaucheria*) a well-marked process of *fertilisation*, in which the cells taking part in the formation of the oöspore showed a distinct difference of sex. In others, however, namely, in *Ulothrix* and *Spirogyra*, union was found to take place between two essentially similar cells, each having an equal share in the act of *conjugation*, with the resulting formation of a zygospore. Both these modes of sexual reproduction are represented also among the Fungi. *Pythium* and *Peronospora* afforded examples of fertilisation; the group to which they belong bears the name of the *Oömycetes*. We are now about to describe a form in which the sexual act is one of *conjugation*; the Fungi in which this process prevails are called the *Zygomycetes*.

A large proportion, though by no means all, of the

Oömycetes are parasites on living plants or animals. Among the Zygomycetes parasitism, though it occurs, is rare ; the majority are of saprophytic habit, and to this group many of the commonest " moulds " belong.

We have chosen as our type a little Fungus which is often found growing on the dung of herbivorous animals. In spite of its disagreeable habitat, *Pilobolus*, and especially the species *P. crystallinus*, is a decidedly pretty object. The part of the plant which alone rises above the surface of the substratum and so meets the eye, consists of the stalks bearing the asexual sporangia. Each of these stalks is about a quarter of an inch high, and swells up near the top into a neat little crystalline globule, surmounted by a kind of black cap, which is the sporangium itself (see Fig. 96, A). These facts will enable us to recognise the plant. We will now consider its structure more in detail.

1. Structure

The mycelium or vegetative thallus of *Pilobolus* and its allies is made up of repeatedly branched hyphæ, which spread in all directions through the substratum. Histologically these hyphæ have the same structure as in the Oömycetes ; they are without transverse walls, at least in the vegetative condition, and their protoplasm contains very numerous small nuclei, the whole plant being a non-cellular but multinucleate organism. Thus this order betrays an unmistakable analogy with Algæ of the non-cellular type, and therefore Oömycetes and Zygomycetes are grouped together in one class, under the name of *Phycomycetes* or alga-like Fungi. The group to which *Pilobolus* belongs has, however, departed much further from the aquatic habit than the Oömycetes have, for the plants are thoroughly adapted to a terrestrial mode of life, and the power of forming zoospores is altogether lost.

2. Reproduction

a. Asexual

The most abundant means of propagation is by asexual spores formed in sporangia. For the purpose of spore-formation certain of the hyphæ grow out into the air, and assume a vertical position. The apex of the aërial hypha enlarges to form a nearly spherical sac, into which most of the protoplasm travels from below. This sac is cut off from the stalk by a transverse cell-wall, and becomes the sporangium (see Fig. 96). Its contents divide up ultimately into a large number of round cells, each of which surrounds itself with a cell-wall and becomes a spore (see Fig. 96, E). A hypha which bears a sporangium is called a *sporangiophore*. We see, then, that in this family the sporangium, instead of forming zoospores, as in the last group, gives rise to motionless spores with cell-walls. This is the typical method of asexual spore-formation in these moulds. We will now consider the special adaptations which are characteristic of our type *Pilobolus*.

The upper end of the stalk just below the sporangium swells up into a bladder much larger than the sporangium itself, and a second bladder is often formed in the lower part of the hypha (Fig. 96, B). The lower of these two bladders is separated by a transverse wall from the rest of the mycelium ; the upper is bounded above by the wall which marks off the sporangium. The whole stalk between the two transverse walls constitutes a water-reservoir, in which a high hydrostatic pressure is set up, so that drops of liquid often exude through the membrane (Fig. 96, A, *st*). Owing to this pressure, the wall bounding the sporangium becomes bulged into its cavity, forming what is here called the *columella*

(Fig. 96, E, *c*). The wall of the sporangium itself is not uniform throughout; the upper part is thickened and

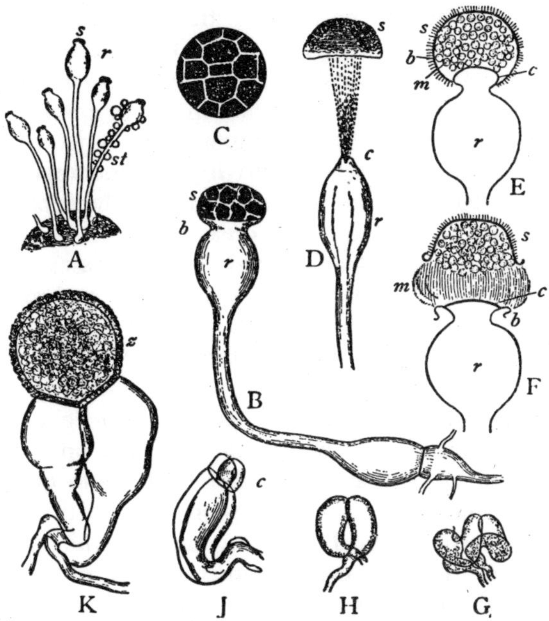

FIG. 96.—*Pilobolus*. A, group of sporangiophores; *st*, stalk; *r*, reservoir; *s*, sporangium. Magnified 5. B, hypha and sporangiophore; *r*, reservoir; *s*, *b*, sporangium. Magnified 15. C, sporangium in surface-view. Magnified 15. D, sporangium (*s*) thrown off by bursting of reservoir (*r*); *c*, columella. Magnified 15. E, reservoir (*r*) and sporangium (*s*) seen in section; *b*, thin part of wall; *m*, mucilaginous layer; *c*, columella. F, the same, when the mucilage swells. Magnified about 25. G and H, mycelial branches preparing for conjugation. Magnified 120. J, later stage; *c*, the conjugating cells. Magnified 120. K, after conjugation; *z*, the zygospore. Magnified 120. (After Zopf and Van Tieghem.)

cuticularised, and bristles with crystals of calcium oxalate, while the lower part adjoining the columella remains thin. Inside the lower part of the sporangium is a mucilaginous layer derived from its protoplasm; when the sporangium is wetted, this mucilage takes up water and swells, bursting the delicate cell-wall, and so freeing the upper portion of the sporangium which contains the spores (Fig. 96, F). The pressure on the upper surface of the columella is thus removed, and no longer balances that of the liquid in the reservoir below; consequently the columella bursts, and a jet of water escapes from the reservoir, driving the sporangium and spores violently before it (Fig. 96, D). The sporangium may thus be hurled to a great distance, amounting it is said to more than a yard in some cases. Hence the name of the plant, which means "a thrower of missiles." The sporangium sticks to any object which it happens to hit, owing to the mucilage which still clings to it. The sporangiophores are positively phototropic, *i.e.* they bend towards the source of light. Thus the sporangia tend to be directed towards open spaces, and in nature they are often shot forth on to grass surrounding the dung. The spores are eaten with the grass by herbivorous animals and become disseminated with the dung. There is evidence that passage of the spores through the body of the animal facilitates germination. When the spores from the sporangium germinate they reproduce the ordinary form of the *Pilobolus* plant.

Other modes of propagation have occasionally been observed. The normal sporangia are only formed in air; if, however, the mycelium is forced to grow in a liquid containing plenty of organic food material, another process takes place. The hyphæ divide up by transverse walls into numerous cells, which may increase in number by budding, each cell putting out a short branch, which becomes separated from the parent. This is called the *oidial* condition. In other cases, namely, when food is less abundant, some of the cells produced by transverse division of the mycelium may acquire thick walls, and pass into a resting condition. These thick-walled cells are called *chlamydospores*, and, like the oidia, may either germinate into a new mycelium or give rise at once to sporangia.

b. Sexual

More important for us is the sexual reproduction, which in *Pilobolus* and many of its relatives takes place but rarely, though in some other members of the group it is a frequent means of propagation. Sexual organs are only formed in *Pilobolus* when some cause hinders the development of the sporangia. It has been found that conjugation can be induced by infecting the aërial hyphæ with a parasitic Fungus (which happens to be a relative, for several members of the group prey upon other members). The parasite hinders the formation of the asexual spores, and the plant is thus led to adopt the other method of propagation, which results in the formation of resting-spores capable of waiting until the bad times are over.

When conjugation is about to take place, two neighbouring hyphæ of the mycelium enlarge, and become club-shaped (Fig. 96, G and H). The swollen portions grow upright, and lay themselves together side by side, accumulating at the same time a large quantity of protoplasm in their interior. The ends of the conjugating hyphæ are next cut off by cell-walls (Fig. 96, J). The terminal parts thus separated, which are the richest in protoplasm, come into close contact, and the cell-walls separating them are absorbed. The protoplasm of the cells now runs together into one mass, the numerous nuclei uniting with one another in pairs, and the two cells completely fuse into a single *zygospore*, which rests upon the two enlarged hyphæ, called the suspensors

(Fig. 99, K, *z*). The zygospore grows to a relatively great size, surrounds itself with a very tough and thick cell-wall, and forms a quantity of oil in its contents. We see that the process is one of perfectly typical conjugation, the two cells concerned taking an exactly equal part in the production of the zygospore. It has recently been shown that in many of the Mucorineæ (to which *Pilobolus* belongs) there is a physiological differentiation of sex, definite sexual "strains" existing, which will alone conjugate with one another, while hyphæ of the same strain will not fuse sexually among themselves. A fungus which exists in the form of two strains sexually differentiated physiologically from one another is said to be *heterothallic*, in contrast to *homothallic*.

After a period of rest, the zygospore, if moistened, germinates. The germination is best known in an allied genus, *Mucor*. If food enough is to be had, it simply grows out at once into a new mycelium; if, however, supplies are scanty, it proceeds without delay to form an asexual sporangium, thus increasing the chances of survival. These differences are quite analogous to those which we found in the germination of the oöspores of *Pythium*.

The Zygomycetes, so far as their sexual reproduction is concerned, stand on a lower level than the last group. On the other hand, they are more fully adapted to a terrestrial mode of life than the Oömycetes. We saw in the case of *Pythium* and its allies how a transition can be traced from the sporangium to a single conidium, germinating directly. A somewhat similar gradation is to be followed among the immediate relations of *Pilobolus*. Some of these produce, in addition to the typical large sporangia, very small sporangioles, containing very few spores, or even only one. In other species sporangioles only are known, and these become detached bodily from the supporting hypha, and behave like single conidia. Thus in the Zygomycetes, as in the Oömycetes, a succession of steps leads from the typical sporangium to the simple conidium—one of the most characteristic forms of reproductive cell in the Fungi.

TYPE XXI

SPHÆROTHECA HUMULI

We now come to the higher Fungi, an immense group of plants, which have become completely adapted to every conceivable variety of parasitic and saprophytic life, and no longer show any clear trace of analogy with the Algæ. While differing among themselves in every other respect, they agree in possessing a septate, multicellular mycelium, with apical growth of the hyphæ of which it is built up.

Our present type is one of the simplest representatives of the great family of the *Ascomycetes*, which are characterised by possessing spore-forming bodies, somewhat like sporangia, of very definite size and form (called *asci*), in the interior of which a definite and usually small number of spores are produced, the number being regularly some multiple of two, usually eight. In this respect, as in many others, they differ from the Phycomycetes, where, as we have seen, the number of spores formed in a sporangium is quite indefinite, and often very large. The whole problem of the real relation of the Ascomycetes to the simpler families of Fungi is still unsolved.

We will proceed at once to the description of our type, which will serve to give us an elementary idea of the main facts in the structure and development of this important and difficult group of plants.

1. Structure

The species of *Sphærotheca* and its nearer allies (forming the family *Erysiphaceæ*, so named after its largest genus *Erysiphe*) are all parasites. One species (*S. Humuli*) is exceedingly common on Hops, and produces a very serious disease, the mildew, which causes great loss to the hop-growing industry. Another species (*S. pannosa*) is equally abundant on the shoots of Roses. All these Fungi are remarkable for being *external* parasites, that is to say, their much-branched mycelium forms a web on the surface of the leaves and other organs of the host plant. The presence of the parasite is quite evident to the naked eye, owing to the white to greyish-brown colour of its mycelium, which obscures the natural green of the leaf. The mycelium is white when young and greyish brown when old. The popular name "mildew" (equivalent to mealy dew) refers to this appearance, the leaves looking as if they had been powdered with flour. It must have been to Fungi of this kind that the name mildew was originally applied, though now it is extended in popular usage to other diseases of plants, presenting quite different symptoms.

The mycelium which spreads over the leaf, when examined microscopically, is found to give rise at intervals on the under surface to short outgrowths that penetrate into the interior of the cells of the epidermis, forming suckers or haustoria, which absorb the organic substances in the cells attacked, and thus supply the whole Fungus with its food, at the expense of the living tissue of the host. The mycelium, as in all the higher Fungi, is multicellular, consisting of branched chains of cells, each of which has a single nucleus, though in some allied Fungi the cells are multinucleate. The mycelium, as already mentioned, forms a dense web on the surface of the leaf; its hyphæ cross and touch each other at many places.

2. Reproduction

It is at the points where two hyphæ cross or come into contact that the fruits originate. Each of the adjoining hyphæ sends out an upright branch; the one

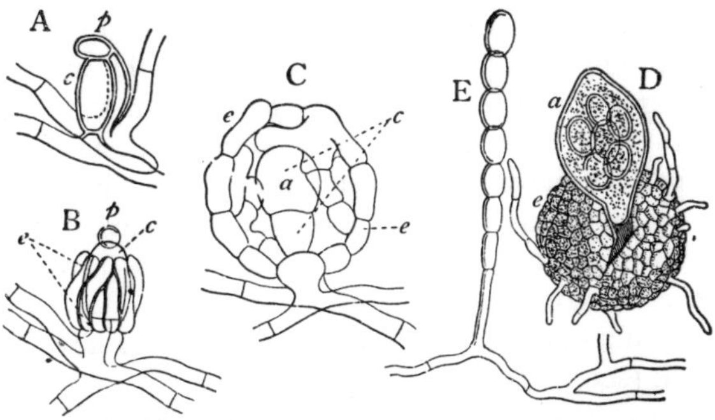

Fig. 97.—A–C, *Sphærotheca Humuli*. A, early stage in formation of fruit; *p*, antheridium; *c*, ascogonium. B, more advanced; *p*, antheridium; *c*, ascogonium; *e*, enveloping hyphæ. C, ripening fruit in section; *c*, ascogonium, from which the young ascus (*a*) is now developed; *e*, *e*, enveloping hyphæ, forming perithecium. D and E, *S. pannosa*; D, ripe perithecium (*e*) bursting to set free the ascus (*a*), in which only six out of the eight ascospores are shown. E, chain of conidia, borne on a vertical branch of the mycelium. A–C magnified 450; D and E less magnified. (After De Bary and Tulasne.)

enlarges and becomes club-shaped, and is cut off by a transverse wall; the other remains more slender, comes into close contact with the former, and grows up with it, soon overtopping it and bending over its apex (Fig 97, A). Two transverse walls are formed in this second branch, one near its base, and the other higher up.

Of these two organs the former, *i.e.* the club-shaped branch, bears the name of the *ascogonium*, for it is from it that the ascus ultimately arises. It corresponds to the oögonium of the lower Fungi. Investigations have proved that the second organ is really an *antheridium*, and that a true act of fertilisation takes place. According to the investigations referred to, fusion takes place between the ascogonium and the terminal cell of the antheridial branch, the cell-walls between them disappearing. Then the nucleus of the antheridium passes over through the opening and unites with the nucleus of the ascogonium. In the case of *Sphærotheca* and some of its allies these facts are now well established, so it is evident that, in many Ascomycetes at any rate, the development of the ascus-fruit is preceded by a sexual process quite comparable to that of the Oömycetes (cf. p. 222). In other Ascomycetes reduced forms of sexual fusion occur, comparable to those observed in apogamous Fern-prothalli (see p. 74).

As regards the subsequent development of the ascogonium, the main facts are clear. After two or three transverse divisions one of the cells of the row (in normal cases the last but one) increases in size, and becomes the ascus. It contains two nuclei, which fuse into one and then divide repeatedly, giving rise to eight daughter-nuclei, around each of which a cell is formed. These eight cells are the *ascospores*. In the meantime the vegetative cell next below the female organ has sent out several branches, which grow up around the ascogonium, completely enveloping it in a double layer of densely crowded hyphæ (Fig. 97, B and C). From the inner cells of the envelope thus formed, short branches filled with very dense protoplasm grow inwards and apply themselves closely to the ascus, probably supplying it with food. The outer cells of the envelope become thick-walled, and form a dense protective layer, completely enclosing the ascus. The ascus-fruit bears the name of the *peri-*

thecium; some of its superficial cells grow out into long hairs (see Fig. 97, D). The ripe perithecia are visible to the naked eye, as little black dots in the mycelium on the surface of the diseased leaf.

The hard perithecium serves to protect the ascus during the winter, for the fruits remain inactive until the following spring on the dead leaves or on the soil to which they may have fallen. When germination takes place, the ascus absorbs water, swells up, and bursts the perithecium (Fig. 97, D), whereupon its own membrane dehisces at the top, and the ascospores are violently shot forth into the air. On germination on a suitable host they at once reproduce the ordinary mycelium of the Fungus. In most of the allies of *Sphærotheca* each perithecium contains several asci (the product of a single ascogonium), instead of one only.

Our plant has another means of reproduction, by conidia (see Fig. 97, E), the formation of which precedes that of perithecia. The conidia are produced on vertical hyphæ (called the *conidiophores*) which produce the conidia at the free end. A whole chain of conidia is formed in basipetal order, the oldest thus being at the top. They are detached and scattered by the wind, germinating immediately under suitable conditions and readily infecting hop leaves and "cones." They produce mycelium like that from which they sprang, and constitute a ready means of propagation during the summer, while, as we have seen, the ascus-fruits are specially adapted for the winter rest. When conidia are being produced abundantly, the mildew assumes a powdery appearance, hence such a fungus as *Sphærotheca* is sometimes called a "powdery" mildew. In general, the conidia of these mildews do not survive the winter. In some allies of *Sphærotheca*, as in the Vine mildew, the conidial fructification was for many years the only one known.

Sphærotheca represents the course of development of an Ascomycete in almost its simplest form. It is not always the case that the ascogonium produces merely a single ascus ; in many Ascomycetes a very large number are formed ; in certain cases it has not been found possible to refer the origin of the asci to any definite ascogonium. There is always a fusion of nuclei in the young asci. In *Sphærotheca Humuli* two nuclear fusions occur in the life-cycle, and this is true of other Ascomycetes. In some of these Fungi, however, the only nuclear fusion which occurs is that in the young asci.

The formation of a completely closed fruit around the ascus or asci is characteristic of the Family (Erysiphaceæ) to which *Sphærotheca* immediately belongs. In the majority of the Ascomycetes, the fruit is more or less open, either having a small pore at the apex, or taking the form of a widely open cup or disc. Our next type will afford an example of the form of fruit last mentioned.

TYPE XXII

Xanthoria parietina (a Lichen)

A very large group of Ascomycetes have their ascus-fruits in the form of an open cup, or even a flat, shield-like disc (*apothecium*). These constitute the Discomycetes. The inside of the cup or the free surface of the disc is coated by the *hymenium*, a name applied in descriptions of the higher Fungi to the layer of spore-producing cells. In the case of Discomycetes the hymenium is made up of a large number of vertical asci, with sterile hairs—the paraphyses—between them.

The particular plant we have chosen to represent this group is a true Discomycetous Fungus as regards its fructification, but it belongs to a set of plants which are so different in habit and mode of life from all other Fungi, that they are sometimes treated as a distinct class of the Vegetable Kingdom. These are the *Lichens*, plants with a definite and often conspicuous thallus, freely exposed to the air and light, very different from the merely filamentous mycelium of ordinary Fungi, which is usually immersed in the substratum. Some Lichens grow on the bark of trees, some on rocks, walls, or roofs, and others on the ground.

1. Structure and Mode of Life

Our example, *Xanthoria parietina*, is extremely common (especially near the sea) on rocks, old walls, and roofs in the country, where it forms a conspicuous and most beautiful object, owing to its brilliant orange colour. The smoky atmosphere of towns, however, is very deleterious to Lichens. We see at once, from the habitat of the plant, that its mode of nutrition must be totally different from that of a typical Fungus. So far from requiring any organic matter, living or dead, on which to feed, *Xanthoria* grows on the most barren and unpromising substratum conceivable. Many Lichens, in fact, thrive for years and even centuries under conditions of drought and apparent starvation, which would be absolutely intolerable to any other plants whatsoever. A Lichen, considered as a whole, is neither a parasite not a saprophyte ; it requires nothing but water and a little mineral food, and can provide itself with carbon from the carbon dioxide of the air, like an ordinary green plant. Lichens therefore can only live in the light, which is not the case with Fungi. We will now proceed to describe the structure of *Xanthoria*, and find out the explanation of its remarkable mode of life.

Xanthoria parietina is described as a foliaceous Lichen

because its thallus has a flat, much lobed, leaf-like form (Fig. 98). New lobes of the thallus arise from just below the tip or growing portion of an older lobe, and grow outwards and upwards. The under surface, which is almost white, is attached to the substratum by means of rhizoids, arranged in dense tufts, which perform the functions of roots.

The anatomical structure of the thallus is shown in vertical longitudinal section in Fig. 99. Towards the upper surface is a dense layer of hyphæ forming closely packed cells at right angles to the surface, with some dead cells on the exterior bearing yellow grains of *parietin* on their outer walls. This tissue is termed *plectenchymatous* because it is formed from closely interwoven hyphæ, and not as in parenchyma. The orange colour is due to crystalline granules of parietin deposited outside the cells, both on the free surface of the thallus and between the hyphæ. The amount of parietin present varies with the light intensity. Such organic pigments as parietin are common in Lichens, and from some of these organisms *Litmus* (used in testing for acids and alkalies) is obtained.

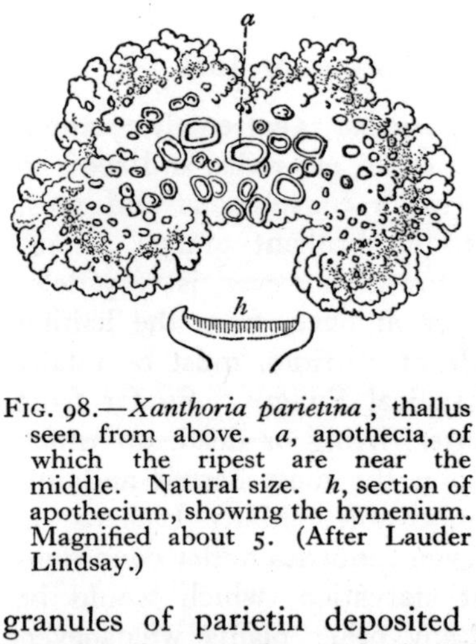

FIG. 98.—*Xanthoria parietina* ; thallus seen from above. *a*, apothecia, of which the ripest are near the middle. Natural size. *h*, section of apothecium, showing the hymenium. Magnified about 5. (After Lauder Lindsay.)

Underneath the dense upper cortex comes the *medulla*, a broad zone of hyphæ with large air spaces between them in certain regions. In the centre of the medulla the hyphæ run longitudinally towards the thallus, forming anastomosing strands, but in other parts of the medulla scattered hyphæ run in all directions.

Below the medulla is the lower cortical layer, resembling that on the upper surface and arising from which are the rhizoids. The latter develop as contiguous outgrowths from the cortical hyphæ. The extremities of the rhizoids broaden out and form flat discs, which secrete mucilage and attach themselves firmly to the substratum.

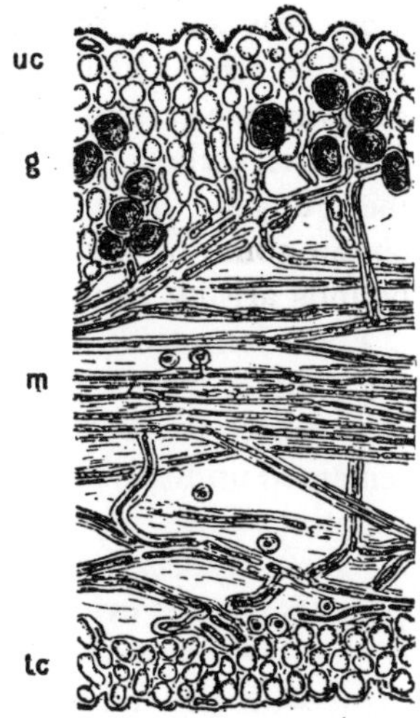

FIG. 99.—*Xanthoria parietina* Vertical longitudinal section of mature thallus. *u.c*, upper cortex; *g*, gonidia; *m*, medulla; *l.c*, lower cortex. Magnified 300. (O. V. Darbishire.)

In the lower part of the upper cortex, and forming a more or less definite layer, are groups of large green cells embedded amongst the hyphæ. The green cells of Lichens bear the name of *gonidia*, and the part of the thallus containing them may be distinguished as the gonidial layer. The gonidia of *Xanthoria* each contain a large, more or less lobed chloroplast with a central pyrenoid, and a nucleus. The gonidia reproduce themselves by the division of the protoplast into a considerable number of separate cells (Fig. 99*), which become free when the wall of the mother cell breaks down. Small suckers or haustoria penetrate the gonidia from the surrounding hyphæ and absorb food from them.

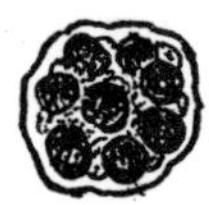

FIG. 99*.—*Xanthoria parietina*. Gonidium showing mode of division. Magnified 600. (O. V. Darbishire.)

Now the most important question as to the thallus of a Lichen concerns the nature of the green cells, or gonidia. They play an essential part in the economy of the plant, for, like other chlorophyll-containing cells, they are able to assimilate carbon from the carbon dioxide of the air or from the carbon dioxide produced by the respiration of the Fungus. It is possible also that the gonidia may obtain some of their carbon from certain organic compounds arising from the metabolism of the Fungus.

For a long time, in fact down to about the year 1868, the gonidia were regarded as forming, like the hyphæ, a constituent part of the thallus. So long as that view remained undisturbed, the Lichens were rightly ranked as a distinct Class, equivalent to the Algæ or the Fungi, and in some respects intermediate between them. Of late years, however, evidence has accumulated, which proves conclusively that the gonidia do not belong to the same plant with the hyphæ, but that they are distinct organisms, identical with definite genera and species of the Algæ. Hence a Lichen is in reality a *compound organism*, made up of two totally different plants, an Alga and a Fungus, living in the closest association, and mutually dependent one on the other for certain essential services.

The evidence on which this striking conclusion is based is of various kinds. In the first place, the so-called " gonidia " of Lichens are always found to agree exactly with certain species of the lower Algæ, which are also well known in a free and independent condition. Thus in our type *Xanthoria parietina*, the algal constituent is *Cystococcus humicola*,[1] a unicellular form nearly related to *Pleurococcus*. Further, the " gonidia " have been isolated from the Lichen-thallus, and are then able to lead a perfectly independent life, growing and multiplying

[1] Sometimes called *Trebouxia humicola*.

on their own account, just like their fellows which have never been in bondage. More recently it has been found possible to raise the fungal constituent of a Lichen by growing its spores in a food solution, which, as Algæ are absent, must of course contain organic food substances. In this way a small thallus can be produced, but it *never contains any gonidia*. Thus the fungal as well as the algal constituent can, under suitable conditions, live by itself.

Most conclusive of all, however, is the actual *synthesis* of a Lichen, that is, the building up of a new plant out of the Fungus and its appropriate Alga. This has been observed in the case of our type, and Fig. 100 represents the process. The ascospores of the *Xanthoria* have been sown among the cells of the Alga, *Cystococcus*. The spore on germination sends out a hypha, which at once begins to branch, and its finer ramifications attach themselves closely to the algal cells (Fig. 100, A). As growth proceeds, more and more of the algal colony becomes involved in the web of hyphæ arising from the fungal spore, and one after another the *Cystococcus* cells are seized upon by the suckers of the Fungus (Fig. 100, B). Soon the filaments of the Fungus, well-fed at the expense of the Alga, are strong enough to build up a thallus. In the middle of Fig. 100 the hyphæ are seen uniting to form a network. If the ascospores of *X. parietina* are sown alone on a nutritive medium the Fungus grows to the extent of forming a rudimentary cortex, but no parietin is produced. If, however, such a culture is then brought into contact with the gonidia, parietin is formed. This indicates the great influence which the gonidia have upon the Fungus.

Observations such as these have removed all doubt as to the compound nature of the organisms called Lichens. What, then, is the real relation between the Alga and the Fungus of which the Lichen is built up? It might be supposed that the case is one simply of parasitism, the Alga playing the part of a mere victim to the devouring Fungus, just as a Cress seedling is preyed upon by *Pythium* or a Hop plant by *Sphærotheca*. This does not, however, seem to be the real condition of affairs. The Alga is not, on the whole, injured when the Fungus

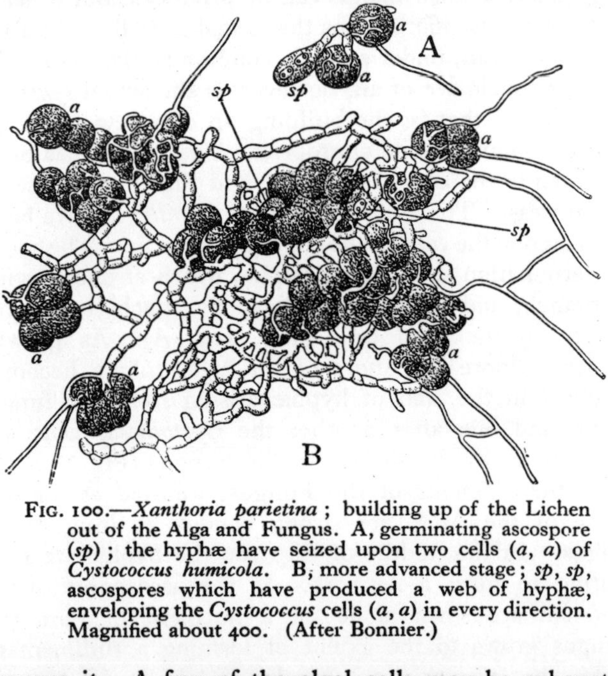

FIG. 100.—*Xanthoria parietina* ; building up of the Lichen out of the Alga and Fungus. A, germinating ascospore (*sp*) ; the hyphæ have seized upon two cells (*a*, *a*) of *Cystococcus humicola*. B, more advanced stage ; *sp*, *sp*, ascospores which have produced a web of hyphæ, enveloping the *Cystococcus* cells (*a*, *a*) in every direction. Magnified about 400. (After Bonnier.)

annexes it. A few of the algal cells may be exhausted and die, but the great majority live and go on multiplying within the Lichen, quite as happily as if living at liberty in the open air. It seems that there are advantages on both sides ; the Alga, by the aid of its chlorophyll-bodies, undertakes the whole duty of the assimilation of carbon, thus providing the Fungus with the organic food which it is unable to manufacture for itself. In many Lichens there are definite pores in the cortex, facilitating the gaseous interchange between the atmosphere and the gonidial layer. On the other hand, the rhizoids of the fungal partner supply water and mineral food, probably in a more effectual way than the Alga could obtain them for itself. At the same time the tissues of the Fungus shelter the Alga and protect it from the weather, and especially from the effects of drought and intense light. It is probable that many unicellular Algæ, when enclosed in the thallus of a Lichen, are able to exist, perhaps for centuries, in places, as, for example, on the surface of exposed rocks, where they could not possibly carry on their life if left to themselves.

Such a relation between two organisms which live in common, and perform certain functions each to the good of the other, is known by the name of *symbiosis*, or *commensalism*, the former word simply calling attention to their living in union, while the latter term means that they share the same table, implying that they mutually help each other to food.

2. Reproduction

The Lichens being, as we have seen, compound organisms, might be described either under the heading "Algæ" or "Fungi." It is usual, however, to take them with the latter class, because the organs of fructification, on which classification is chiefly based, belong entirely to the fungal partner. The captive Algæ go on increasing by division, but rarely produce any characteristic reproductive organs, so long as they form part of the Lichen. Nearly all Lichen-Fungi are Ascomycetes, and the majority belong to the group Discomycetes, in which the hymenium is exposed to the air when mature. So far as the fructification is concerned, there is no essential difference between Lichens and other Fungi of the same group, which lead an ordinary parasitic or saprophytic existence.

The ascus-fruits of *Xanthoria parietina* are conspicuous to the naked eye as flat, shield-like discs on the upper surface of the thallus, generally of a rather deeper orange colour than the rest of the plant (see Fig. 98). These open fruits of the Discomycetes bear the name of *apothecia*. In the mature condition there is a rim of sterile tissue at the edge of the apothecium, the whole disc within this rim being covered by the hymenium. Below the hymenium is a dense mass of closely interwoven hyphæ forming the *hypothecium*. The hymenium itself consists of elements of two kinds,—the asci and the paraphyses. The asci, of which a great number are present in each fruit, are stout, club-shaped cells set vertically to the surface of the apothecium, each ascus when ripe containing eight ascospores. The paraphyses are sterile hairs rising to a greater height than the asci, both being closely packed together, so that the hymenium has a smooth, continuous surface (see Figs. 98 and 101).

At an earlier stage of development the apothecium is closed, and consists of a mass of hyphæ surrounded by a cortical layer. The paraphyses which arise from the hypothecium are the first elements of the hymenium to be developed. The asci, which in many cases have been observed to arise from the branches of distinct hyphæ, differing from those which produce the paraphyses, are developed relatively late. They grow up among the paraphyses, insinuating themselves between them until they attain nearly the same height. At the same time the envelope of the fruit is opened at the apex, and the edges gradually pushed back as the apothecium expands.

The development of the apothecium may go on for a very long time, even for years in some cases, new asci arising towards the exterior margin. Each ascus at a certain stage of development contains a single nucleus (the product of a fusion), which subsequently undergoes repeated division, into two, four, and eight. When the full number is attained, a cell is formed around each nucleus, and these cells become the eight ascospores;

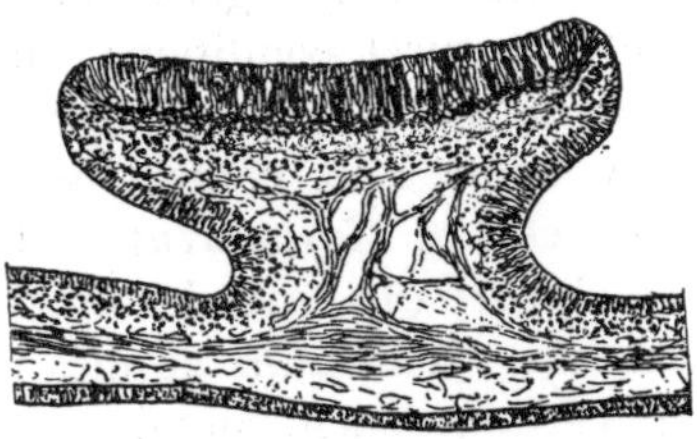

FIG. 101.—*Xanthoria parietina.* Vertical section through a young apothecium. Magnified 50. (O. V. Darbishire.)

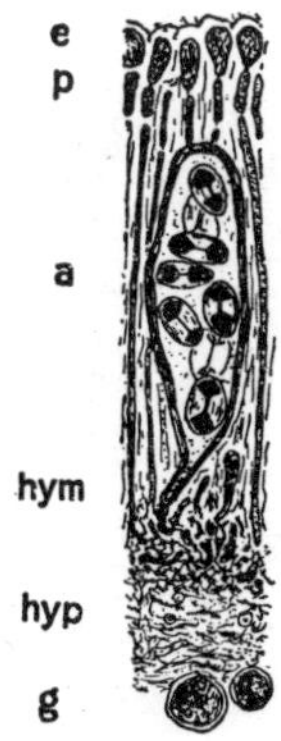

FIG. 101*. — *Xanthoria parietina.* Vertical section through a mature ascus, etc. *e*, epithecium; *p*, paraphyses; *a*, ascus with ascospores; *hym*, hymenium; *hyp*, hypothecium; *g*, gonidia. Magnified 300. (O. V. Darbishire.)

each ascospore at maturity consists of two cells (Fig. 101*). The contents of the ascus are not, however, completely used up in the process of spore formation; a certain part remains over, lying between the spores; this unused substance becomes gelatinous, and in taking up water tends to swell, and so to burst the ascus. The dissemination of the spores, however, is not entirely due to the pressure from within the asci. The whole apothecium, including the paraphyses, endeavours to expand laterally when wetted, and this expansion is resisted by the rim of tissue at the edge. Hence the asci are subjected to very considerable pressure, the result of which is that those which are ripe dehisce at the top, expelling all the eight spores with considerable force, so that they are shot up as much as a centimetre into the air. The asci open successively as they become mature, the dehiscence taking place whenever wet weather occurs.

The ascospores of *Xanthoria*, we have seen, can only complete their germination under natural conditions, and form a new Lichen-thallus, if they come into contact with the cells of *Cystococcus*, with which they can enter into partnership. The same applies to all Lichens, each containing a particular Alga.

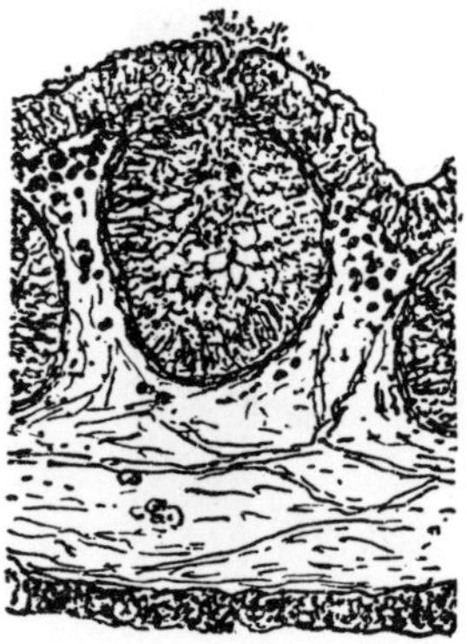

FIG. 102.—*Xanthoria parietina*. Vertical section through a spermogonium. Magnified 100. (O. V. Darbishire.)

There is another form of fructification consisting of small flask-shaped receptacles called *spermogonia* (Fig. 102), which, in our type and its near allies, are formed in groups on rounded prominences of the thallus. The opening at the top of the spermogonium is called the *ostiole*. The interior of the spermogonium is occupied by a net-work consisting of sterile and fertile cells. From the latter, excessively minute cells, the *spermatia*, are budded off (Fig. 103). Each spermatium contains one nucleus. The nature of these spermatia has been much disputed; there appears to be weighty evidence for regarding them, at least in certain cases, as the male cells of the Lichen.

In most species investigated the apothecium has been

found to arise from a special cellular filament, the *ascogonium*. Its lower end, which usually forms a coil, is embedded in the medullary tissue of the thallus, while the upper portion projects beyond the surface of the cortex, terminating in a long cell, the *trichogyne* (see Fig. 104). The trichogyne has a gelatinous cell-wall, to which numerous spermatia are found adhering. There is reason to believe that only those ascogonia, of which the trichogynes have been associated with spermatia, develop into ascus-fruits. The asci spring from cells of the coil, while the paraphyses arise lower down. It appears, then, that in these Lichens there may be a sexual process somewhat resembling that in the Florideæ and in certain Fungi. The details of fertilisation, however, still require to be worked out. In some cases, on the other hand, the spermatia have been observed to germinate, so their function as male cells seems not to be constant throughout the group. The illustrations (Fig. 103–104) are taken from a Lichen (*Physcia pulverulenta*), closely allied to our type. In certain free-living Discomycetous Fungi there is undoubtedly a process of fertilisation in the ascogonium.

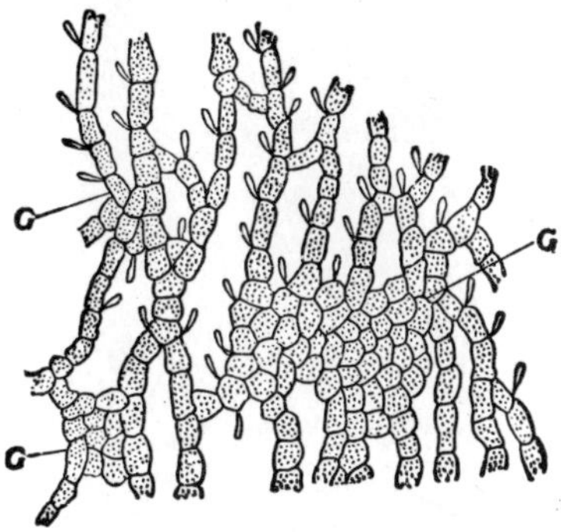

FIG. 103.—*Physcia pulverulenta*. Portion of the tissue inside a spermogonium, showing the groups of sterile cells (*G*), and the cells bearing spermatia. Highly magnified. (After Darbishire.)

The ascospores, as we have seen, reproduce only the fungal element of the Lichen; they must meet with algal cells in order to form a perfect Lichen-thallus.

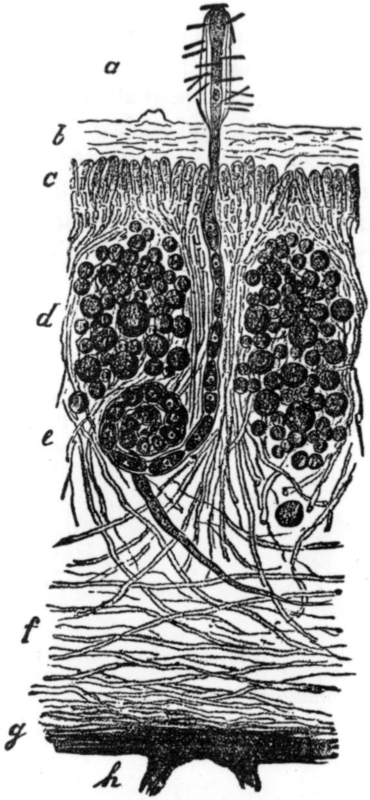

FIG. 104.—*Physcia pulverulenta*. Portion of a vertical section of the thallus, showing a complete ascogonium. *a*, projecting trichogyne, with swollen gelatinous cell-wall, in which numerous spermatia are caught; *e*, coiled part of the ascogonial filament; *b*, dead layers of upper cortex; *c*, living cortical hyphæ; *d*, groups of gonidia; *f*, medullary layer; *g*, lower cortex; *h*, rhizoids. Magnified about 1100. (After Darbishire.)

In many members of the group, however, though not in our type, there is a special provision for the reproduction of the compound organism as a whole. In these Lichens we find some of the apothecia replaced by patches of a powdery substance, each grain of the powder consisting of a few algal cells invested by fungal hyphæ. These little groups become isolated, and are dispersed by the wind or rain, or by the agency of insects. They are called *soredia*, and serve to reproduce at once the algal and fungal constituents of the Lichen. The group of soredia is here homologous with an apothecium; in other Lichens similar bodies are formed on other parts of the thallus.

Xanthoria then has served to illustrate, on the one hand, a highly organised Ascomycete, with a complex fruit containing a large number of asci ; while, on the other hand, it has made us acquainted with the remarkable phenomenon of *symbiosis*, or the associated life of two distinct organisms, each partner performing certain physiological functions to the benefit of the other. Of this widespread manner of life, the Lichens afford the most striking example among plants. Lichens live their lives as if they were single organisms, so perfectly are the two constituents of the thallus attuned to one another. Lichens, therefore, possess an individuality of their own. The association of Alga with Fungus enables both organisms to flourish in situations in which neither by itself could survive, and results in the formation of novel types of plant body not attained by Algæ or Fungi separately.

TYPE XXIII

SACCHAROMYCES

Saccharomyces, or Yeast, is a very simple type of fungus which belongs to the Ascomycetes, although its relationships to other members of the group is somewhat obscure. The genus Saccharomyces is a large one, but the species are only distinguishable from one another with great difficulty.

Yeasts occur naturally on the surface of fruits, *e.g.* grapes and apples, and on other plant tissues. Some forms have been " domesticated," notably those used in the manufacture of beer and in bread-making. Yeast cells are small and light, and are readily carried about by wind. When seen in mass, yeast cells are usually white or brownish white, but some species are pink or otherwise coloured.

If a small portion of Baker's Yeast is mounted in water and examined under the microscope, it will be seen to consist of innumerable hyaline, spherical or oval cells, about 10–15 μ in diameter (Fig. 104*). In the centre of each cell is a large "vacuole," which is considered by some authorities to be part of the nucleus. It is claimed that the "vacuole" is traversed by chromatinic fibrils demonstrable by careful staining; otherwise the vacuole is filled with sap. On the margin of the "vacuole" is a large dense granule, the nucleolus,[1] but this is only visible after staining. The rest of the cell is filled with cytoplasm containing granules of reserve food, including glycogen, a substance allied to starch which is commonly found in Fungi.

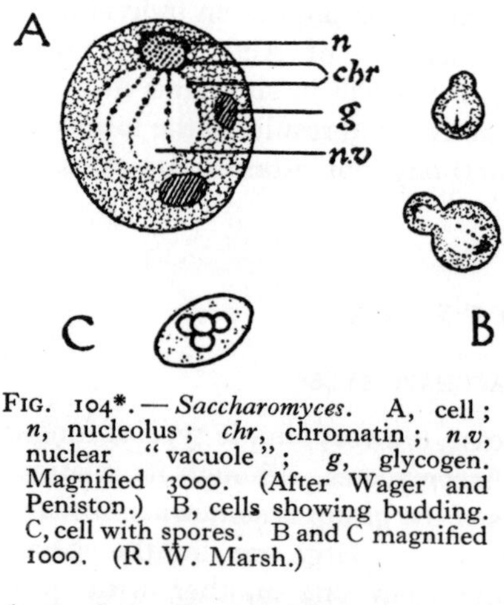

FIG. 104*.—*Saccharomyces*. A, cell; *n*, nucleolus; *chr*, chromatin; *n.v*, nuclear "vacuole"; *g*, glycogen. Magnified 3000. (After Wager and Peniston.) B, cells showing budding. C, cell with spores. B and C magnified 1000. (R. W. Marsh.)

If yeast cells are kept in a nutritive solution they reproduce themselves rapidly by a peculiar process known as *budding*. In this, a cell gives rise to a small protuberance or bud, which gradually grows until the bud is nearly as large as the mother-cell. The bud then becomes separated from the mother-cell, and the process may be repeated indefinitely. Sometimes, however, the first bud proliferates again before abstriction from the mother-cell, and the second or third buds may behave similarly,

[1] Other authorities consider the "nucleolus" to be the entire nucleus, and that the "vacuole" is in the cytoplasm.

so that considerable colonies of cells loosely joined together may be seen. Sooner or later, however, the colonies break asunder into their constituent cells. At the commencement of budding, the nucleus divides into two, one of the daughter nuclei passing into the bud. In most yeasts hyphæ are never formed, but in some of them the cells may either grow out into short hyphæ, or produce buds.

In many yeasts "budding" is the only form of reproduction. In some, however, spores are formed under certain conditions, especially when the food in the surrounding medium is becoming exhausted; or when there is danger of desiccation. In the process of spore formation the protoplasm divides, usually into four portions, each of which becomes surrounded with a comparatively thick wall (Fig. 104*). When four spores are formed they are usually arranged tetrahedrally in the mother-cell. After a time the spores are liberated. By virtue of their thick walls the spores can remain alive under adverse conditions. On germination the spores bud and give rise to colonies of cells. It is in consequence of this method of spore formation that *Saccharomyces* and allied genera are included in the Ascomycetes, for the mother-cell, which gives rise to a definite and limited number of spores corresponds to an ascus. In some yeasts spore formation is preceded by a sexual fusion or conjugation of two ordinary cells. Subsequent to the union of the cells and the fusion of the two masses of protoplasm two spores are usually formed in each of the conjugating cells.

Like other Fungi, yeast cannot assimilate the carbon dioxide of the air. It can, however, utilise for food the carbon in comparatively simple organic substances, *e.g.* ammonium tartrate $[(NH_4)_2C_4H_4O_6]$. It also requires for its growth nitrogen (present in ammonium tartrate), sulphur, phosphorus, potassium, magnesium and calcium, as well as hydrogen and oxygen. A convenient nutritive solution for growing yeast consists of potassium hydrogen phosphate [K_2HPO_4], magnesium sulphate [$MgSO_4$], calcium phosphate [$Ca_3(PO_4)_2$], and ammonium tartrate, dissolved in water. In this solution, yeast grows fairly well and gives rise to large numbers of new cells. If, however, glucose [$C_6H_{12}O_6$], a kind of sugar, is added to the nutritive medium, the yeast grows much faster and produces larger quantities of carbon dioxide than when grown in the simpler solution. In addition, a considerable quantity of alcohol [C_2H_5OH] will be found to accumulate in the medium as time goes on.

When supplied with free oxygen, yeast respires partly in the usual way, but it also liberates carbon dioxide as the result of the decomposition of glucose without oxidation, alcohol being formed at the same time as follows :

$$C_6H_{12}O_6 \rightarrow 2CO_2 + 2C_2H_5OH$$

As with normal respiration this process of decomposition of glucose without oxidation (known as *intramolecular respiration*) results in the liberation of energy which is utilised for carrying out other vital activities. Even in the entire absence of oxygen, yeast can grow rapidly owing to its ability to decompose glucose into carbon dioxide and alcohol. When the alcohol reaches a certain concentration, the yeast ceases to grow. The giving off of large quantities of carbon dioxide by yeast in the complete or partial absence of free oxygen is known as *fermentation*, and the alcoholic fermentation of sugar by yeast is the best known process of this kind. Many other micro-organisms, especially certain bacteria, ferment organic compounds, and gases, other than carbon dioxide, and substances, other than alcohol, are often produced in these fermentation processes.

It has been shown that the decomposition of glucose into carbon dioxide and alcohol by yeast is due to an enzyme, *zymase*, contained in its cells. By pounding up yeast cells the zymase can be extracted with difficulty from the protoplasm, and it will then decompose glucose precisely as the living yeast does. Zymase, like other enzymes, is a lifeless substance of very complex composition, which is formed by the protoplasm (see *Structural Botany*, Part I, p. 60). The yeast cell contains many other enzymes beside zymase. If yeast is grown in a solution of cane sugar (sucrose) the enzyme *invertase* first transforms the sucrose into lævulose and glucose before the latter is decomposed by zymase. The metabolism of yeast is extremely complicated, and only an outline of it has been presented. In particular it should be borne in mind that small quantities of several other substances besides alcohol are formed during the growth of yeast, these depending upon the particular kind of yeast and upon the constitution of the nutritive medium.

In brewing beer, wort, an aqueous infusion of germinated barley grains (malt) is fermented by yeast. Sometimes the yeast is added in the form of a pure culture of a known variety in order to prevent adverse developments in the beer, which may be caused by the inclusion of wild yeasts. When the barley grains germinate, the enzyme *diastase* (see Part I, p. 60) in them converts the starch into sugar, so that there is a large quantity of this substance in the wort for the growth of the yeast. Hops are added to the wort during brewing to flavour the beer. In making cider, the juice is extracted from apples by pressure and is fermented by yeasts present on the surface of the fruits or in the air. Wine is similarly made by the spontaneous fermentation of grape juice. Yeast also plays an important part in bread making, causing it to "rise" and thereby increasing its palatability. In making bread a special kind of yeast, called Baker's Yeast, is used. This is produced on a large scale by the fermentation industries, and is pressed together into blocks for convenient handling. The baker incorporates some of this yeast with the dough, which is kept warm for some hours to promote the rapid growth of the yeast. There are small quantities of sugar in the dough produced by the action of diastase on the starch in the flour. The yeast ferments the sugar, and the bubbles of carbon dioxide cause the dough to rise and the bread to become porous. When the dough is baked, the bubbles of carbon dioxide burst and the gas is dissipated as well as the alcohol formed during the fermentation.

TYPE XXIV

Puccinia graminis

The group of Fungi (*Uredineæ* or Rust Fungi) represented by this type is a comparatively large one, but it shows a narrow range of diversity as compared with a great group like the Ascomycetes. The plants, however, are of much interest, for they afford one of the very best examples of typical parasitic Fungi, which have adapted themselves exclusively to life at the expense of other plants. Some of them, and especially that species which we have chosen as our type, are extremely injurious to important crops, and so possess a very considerable practical interest. The life-history of these parasites is singularly complicated, at least in some of their representatives.

Puccinia graminis is the cause of the Black Rust or "mildew" of Wheat and other cereals. Other species of *Puccinia* cause different rusts of cereals, some of which are much commoner in Britain than is *P. graminis*.

P. graminis occurs in summer on the leaves and stems of Wheat, Barley, Rye, and Oats, as well as on various wild grasses. In its early stage on these hosts the parasite is easily recognised, for it forms conspicuous rusty brown streaks between the veins of the leaf or along the surface of the stem, but it is usually much more abundant on the stems and leaf sheaths than on the laminæ. When the rust is mature, we see that these streaks are made up of a fine powder, bursting out through the epidermis of the host-plant, the powder consisting of the *conidia* or *uredospores* of the Fungus. Thus it is only the fructification of the parasite which is visible externally.

The vegetative part or mycelium is hidden in the tissues, and requires very careful microscopic examination for its detection. It does not spread throughout the whole plant, but is limited to isolated patches of the particular organs attacked. The mycelium consists of a dense web of excessively fine hyphæ, growing luxuriantly between the cells of the parts affected, and also sending out haustoria, which penetrate the cells themselves. The mycelium is multicellular, the transverse septa, however, only occurring at long intervals. The elongated form of the rust-streaks, which is most conspicuous on the stems, is due to the fact that the Fungus attacks the soft tissues, lying between the longitudinal bands of fibres, which accompany the vascular bundles.

The fructification characteristic of the rust-stage of the Fungus is produced in great quantities during the summer months. Preparatory to its formation, certain of the more superficial hyphæ pack themselves closely together, forming a dense layer just below the epidermis of the host plant. From this layer the conidia are formed. Each conidium is a single cell, borne at the end of a vertical unicellular stalk (see Fig. 105, E, *u*). They arise in great numbers close together, forming large groups or

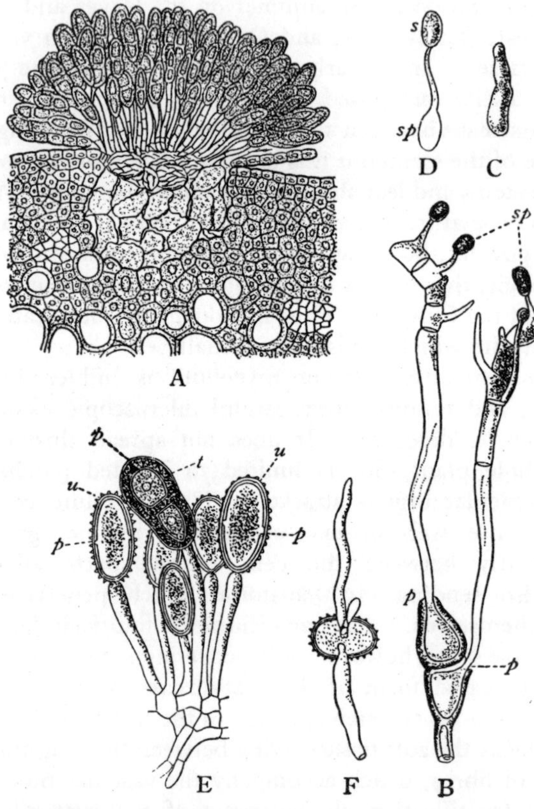

FIG. 105.—*Puccinia graminis*. A, section of the cortex of a wheat-stalk, showing a pustule of teleutospores, with the mycelium below. Magnified 150. B, teleutospore germinating; *p*, *p*, germ-pores; *sp*, sporidia borne on sterigmata arising from the promycelium. Magnified 230. C, sporidium germinating directly. D, sporidium (*sp*) germinating indirectly, forming a secondary sporidium (*s*). Magnified 370. E, part of a pustule, showing several uredospores (*u*) and a teleutospore (*t*); *p*, *p*, germ-pores. Magnified 300. F, uredospore germinating. Magnified 300. (After Von Tavel, Tulasne, and De Bary.)

pustules ; in each pustule the development begins near the middle of the mass and spreads centrifugally. As the pustule develops, the epidermis lying above it is burst, and the ripening conidia are exposed to the air.

The single conidium, as it matures, acquires a rather thick cell-wall, consisting of two layers, the exospore and endospore, the outer of which is of a brownish colour and is covered with short spines, while the inner is colourless. There are usually four *germ-pores* or thin places in the cell-wall around the middle of the spore (see Fig. 105, E, *u*). In the cell-contents a quantity of oily, brownish-coloured pigment is present.

These conidia are called the *uredospores*, because they were formerly regarded as belonging to a distinct genus—*Uredo*—the species of which are now known to represent merely a particular form of fructification of the *Puccinia*. These uredospores become detached from their stalks, and are scattered by the wind and possibly also by the agency of insects. They are capable of immediate germination, and give rise to the same form of the Fungus as that which produced them, growing on the same kind of host plant, or at least on an allied species. The hyphæ grow out from the germinating spore through the germ-pores (see Fig. 105, F). If germination takes place upon a Wheat plant or other suitable host of the Grass family, the hyphæ grow along the surface of the epidermis until a stoma is reached, through which an entrance into the intercellular spaces of the host is effected. Thenceforth the hyphæ at once proceed to develop a new mycelium, from which new crops of uredospores arise. This form of fructification therefore serves for rapid propagation during the summer, though, as we shall see, it is not of necessity limited to that season.

Later in the summer another kind of spore, borne on

the same mycelium with the uredospores, begins to make its appearance. The external sign of the change of fructification is a change in the colour of the pustules, from reddish-brown to nearly black. This is due to the development of the *teleutospores*, which owe their name (meaning *final* spores) to the fact that they appear at the end of the season of growth. It is in consequence of the black streaks of teleutospores on the stems that *P. graminis* is often called " Black Rust."

It was to the teleutospore form of fructification that the name *Puccinia* was applied in the first instance, before the life-history was completely understood ; for this Fungus was originally put in three distinct genera, which are now known to represent stages in the development of one and the same plant. The teleutospore condition used to be known as the " *Mildew* " of wheat and other cereals. The teleutospores are produced in just the same way as the uredospores ; in fact both kinds of spore are often found in the same pustule (Fig. 105, E) during the intermediate period while the one fructification is being gradually replaced by the other. Later in the season we find pustules consisting of teleutospores only (Fig. 105, A).

The teleutospore is borne on a stalk like the uredospore, but is quite different from it in structure. The membrane is excessively thick, consisting of a stout outer coat of a dark-brown colour, and an inner colourless layer. The spore is made up of two cells, separated by a comparatively thin transverse septum. Within the cytoplasm of each young cell are two nuclei, which subsequently fuse into one, and there is also a vacuole containing oil. The surface of the cell-wall is smooth, unlike that of the uredospores. Each cell has a germ-pore, that is to say, a deep pit in its membrane ; in the upper cell this pit is situated at the apex, while in the lower it lies on one side, just below the septum (Fig.

105, B). These teleutospores represent the resting-stage of the Fungus, in which it passes through the winter.

This completes the history of the parasite, so far as its life on the Wheat or other gramineous [1] host is concerned. The damage which it does when it attacks the crop before the grain is set is very serious, though its immediate effect is only local. The chlorophyll of the part attacked is destroyed, and the tissues thus rendered useless for assimilation, while the cells affected ultimately become exhausted and die. Thus, if the seats of infection are numerous, the plant may gradually lose almost the whole of its effective assimilating tissue, and thus become starved and quite incapable of producing good grain. When the stems are severely affected the straw sometimes falls to the ground. Fortunately, early attack of cereals by *P. graminis* is now rare in Britain, but in N. America and certain other countries grave damage is caused by it in certain seasons.

The germination of the teleutospores in this species takes place in the following spring, and occurs while the teleutospores are still attached to the straw. Each cell sends out a hypha, which starts from the germ-pore, as shown in Fig. 105, B. These hyphæ do not develop into a normal mycelium, but are of limited growth, forming what is called a *pro-mycelium* (see Fig. 105, B) or *basidium*, which divides by transverse walls, cutting off a row of four cells from its terminal portion.[2] Each of these cells sends out a slender lateral outgrowth, which swells up at the end to form a small spore-like cell (Fig. 105, B). These cells are the *sporidia* or *basidiospores* ; the stalks on which they are borne are called the *sterigmata*. Hence we see that the teleutospore is incapable of directly repro-

[1] The family Gramineæ, or Grasses, includes all cereals.

[2] In water, however, the pro-mycelium may grow to a considerable length before producing sporidia.

ducing the typical form of the Fungus, for it only gives rise to a rudimentary mycelium, which proceeds at once to form yet another kind of spore. The sporidia become detached from their stalks, and are capable of direct germination ; but if they do not happen to be carried by the wind to their proper host plant, they form only a very short hypha, which at once gives rise to a secondary sporidium (Fig. 105, D), thereby gaining another chance of successful dissemination.

The sporidia, whether primary or secondary, are quite incapable of infecting any plant of the Grass Family. They are dependent upon a totally different kind of host, namely, the Barberry (*Berberis vulgaris*) or some of its allies. The sporidium, if it germinates on a young leaf of a Barberry bush, sends out a hypha which penetrates the cuticle instead of obtaining entry into the tissues by way of a stoma. In this respect the germinating sporidium differs from all the other forms of spore in this Fungus. When the mycelium is once started, it spreads in the tissues of the leaf, establishing haustoria in the cells, just as it did in the Wheat. The fructifications produced on the Barberry, however, are of a totally different kind from any of the forms already described.

During the spring the Barberry may show signs of disease, consisting in the appearance of swollen discoloured patches on its leaves. When the disease has advanced further we find on the under-side of the leaf, seated upon the swollen place, clusters of exceedingly pretty little yellow cups containing spores (Fig. 106, A). This is the *Æcidium* form of the parasite, and, like the *Uredo* and *Puccinia*, was long described by botanists as belonging to a distinct genus.

When an *Æcidium*-fruit is to be produced, a group of hyphæ becomes densely felted together in an intercellular space of the leaf ; the inner hyphæ of the group enlarge their cells, so as to give rise to a little nest of apparently parenchymatous tissue, surrounded by a

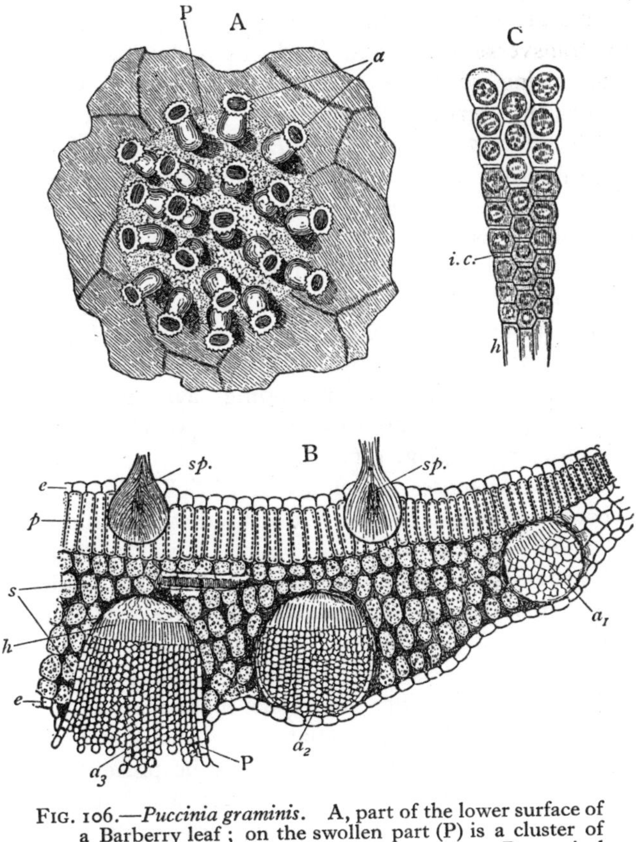

FIG. 106.—*Puccinia graminis*. A, part of the lower surface of a Barberry leaf; on the swollen part (P) is a cluster of *Æcidium*-cups (*a*). Magnified about 10. B, vertical section of the diseased part of the leaf; *e*, *e*, epidermis; *p*, palisade tissue; *s*, spongy tissue; *sp*. *sp*, spermogonia of the parasite; a_1, a_2, a_3, three stages in the development of the *Æcidium*-fruit; *P*, peridium; *h*, hymenium. Magnified 40. C, three chains of æcidiospores; *h*, hymenial cells; *i.c.* intercalary cells. Magnified about 200. (After Zopf and Kny.)

web of ordinary mycelium ; at the base of this mass a now of vertically elongated cells—the *hymenium* (Fig. 106, B, *h*)—is formed, and it is from these cells that the spores are formed. Each cell of the hymenium divides by transverse walls, and produces in basipetal order a long string of spores, separated from each other by intercalary sterile cells (Fig. 106, C). In this way the whole interior of the young *Æcidium* becomes filled up by numerous parallel chains of spores, which, as they grow, completely displace the cellular tissue by which the space was at first occupied. The wall or *peridium* of the cup is built up of vertical rows of sterile cells resembling the chains of spores, but connected together into a permanent tissue. This peridium at first completely encloses the fruit, but as the spores within increase in number, the enveloping layer is burst and thrown open, showing a toothed margin where its edges were torn apart (Fig. 106, B).

The æcidiospores, which are of a bright yellow colour, become separated by the breaking down of the sterile cells between them. The spores have a somewhat polygonal form, owing to mutual pressure while enclosed in the peridium, but they become approximately spherical on liberation. Their walls are thick, and each spore possesses six germ-pores or pits, through which, on germinating, the hyphæ make their exit.

The æcidiospores germinate within a few hours of their discharge, if sufficient moisture be present. They do not, however, infect the host on which they were produced, but are only able to form a mycelium if conveyed by the wind or rain on to the leaves of some member of the Gramineæ, such as the Wheat or Rye. In this case a hypha is sent out through one or more of the germ-pores. The hypha receives the protoplasm from the spore and goes on growing, bending first in one direction and then in another until its tip lights on a stoma. Then the hypha turns in through the pore of the stoma, and so makes its way into the intercellular spaces of the host, where it develops a mycelium from which uredospores are soon produced. Thus the cycle of the parasite's existence is completed.

There remains, however, yet another form of reproductive structure to be considered before we proceed to sum up the life-history. Accompanying the *Æcidium* on the Barberry, but usually on the upper surface of the leaf, are minute bodies called the *spermogonia*,[1] which are visible to the naked eye merely as little yellowish specks. They make their appearance before the cups on the opposite side of the leaf fully develop. Each of these spermogonia, when observed in a section vertical to the surface of the leaf (see Fig. 106, B, *sp*), is found to be a little flask-shaped body, consisting of a sheath of slender converging hyphæ, leaving a cavity in the middle. The spermogonium arises from the mycelium below the epidermis of the host, but ultimately breaks through it, so that the neck of the flask reaches the surface (Fig. 106, B). The hyphæ which project into the cavity form minute cells at their ends, a little row of such cells being formed on each filament. These minute cells—the *spermatia* [2]—are very much smaller than any other form of spore in the Uredineæ, their average diameter being less than one two-hundredth of a millimetre. They are produced in enormous numbers, and sugary substances (" nectar "), secreted by the spermogonium, accompany their exudation through the opening (*ostiole*) at the top. Each spermatium contains a large nucleus and a little cytoplasm.

It used to be thought that the spermatia were functionless male cells, but it is now known that they play a very

[1] Sometimes called *pycnidia*. [2] Sometimes called *pycnospores*.

important part in the life-history of *P. graminis*. Like certain other Fungi (see p. 232) *P. graminis* is heterothallic, *i.e.* it exists in two sexual strains physiologically differentiated from each other. Unless these two strains are united by a kind of sexual process, æcidia and æcidiospores do not normally develop. Let us denote the two sexual strains of this Fungus as " A " and " B." It will be remembered that when the teleutospores germinate, each cell produces a pro-mycelium or basidium that forms four spores. Two of these spores belong to the sexual strain " A " and two to " B." Both " A " and " B " sporidia can infect Barberry leaves and give rise to mycelia and spermogonia. There will, however, be two kinds of sexually differentiated mycelia and spermogonia, " A " and " B " respectively, although they look alike. Unless the two kinds of mycelia become united, or unless " A " spermatia or their derivatives come into contact and fuse with hyphæ of the " B " type, or *vice versa*, thus leading to the formation of a mycelium having both " A " and " B " characters, æcidia do not fully develop. The exact means by which an " A + B " mycelium becomes established is not yet clearly understood. It has recently been shown that flies and other insects visit the spermogonia on the Barberry leaf to obtain the " nectar," in which there are large numbers of spermatia. In this way the " A " and " B " spermatia probably become intimately associated with hyphæ of opposite sex. Often, however, in nature the " A " and " B " types of mycelium develop in close proximity so that, *e.g.*, " A " spermatia have abundant opportunity of coming into contact with " B " hyphæ. It may be that the nucleus of an " A " spermatium enters a cell of a " B " hypha, thus forming a cell with " A + B " nuclei. By nuclear division and migration other " A + B " cells may arise, and thus an " A + B " type of mycelium may become established. As a consequence the æcidia develop and produce æcidiospores. The function of the spermatia is undoubtedly to give opportunity for the establishment of mycelia of the " A + B " type.

Before the discovery of the function of the spermatia it was known that the cells which give rise to the chains of æcidiospores were bi-nucleate and that each æcidiospore was also bi-nucleate. It is clear that one of each pair of nuclei is of the " A " type and the other of the " B " type. The " A " and " B " nuclei, however, do not fuse together at this stage. When the æcidiospores germinate the " A " and " B " nuclei divide simultaneously, and mycelia are formed consisting of bi-nucleate cells. The uredospores are also bi-nucleate, and so is each cell of a young teleutospore. The " A " and " B " nuclei, however, fuse together as the teleutospores mature. The " A " and " B " sexual characters again become separated from each other during the germination of the teleutospores. Thus processes of an undoubtedly sexual nature play an important part in the life-cycle of the Rust Fungi.

We are now acquainted with the full normal life-history of *Puccinia graminis*, which affords a typical instance of the phenomenon known as *heterœcism*, this term implying that the parasite at different stages of its career necessarily inhabits two distinct hosts. In this case we have seen that on the Wheat or other members of the Grass Family, two forms of fruit—the uredospores and the teleutospores—are produced. The latter on germination give rise to sporidia which infect the other host, namely, the Barberry. It is only on the Barberry that the *Æcidium* fructification and the spermogonia are developed. The æcidiospores once more infect the Wheat or some allied plant, and the cycle is complete.

The fact that the Barberry has something to do with the appearance of Black Rust in Wheat was well known to practical farmers, long before botanists found out the scientific explanation, or even allowed the truth of the observation. During the eighteenth century a vast amount of evidence was accumulated showing that Barberry bushes acted as centres of infection, from which rust spread over the cornfields. So strong was this conviction among agriculturists, that in the year 1760 a " Barberry Law " was enacted in the province of Massachusetts in North America, ordering the rigorous extirpation of Barberry bushes throughout the province. The preamble to the Act runs thus : " Whereas it has been found by experience that the Blasting of Wheat and other English Grain is often occasioned by Barberry Bushes, to the great loss and damage of the inhabitants of this province . . . " etc. The almost complete eradication of Barberry bushes from the vicinity of arable fields in Britain accounts for the comparative rarity of the Fungus nowadays. The true explanation " that the parasitic Fungus of the Barberry and that of Wheat are one and the same species," was first suggested by Sir Joseph Banks in 1805, and fully confirmed a few years later by the independent experiments of a Danish schoolmaster named Schoeler. Botanists, however, were still unwilling to accept the fact, because the *Puccinia* of the Wheat had quite different characters from the *Æcidium* of the Barberry. It was not till 1865 that the complete demonstration of all stages of the life-history of the parasite was accomplished by the German botanist De Bary. We now know of a great many other cases of heterœcism among allied Fungi.

We must not suppose, however, that the change of host is absolutely necessary for the perpetuation of a heterœcious parasite such as *Puccinia graminis*. In Australia, for example, Black Rust is prevalent on Wheat to a serious extent, though there are no Barberry bushes nor any other plant on which the Æcidium form naturally develops. In such cases it is evident that the *Uredo* form must persist through the mild winter. Under such conditions, the teleutospores are useless, for their sporidia can only infect the other host, and not the Gramineæ.

There is another interesting feature about the parasitism of *Puccinia graminis*. Although this Fungus occurs on all the cereals, the uredospores taken from wheat cannot infect oats, or *vice versa*. It is evident therefore that *P. graminis* exists in several distinct *biologic forms*, physiologically, but not morphologically, distinguishable from one another. This kind of parasitism is known as *specialised parasitism* ; it is of common occurrence in many parasitic Fungi, especially in Mildews and Rusts.

Some other members of the *Uredineæ* have a very simple life-history compared with *Puccinia graminis*. In many of them all stages of the Fungus are passed through on the same host plant (*autœcism*), while in others certain of the stages are missing altogether.

TYPE XXV

THE MUSHROOM (*Psalliota campestris*)

The Mushroom, which to most people is the best known of all Fungi, represents a group of great extent, including more than ten thousand species. The Mushroom and its near allies (most of which are commonly called " Toadstools ") are among the most highly organised of the Fungi. What is known in ordinary language as the Mushroom is simply the fructification ; for the vegetative part of the plant, or mycelium, is very inconspicuous, and remains hidden in the soil. The " mushroom spawn " of gardeners, from which the

Fungus is often raised in cultivation, consists of blocks of richly manured soil permeated with the mycelium.

The vegetative structure is simple enough, the mycelium consisting of long, branched, multicellular hyphæ, which traverse the substratum in every direction. The individual hyphæ are usually not isolated, but woven together into strands. Fusions of the cells are very common, and take place both between neighbouring cells of the same hyphæ and between those of adjacent hyphæ. Such fusions have nothing to do with reproduction, and so far as we know serve no other purpose than to facilitate nutrition. Each cell of the mycelium contains numerous small nuclei in its cytoplasm.

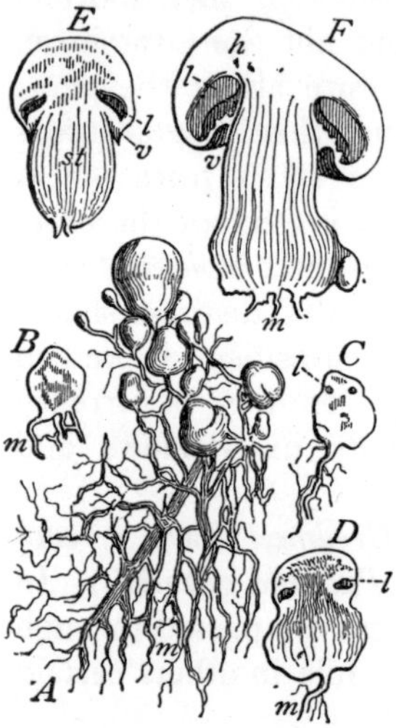

FIG. 107.—Development of a Mushroom. *A*, mycelium (*m*) giving rise to a number of young fructifications. *B*, very young mushroom in section; *m*, mycelium. *C*, slightly older; *l*, the gills just appearing. *D*, still older; *l*, gills; *m*, mycelium. *E*, older again; *l*, gills; *v*, velum; *st*, stipe. *F*, nearly ripe; *h*, pileus; other letters as before. Reduced. (After Sachs.)

The matured fructification consists, as everyone knows, of a thick stalk (the *stipe*) swollen at the base, supporting a hat-like expansion (the *pileus*), on the under-side of which are an immense number of radiating *gills* or *lamellæ*, pink when young, but of a rich brown colour when mature. If we pull up a Mushroom entire we can see, hanging on to the base of the stalk, remains of the strands of mycelium from which it arose. In Fig. 107, *A*, is shown a large piece of the mycelium made up of the thick branched bundles of hyphæ, and bearing a number of young fructifications.

The fruit itself, like every other fungal organ, is entirely built up of hyphæ. In the stalk these filaments are closely packed towards the outside, forming an apparently parenchymatous cortex. Towards the middle they are more loosely arranged, so that the separate strands are easily distinguished, and large air-spaces are left between them. The multinucleate cells of which the hyphæ are composed communicate with each other by means of pits, one of which is present in the middle of each transverse wall.

On the stalk of a ripe Mushroom, rather more than half-way up, is a membranous ring or *annulus*, formed of the remains of the *velum*, which at an earlier stage covered in the lower surface of the pileus, as shown in Fig. 107, *E* and *F*.

The tissue of the pileus is like that of the stalk, but rather denser. The gills on the under-surface are formed by an extension of the hyphæ of the pileus. If we cut a tangential section of the pileus, we see the gills or lamellæ in transverse section, and can make out their structure (see Fig. 108). The middle part of each lamella is formed of hyphæ coming down from the pileus, and following on the whole a longitudinal course, their lateral branches, however, diverging towards the two surfaces. This central tissue of the lamella is called the *trama* (Fig. 108, *B*, *C*, *t*). Towards the free surfaces the cells of the diverging hyphæ are shorter and more closely packed, forming the *sub-hymenial layer* (*sh*), and beyond this again we come to the hymenium itself, which is thus composed of the terminal cells of the same hyphæ which constitute the trama and sub-hymenial layer (Fig. 108). In this last part of their course the filaments have diverged from their original direction to such an extent that they now stand at right angles to the surface of the lamella.

The hymenium consists of a palisade-like layer of club-shaped cells rich in protoplasm. A few of these are somewhat more slender than the rest, and remain sterile, bearing the name of *paraphyses*. The others are of stouter build at maturity, and are the spore-producing elements, here called *basidia*. Each basidium gives rise at its free end to from two to four minute peg-like outgrowths (the *sterigmata*), each of which enlarges at the tip to form a *basidiospore* (see Fig. 108, *C*, *s*). The wild Mushroom has usually four spores on each basidium and the cultivated form two. The spores when young are pink, but when ripe are purplish black; they contain oil, and have each four nuclei. These are derived from the basidium, which at an earlier stage possesses a single nucleus formed by the fusion of the two nuclei which it originally contained. The fused nucleus of the

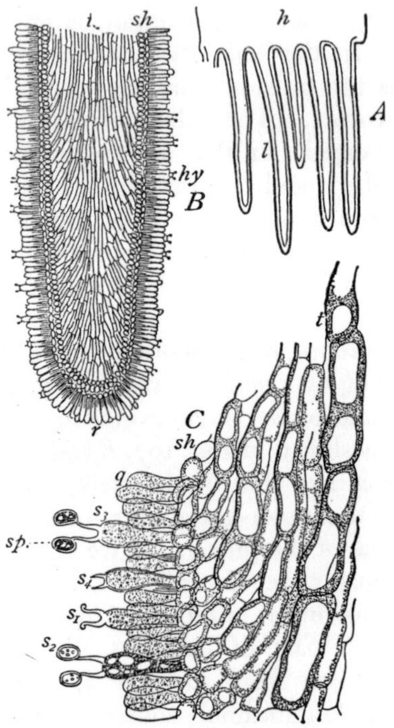

FIG. 108.—Gills of Mushroom. *A*, part of tangential section of pileus (*h*), showing gills (*l*). Slightly magnified. *B*, single gill in section; *t*, trama; *sh*, sub-hymenial layer; *hy*, hymenium; *r*, lower edge of gill. Magnified about 80. *C*, part of *B* enlarged; *t*, cells of trama; *sh*, sub-hymenial layer; *q*, paraphyses and very young basidia; s_1-s_4, stages in development of basidia; *sp*, basidiospores. Magnified 230. (After Sachs.)

basidium divides into four, and the daughter-nuclei pass over into the basidiospores, one or two into each. They again divide, so that the mature spore has four nuclei. If the surface of a young gill is examined it will be seen to be faintly mottled in an irregular manner, some areas being pink and others nearly white. This is due to the fact that the basidia do not develop simultaneously on all parts of the gills. In the pink zones the basidia are nearly ready to shed their spores, whereas in the paler zones the basidia are only just beginning to form their spores. At maturity the spores are violently shot forth from the sterigmata. After spore discharge the basidia collapse, and this part of the gill again becomes pale in colour. Subsequently, the same area once more becomes pink owing to the development of another crop of basidia. At a later stage the whole of the gills become brownish or purplish black on account of a darkening of the cells in the matrix of the gills. The wave-like ripening of the basidia can be seen by examining the surface of a young gill under the microscope.

This mode of fructification, consisting of basidia bearing spores on sterigmata, is universal throughout the great group to which *Psalliota* belongs, hence called the *Basidiomycetes.* The subdivision of this group, represented by *Psalliota,* is characterised by the hymenium being exposed to the air when ripe, and bears the name of the *Hymenomycetes.* The basidium, when it has once produced its two or four spores, is exhausted, and does nothing more ; but for a time new basidia may arise, growing up between the old ones.

An immense number of spores are produced from the gills of a Mushroom. Some idea of their multitude may be obtained by cutting off the pileus of a Mushroom and laying it, gills downward, on a sheet of white paper. If it be removed after a time an exact print of the gills will be found on the paper, in the form of a fine powdery deposit of spores which have fallen from them.

Until recently, nothing satisfactory was known as to the germination of the spores of the Mushroom. Of late years, however, Mushrooms have been successfully raised from spores; the entire development, up to the formation of ripe fructifications, takes from six to seven months. As a rule they are raised from the mycelium or "spawn."

The basidial fructification is quite distinct from that of any other group of Fungi which we have described. There are, however, a number of doubtful forms among the lower Basidiomycetes which appear to connect that group with the Uredineæ, and the basidium of the Mushroom group is undoubtedly homologous with the pro-mycelium produced from the teleutospores of the Uredineæ, the basidiospores corresponding to the sporidia. There are some Uredineæ, such as the *Puccinia* so common on hollyhocks (*P. Malvacearum*), in which the teleutospores germinate *in situ*, *i.e.* while still in the pustule and attached to the mycelium. In this case the resemblance of the pro-mycelium to the basidia of some of the simpler Basidiomycetes is very striking.

In Fig. 107 the development of the Mushroom-fruit is illustrated. The young Mushroom arises from a tangle of hyphæ borne on a strand of mycelium. The intertwined hyphæ group themselves into a tissue, thus forming a little oval tubercle. At first the Mushroom is all stalk; soon, however, the pileus begins to appear at the top. In the earlier stages there is no separation between pileus and stipe (Fig. 107, *B*, *C*). The gills are developed endogenously, while enclosed on all sides by continuous tissue (Fig. 107, *D*). Later on the pileus begins to spread out laterally (Fig. 107, *E*), but its under-side is still closed in. The tissue which connects the edge of the pileus with the stalk, and thus encloses the gills from below, is called the *velum* or veil (Fig. 107, *F*). At last this becomes ruptured as the pileus expands, and its torn remains adhere to the stipe, forming the *ring*, which we mentioned in describing the ripe fructification.

Mushrooms develop normally in complete darkness, and in fact they are often cultivated in cellars and caves. The fruit bodies are sensitive to gravity, the stipe being negatively, and the gills positively, geotropic. These tropisms lead to the correct orientation of the fructifications for spore discharge. After being hurled from the sterigmata the spores fall downwards in the spaces between the gills and are then caught up by convection currents in the air and dispersed. If a ripe Mushroom be suspended in a dimly lighted room a faint cloud of spores can be seen arising from the lower part of the pileus. The occurrence of the hymenia on the under surface of the fruit body protects the basidia from rain and from excessive loss of water in dry weather, and the erection of the pileus above soil level facilitates the dispersal of the spores by air currents after discharge from the basidia. The fructification of the Mushroom is a most efficient mechanism for the formation and liberation of large numbers of spores.

The Mushroom is a saprophyte growing in pastures and in richly manured soil, but some of its near relations are parasitic on trees to which they do great damage. Some Toadstools are poisonous and even fatal to human beings if eaten, hence great care should be taken in the selection of mushroom-like Fungi for food.

Although there are no true sexual organs in the Mushroom and its allies, the invariable fusion of nuclei in the young basidia is undoubtedly a kind of sexual process. Some Fungi closely related to the Mushroom are heterothallic, the spores being of two sexual types, "A" and "B," as in the sporidia of *Puccinia graminis*. Unless a hypha from an "A" spore fuses with that from a "B" spore, thereby establishing an "A + B" type of mycelium, fertile fruit bodies are not formed. In such a heterothallic type one of the nuclei in the young basidium is of the "A," and the other of the "B," kind.

In the Mushroom no other form of spore than the basidiospores has so far been discovered. In some nearly allied Fungi, however, additional forms of spores, such as conidia, are produced on the mycelium.

We have now finished our series of types of Fungi. It has only been possible to consider a very few representatives, and many important groups have been left altogether untouched. We have gained, however, some slight idea of the great range of structure which the class presents, and in our later types we have seen how very far the higher Fungi have diverged from the simple forms with which we started our survey.

CHAPTER V

THE BACTERIA

THE Bacteria, which in these days are familiar, by name at any rate, to everyone, are an extensive group of organisms of the most minute size, and, so far as we know them, of the most simple structure. Most Bacteria are unicellular. In their mode of life they bear a general resemblance to Fungi, for, though sometimes pigmented, they are destitute of chlorophyll, and as a rule adapted either to a parasitic or saprophytic existence, obtaining both their carbon and nitrogen supplies usually from organic sources. They are, however, as we shall see, quite different from any known Fungi in structure and development, and constitute a wholly distinct class of organisms.

Both as parasites and as saprophytes, the Bacteria play an enormously important part in the world. Parasitic Bacteria are now known to be the cause of very many of the infectious diseases of man and animals, *e.g.* tuberculosis and cholera, and in many cases the actual species to which the different diseases are due have been strictly determined. Other Bacteria cause important plant diseases. As saprophytes, Bacteria are the great agents of decay of all kinds, owing to the fact that they set up rapid and profound chemical transformations in the organic substances on which they feed. Thus when milk turns sour, or when wine is converted into vinegar, or protein substances, such as meat, undergo putrefaction, the change is in each case due to the action of a definite species of the Bacteria. On the same power of initiating far-reaching decompositions in the bodies which they inhabit, depends the fatal efficiency of the parasitic Bacteria in producing disease. The whole subject of the fermentations set up by these organisms has become in recent years of the greatest possible practical importance in relation both to medicine, as regards the parasitic forms, and to innumerable branches of industry, as regards the saprophytes. A new science, Bacteriology, has grown up on these subjects, which lie beyond the province of the present Introduction.

TYPE XXVI

BACILLUS SUBTILIS

This is one of the commonest and best-known forms of Bacteria. It occurs constantly in hay, and can be obtained with certainty by soaking or boiling hay in water. In the latter case the appearance of the *Bacillus* depends on the extraordinary resistance to heat shown by its spores, which can stand an hour's boiling with impunity. After a little time the whole of the liquid simply swarms with the cells of the *Bacillus*, which in its active vegetative condition is a strictly unicellular organism, the isolated cells having the shape of short rods rather more than $\frac{1}{1000}$th of a millimetre (1 μ) in diameter and from $\frac{5}{1000}$ to $\frac{8}{1000}$ mm. (5–8 μ) in length. The cells are thus far more minute than those of any plant we have hitherto considered. The excessive smallness of the cells has placed great difficulties in the way of their investigation, and the cell-structure of Bacteria is still very little understood. So far as we know at present, however, this structure appears to be very simple. There is a definite membrane which, however, does not consist of cellulose, but seems to be chiefly of a protein nature. The cells move actively, and their movements are now known to be due to cilia (flagella) which are attached to the protoplasm and penetrate the wall (see Fig. 109, *a*, *d*).

Other Bacterial cells are spherical (*Micrococcus*), while some are kidney-shaped (*Vibrio*) or are in the form of loose spirals (*Spirillum*). Some Bacteria form filaments which are spuriously branched (*e.g. Cladothrix*). Many Bacteria have no cilia and are non-motile.

The whole interior of the cell is usually occupied by the protoplasm. No definite nucleus has been demonstrated, though a substance similar to that of which nuclei are composed is present scattered throughout the protoplasm.

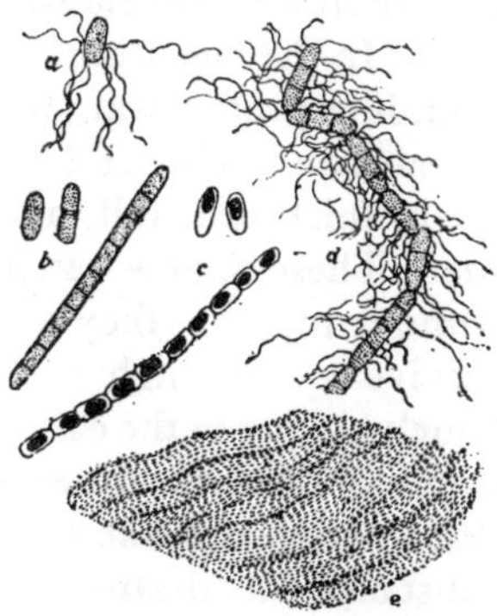

FIG. 109. — *Bacillus subtilis*. *a*, *d*, ciliated motile cell and filament; *b*, non-motile cells and filament; *e*, zooglœa, during spore-formation; *c*, cells and filament with endospores, fıom the zooglœa. *a-d*, ×1050. *e*, ×175. (After A. Fischer.)

For some time the *Bacillus* continues in the actively swarming condition, multiplying abundantly by the repeated transverse division of the cells. After some days the individuals begin to seek the surface of the liquid, where they pass into a resting condition. At this stage the cells remain connected together in long filaments, and their outer cell-walls become very gelatinous. This is called the *zooglœa* condition, and is easily recognised by the gelatinous iridescent film which the colonies of the organism form on the surface.

Lastly, when the food in the liquid is getting exhausted, the spores begin to form. This takes place after locomotion has ceased and the *Bacillus* has entered the filamentous condition (see Fig. 109). The spores in this species and in a very large group of allied Bacteria, are *endospores*, one spore being produced in the interior of each cell. A new wall appears round a portion of the contents, in which the nuclear substance is included. The young spore absorbs the remaining protoplasm, and it becomes elliptical in form, and increases sufficiently in bulk for its walls to touch those of the mother-cell. In the meantime it has completely used up the surrounding protoplasm, and now lies within a mere empty membrane. The endospore itself acquires a comparatively thick cell-wall, and is extraordinarily tenacious of life. These spores can bear being completely dried up without injury; they are little affected by poisons, and survive a very high temperature, withstanding even an hour's boiling in the case of the hay *Bacillus*. Hence spore-forming Bacteria are extremely difficult to extirpate, so that in order to make sure of effectually "sterilising" any substance (*i.e.* destroying any living things which it contains) it is often necessary to expose it to a temperature considerably above the boiling-point of water, or, if that be impracticable, at least to continue boiling for some hours.

The spores germinate when brought into a suitable food-solution. The outer membrane splits across, and the entire contents escape as an ordinary bacterial cell, which at once begins to move about by means of cilia (see Fig. 109, *a*).

Fig. 110 shows very completely the stages in the formation and germination of the spores in another *Bacillus*, called *B. megatherium*, because for one of the Bacteria it is quite a monster, though its cells are only about $\frac{1}{400}$th of a millimetre ($2{\cdot}5\,\mu$) in diameter. This species was originally found in boiled cabbages, and was afterwards cultivated by its discoverer in solutions of glucose, to which a little extract of meat had been added. The formation of endospores characterises one great group of Bacteria, and distinguishes them from similar unicellular organisms. Many other Bacteria, however, are devoid of spores.

Bacillus subtilis, like most other living things, requires plenty of atmospheric oxygen in order to flourish. Some of the other Bacteria, however, have the remarkable peculiarity that they thrive best in the *absence* of free oxygen. This is the case, for example, with *Bacillus butyricus*, the organism to which the formation of butyric acid by the fermentation of sugar is due. In this case the oxygen necessary for respiration is not absorbed in the free state, but obtained from the breaking down of the organic substance in which the organism lives.

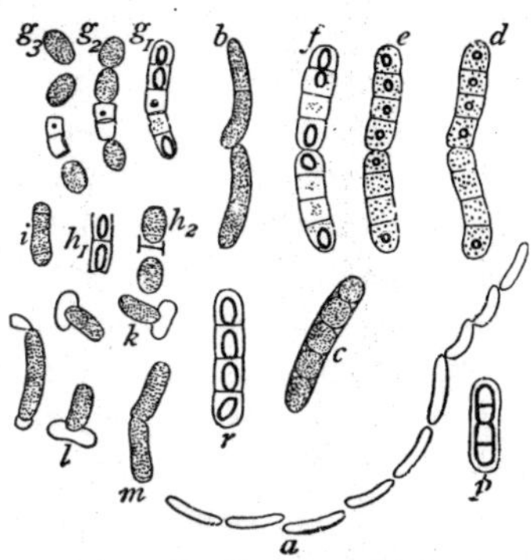

FIG. 110.—*Bacillus megatherium*. *a*, chain of vegetative rods, each consisting of two or more cells, but septa not shown. Magnified 250. *p*, four-celled rod, after treatment with alcoholic solution of iodine; *b*, vegetative rods; *c*-*f*, rods, showing the formation of endospores; *r*, four-celled rod with ripe spores; g_1-g_3 and h_1, h_2, spores swelling before germination; the mother cell-walls disappear; *k*-*m*, germination of spores. All figures except *a* magnified 600. (After De Bary.)

It may be mentioned here that numerous experiments have proved that bright light has a very unfavourable effect on Bacteria, completely stopping their growth and multiplication in many cases, and even, when intense enough, killing the cells outright. It is the rays towards the violet end of the spectrum and the ultra-violet rays which exercise the greatest retarding effect on the growth of these creatures. The action of light in checking the increase of these agents of decomposition and disease is evidently a fact of great practical importance.

It was stated in Part I. (p. 206) that plants of the Pea and Bean kind, unlike ordinary green plants, are able, by the help of certain Bacterial companions, to obtain their nitrogenous food from the free nitrogen of the atmosphere. The plants in question, including most if not all of our native Leguminosæ, invariably have swellings or tubercles on their roots. These tubercles are inhabited by a symbiotic bacterium, the entrance of which into the root is the cause of the first formation of the tubercle. It has been proved conclusively that it is only when this organism is present in the soil that the tubercles develop on the roots, and only when the tubercles are formed that free nitrogen can be assimilated. If the plants are grown in sterilised soil, *i.e.* soil which has been heated sufficiently to kill all living things contained in it, then no tubercles develop, and no nitrogen is absorbed from the air. When the tubercles are present, however, great quantities of nitrogen are assimilated, and the plant can thrive even if nitrogenous compounds be quite absent from the soil. A very important result of this fact is that leguminous crops actually enrich the soil in nitrogen. In Germany one sees whole fields of Yellow Lupine grown for no other purpose than to be ploughed in and so enrich the soil for other crops. In this country clover is often grown before wheat in the farm rotation of crops. The remains of the clover on being ploughed into the soil enrich it in nitrogen for the growth of the wheat.

The subject is mentioned here because the organism to which this assimilation of gaseous nitrogen is due, is a *Bacillus*, and has been described under the name of *B. radicicola*. Moreover, the relation of this creature to its leguminous host seems to be one of symbiosis, or mutual service, rather than of one-sided parasitism, for the bacterium obtains its supplies of carbon from the green plant.

B. radicicola is only one of many kinds of Bacteria that play an essential part in the circulation of nitrogen in Nature. Certain Bacteria convert urea into ammonium compounds, which other soil Bacteria (the nitrifying Bacteria) change first into nitrites and then into nitrates. These nitrifying Bacteria have the remarkable property of assimilating carbon dioxide, although devoid of chlorophyll. This is done with the aid of the energy liberated in the processes of oxidation that lead to the formation of nitrites and nitrates. Still other soil Bacteria utilise atmospheric nitrogen in their own metabolism, and in this way " fix " it for the future benefit of the green plant. It is not too much to say that green plants are ultimately dependent on Bacteria for their nitrogen, for, in general, green plants can only absorb nitrogen from the soil in the form of nitrates or sometimes ammonium compounds. Without these Bacteria, green plants would cease to exist.

Before leaving the subject of Bacteria it may be mentioned that as a group they show an extraordinary range of different types of metabolism. It has already been implied that the nitrifying Bacteria obtain the whole of their food supplies from inorganic sources. In certain other Bacteria, carbon dioxide is decomposed under the influence of light, with the help of a purple pigment which seems to have similar properties to those of chlorophyll. In the curious group of the " sulphur bacteria " sulphuretted hydrogen is oxidised to sulphur and energy thereby gained, the sulphur being deposited as granules in the cells.

CHAPTER VI

THE MYXOMYCETES[1]

THE present type represents a group of organisms lying on the borderland of the animal and vegetable kingdoms. It may be doubted whether they have any right to a place in a book on Botany, but we give them the benefit of the doubt because of their great scientific interest; for in them we can study living protoplasm and its behaviour on a greater scale than in any other creatures. Myxomycetes, unlike Fungi and Bacteria, are of no practical importance, apart from a few which are parasitic on plants, *e.g. Plasmodiophora Brassicæ*, the cause of Finger and Toe or Club Root disease of turnips and cabbages. They are probably known to very few people except naturalists; yet they are common enough, easily visible to the naked eye, and in some conditions extremely conspicuous.

In the vegetative state a typical Myxomycete consists of a mass of naked protoplasm, sometimes several inches in extent, which creeps slowly about, on the surface of dead leaves or bark or wood. Such immense aggregations of living matter in so simple a form are quite unknown in any other group of organisms.

When reproduction is about to take place, the creature completely changes its character, gradually ceases to be active, and converts itself into a collection of fruits of rather complex structure, in which the microscopic spores are produced. The spores on germination give rise to

[1] Also called *Mycetozoa*.

swarm-cells, which unite together to build up the great protoplasmic body with which we started. Such are the rough outlines of a Myxomycete's career. We will now proceed to study a particular example more in detail.

TYPE XXVII

BADHAMIA UTRICULARIS

1. THE PLASMODIUM

This Myxomycete is common in some years, though rare in others; it occurs on the bark or wood of fallen trees, on old garden seats, and in fact in all places where timber is left exposed to damp and decay. In its ordinary state the organism forms irregular flat gelatinous masses of a deep chrome-yellow colour, spreading and creeping over the surface of the rotting wood, especially in damp weather. These creeping masses are called *plasmodia*. A small plasmodium is shown in Fig. 112. This specimen was only about half an inch across, but much greater dimensions are usually attained, the area covered by one plasmodium sometimes amounting to as much as six square inches. The plasmodium is not uniform in thickness throughout, but is traversed by thicker veins, which unite together to form a kind of network. The thinner protoplasmic layer between the veins is sometimes interrupted so as to leave some of the meshes empty. The whole plasmodium is a mass of living protoplasm which is in constant movement. The movement is of two kinds—(1) an advancing locomotion of the whole plasmodium, and (2) an internal circulation of the protoplasm, especially in the veins. The locomotion is a slow, creeping movement; the advancing edge of the mass (*a*, *a* in Fig. 111) is constantly putting out feelers (*pseudopodia*, as they are called), which are sometimes withdrawn again, but more often maintain their position, and are increased by the flow of protoplasm from behind. The outer layer of the whole plasmodium is clear and transparent ; the inner mass is very granular, and the granules are especially abundant in the veins. Many of the granules consist of lime (calcium carbonate), and it is around these lime-granules that the yellow colouring-matter is chiefly

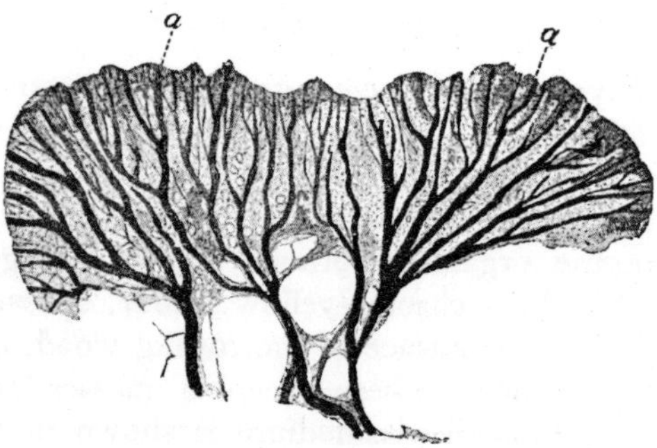

FIG. 111.—*Badhamia utricularis*; plasmodium, from a stained specimen, prepared by the late Mr. A. Lister, F.R.S. *a*, *a*, advancing margin. Magnified about 5. (R. S.)

deposited. The clear part of the protoplasm is colourless. When a pseudopodium is first extended it consists of the clear part (*hyaloplasm*) only ; subsequently the inner granular substance flows into it and increases its mass.

The internal movement along the veins is extremely active, and can be followed with ease under the microscope by means of the granules which are swept along with the current. The flow is curiously rhythmical. In each vein the current sets steadily in one direction for from one and a half to two minutes, then it slackens and stops altogether for an instant, only to recommence with equal energy in the opposite direction. This internal flow is closely related to the locomotion of the whole plasmodium, for it is found that the current lasts longest in that direction in which the plasmodium is advancing.

The movements do not go on at random, but take a definite direction in accordance with the needs of the organism. Thus an active plasmodium, if the wood be wetted on one side of it more than on the other, will move towards the damper side; if, however, the Myxomycete be about to form spores (for which moisture is not favourable) it will move the opposite way, in the dry direction. Generally speaking, a plasmodium will try to avoid intense light (which no doubt has a bad effect, as in the case of the Bacteria); for the purpose of spore-formation, however, it will leave any dark recess of the wood in which it may be hidden, and seek the light.

This particular Myxomycete, *Badhamia utricularis*, feeds on living Fungi, especially on certain members of the Hymenomycetes, which grow on decaying wood. If a piece of one of these Fungi be placed in its way, the advancing margin of the *Badhamia* at once begins to flow over it, and the whole plasmodium will turn aside in the direction of the prey. Individual hyphæ, or small pieces of the fungus, or even Bacteria living on the surface, become enclosed in vacuoles in the protoplasm of the *Badhamia* and digested, their useless remains being afterwards disgorged and left behind on the track. This way of feeding, by taking solid food into the body, and then digesting it, is characteristic of animals, and is not known among true plants. The Myxomycetes generally have this power, but in most cases they live on dead and decaying substances, such as fragments of bark and wood, and to some extent their mode of nutrition may be saprophytic. As a reserve food material the plasmodium forms glycogen. Our type is exceptional in preying on living tissues. Its frequency in any particular season depends on the weather and the abundance of the Fungi which it grows on.

The plasmodium contains an immense number of nuclei, which are only absent from the clear external portion. The nuclei (see Fig. 112) have the same structure as those of higher organisms, possessing a nucleolus and a framework of delicate threads. They increase in number by division, as the plasmodium grows.

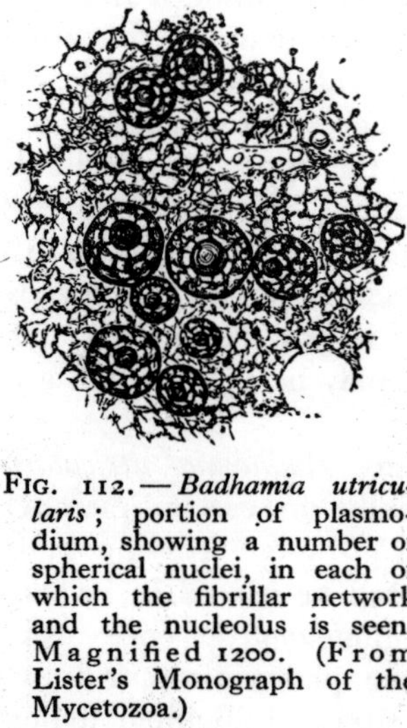

FIG. 112.—*Badhamia utricularis*; portion of plasmodium, showing a number of spherical nuclei, in each of which the fibrillar network and the nucleolus is seen. Magnified 1200. (From Lister's Monograph of the Mycetozoa.)

We see, then, that the plasmodium of a Myxomycete is a typical example of non-cellular structure, consisting of a perfectly continuous protoplasmic body of large size, containing vacuoles and numerous nuclei, but entirely destitute of any cell-wall.

Sometimes the plasmodium passes into a resting condition, a change which often, though not always, happens in consequence of drought. The movements cease, and the protoplasm becomes partitioned into a number of irregular *cysts* or cells, each containing about ten or twenty of the nuclei. The cysts are separated from each other by firm walls, which are hardened portions of the protoplasm. In this resting stage, which is known as the *sclerotium*, the plasmodium may remain alive for as long as three years. The external appearance of the sclerotium of our type is that of a dry, horny, irregular mass, of a brick-red colour. When moistened, it revives, the walls of the cysts become absorbed, and the contents reunite and recommence the movements characteristic of active life.

2. The Sporangia and Spores

When a Myxomycete fructifies it completely changes its appearance. The whole of the active protoplasm is used up to form a sorus of sporangia in which the spores are contained. The sporangia are often brightly coloured; in *B. utricularis* they are ashy-grey. In Fig. 113 a cluster of sporangia (from an allied genus) is shown. The ripe sporangium is a rounded hollow case, borne on a stalk; it has a firm external wall, and its interior is traversed by a network of threads, among which lime is deposited (see Fig. 113, *b*). In the meshes of the network are contained the numerous spores.

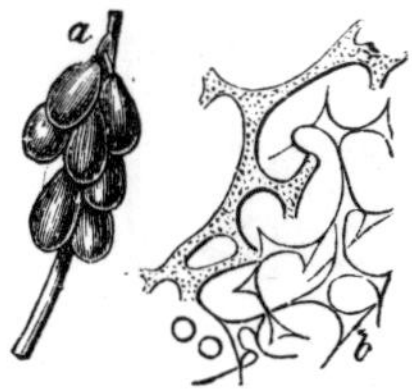

Fig. 113.—*Leocarpus vernicosus*. *a*, group of sporangia on a fragment of dead leaf. Magnified 2½. *b*, portion of capillitium with spores. Magnified 120. (From Lister's Monograph of the Mycetozoa.)

When fructification is about to take place, the protoplasm accumulates at certain points, corresponding to the position of future sporangia. At each of these points the protoplasm heaps itself up to form a projecting mass; a portion of this hardens and becomes the stalk, while the living part continues to creep upwards, and constitutes the sporangium itself at the top. The outer layer of the terminal mass of protoplasm forms itself into the firm outer wall, while the interior part builds up a network of hollow branched threads (the *capillitium*), which traverse the cavity in all directions (see Fig. 113, *b*). Between these threads masses of living protoplasm remain. After the wall and capillitium are completed, the formation of spores takes place. The protoplasm in the meshes breaks up into distinct masses, the nuclei of which all undergo division. It may be mentioned here that nuclear division, both in the sporangia and in the plasmodium as well, takes place in just the same complicated way as in the tissues of the higher animals and plants. Ultimately the whole of the living protoplasm in the sporangium further divides up into spherical spores, each of which includes a single nucleus.

FIG. 114.—*Comatricha obtusata*. *a*, group of sporangia. Natural size. *b*, empty sporangium, showing capillitium. Magnified 16. (From Lister's Monograph of the Mycetozoa.)

The cell-wall of the spores, and also the substance of the sporangial wall and the capillitium, resemble the cuticularised membrane of vegetable cells. In a few cases cellulose has been found. It will be noticed that complicated as the structure of the sporangia is, there is no formation of distinct cells until the spores themselves are developed. In Fig. 114 the sporangium of another Myxomycete (*Comatricha obtusata*) is figured. Here the sporangial wall soon disappears, so that the whole capillitium in connection with the stalk becomes visible.

The sporangia of *Badhamia* open by the breaking down of the membrane, and the spores are exposed. They hang for a time on the threads of the capillitium, which acts as a supporting scaffolding, and are gradually scattered. The spores are brownish, spherical, and 9–12μ in diameter. The spores can be kept dry for a long time in a living condition, and germinate readily when wetted (see Fig. 115, from another Myxomycete). The membrane splits, and the whole contents become free. The protoplasmic body at first shows "amœboid" movements, changing its form by putting out and again withdrawing pseudopodia. After a few minutes a single cilium is developed at one end, and now the pear-shaped swarm-spore is fully formed (Fig. 115, *d*). It contains one nucleus, placed near the thin end, and its cytoplasm is vacuolated; one of the vacuoles is contractile, expanding and contracting at regular intervals. The swarm-spore swims through the water with a dancing movement, or it creeps along the surface of any solid body like a snail.

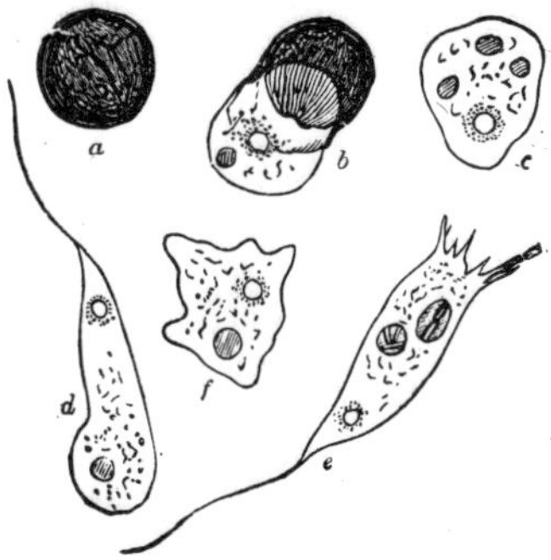

FIG. 115.—*Didymium difforme*. *a*, spore; *b*, swarm-cell escaping from the spore-membrane; *c*, newly-hatched swarm-cell, containing a nucleus and three vacuoles; *d*, ciliated swarm-cell; *e*, swarm-cell, with two vacuoles containing bacteria—another bacterium is just caught by the pseudopodia; *f*, amœboid swarm-cell. Magnified 720. (From Lister's Monograph of the Mycetozoa.)

These swarm-spores, like the plasmodia, can take in their food in the solid state, but they may also feed in a saprophytic manner. They catch minute objects in the water by means of pseudopodia put out at the broad posterior end. They are particularly fond of Bacteria, which are often caught in this way, the pseudopodia laying hold of the microbe, and in spite of its struggles dragging it in until it is enclosed in a vacuole of the cytoplasm, and ultimately digested (Fig. 115, *e*).

The swarm-spores multiply repeatedly by division

into two, the movement ceasing before the division takes place, and starting again when the daughter-cells are formed. The division of the nucleus precedes that of the cell. The swarm-spore may also become encysted, surrounding itself with a cell-wall, from which it afterwards escapes, resuming its active career. The encysted swarm-cells bear the name of *microcysts*.

From the swarm-spores the plasmodium is built up. Before this happens they withdraw their cilia, and henceforth confine themselves to creeping amœboid movements. When two creeping *myxamœbæ* (as they are called at this stage) meet, their protoplasm flows together into a single mass. According to recent observations on other Myxomycetes the nuclei of these *myxamœbæ* also fuse, so that the act of union appears to be a sexual conjugation. Then other swarm-spores, *microcysts* and *myxamœbæ* coalesce with the two cells that have fused sexually into one, and so the beginning of a new plasmodium is made (Fig. 116). The nuclei of the accessory swarm-spores, etc., are digested by the young plasmodium. The nuclei of the latter arise solely by division of the single nucleus formed by the sexual fusion of the first two *myxamœbæ* to coalesce. In some Myxomycetes sexual fusion occurs between motile swarm-spores instead of between *myxamœbæ*.

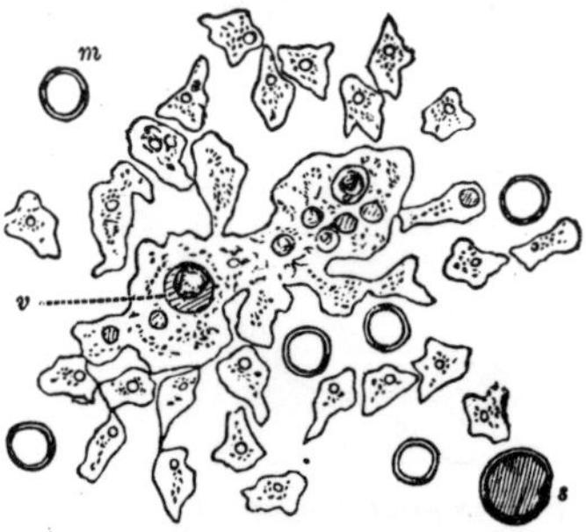

FIG. 116. — *Didymium difforme*; young plasmodium with attendant myxamœbæ. *m*, microcyst. One microcyst is being digested in a vacuole (*v*). *s*, empty spore-membrane. Magnified 470. (From Lister's Monograph of the Mycetozoa.)

Thus we see that a plasmodium is a compound structure, built up in the first instance by the union of a number of distinct protoplasmic bodies. As there is no fusion of nuclei in the later stages, only the first union can be regarded as sexual.

When a plasmodium has been started in this way it continues to grow, and the number of its nuclei increases by division to keep pace with its growth,—a fact which is now well ascertained.

Such are the outlines of the life-history of these extraordinary creatures, which in their whole structure and mode of development differ widely from any other organisms. We can study in them, better perhaps than in anything else, the behaviour of living protoplasm when untrammelled by the bonds of cellular structure. We see how closely movement and growth are connected ; and when the period of fructification comes on, we can observe how the protoplasm by its active exertions literally builds up the new structure out of its own substance. It has been well said that in the plasmodium of a Myxomycete we have a type of the organisation of all plants, for we see in these organisms, freely exposed to view, the same movements and the same constructive activity of the living matter, on which the growth and development of the highest plants depend. In the latter, however, the living agent is concealed within the framework of the cells, and its successive changes of form are stereotyped by the rigidity of the structures which it has itself built up.

CHAPTER VII

THE FLAGELLATA

THE Flagellata are a large group of motile, unicellular organisms, some of which possess wholly animal characteristics, while others, like *Euglena viridis*, show a mixture of plant and animal characters. The Algæ may have evolved from pigmented forms like the Flagellata, and the Fungi from colourless members of this group or allied groups of the Protozoa. It is not possible to draw a hard-and-fast line between the Flagellata and certain unicellular Algæ (see p. 180).

TYPE XXVIII

EUGLENA VIRIDIS

Our type, *Euglena viridis*, is sometimes found in great abundance in shallow stagnant water, especially in water containing considerable quantities of nitrogenous substances, such as pools in farmyards and around manure heaps. In such situations *Euglena viridis* may be so plentiful as to colour the water a brilliant green.

Under the microscope the organism is seen to be actively motile, as in *Chlamydomonas*, but it may also exhibit a slow, crawling amœboid movement when not progressing rapidly. In the latter case *Euglena viridis* changes its shape gradually as it slowly moves. At one moment an individual may be pear-shaped or spindle-shaped with a pointed posterior extremity, at another it may be nearly spherical. These changes of form are due to the fact that the organism is not bounded by a rigid cell-wall as in *Chlamydomonas*, and it is the absence of a cell-wall in the motile condition which is one of the most important differences between the Flagellata and the unicellular Algæ. The peripheral region of the protoplast of the motile *Euglena*, called the *periplast*, is somewhat more rigid than the interior and shows a faint spiral striation.

Euglena viridis moves rapidly through the water by means of a single protoplasmic thread, the cilium or flagellum (Fig. 117), which arises from the gullet and which by its own movements propels the organism onwards. The flagellum is about the same length as the cell. It is comparable with the cilia of *Chlamydomonas*, but is considerably coarser. The flagellum can generally be seen easily, although by running a little iodine solution under the coverslip it may be made clearer.

Within the cell is a bright green chloroplast of stellate form (Fig. 117), which enables the organism to carry out photosynthesis of the carbon dioxide dissolved in the water. The visible product of photosynthesis, however, is not starch, but an allied substance, paramylon, which does not stain blue with iodine. Grains of paramylon can be seen clustered round the "pyrenoid" in the central part of the chloroplast, and also in other parts of the cell. In a median position in the lower part of the cell is a relatively large nucleus containing a conspicuous nucleolus (Fig. 117). The nucleus is usually only visible after staining with iodine or other reagent. Towards the anterior part of the organism is a red eye-spot, which, as in *Chlamydomonas*, there are reasons for thinking renders the organism sensitive to varying light intensities. *Euglena* tends to move towards light of moderate intensity. Also in the front of the organism and at one side is a large contractile vacuole, which is sometimes accompanied by subsidiary ones (Fig. 117). The contractile vacuole expands and contracts, and empties itself into the reservoir which connects with the exterior by a very narrow channel, the gullet. In many colourless Flagellata the gullet is the channel by which solid food is taken into the organism and digested in an animal-like way.

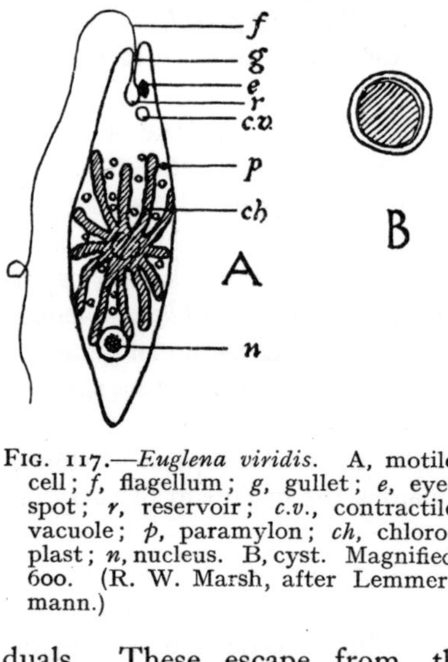

FIG. 117.—*Euglena viridis.* A, motile cell; *f*, flagellum; *g*, gullet; *e*, eyespot; *r*, reservoir; *c.v.*, contractile vacuole; *p*, paramylon; *ch*, chloroplast; *n*, nucleus. B, cyst. Magnified 600. (R. W. Marsh, after Lemmermann.)

Division takes place in *Euglena* after the organism has come to rest and has discarded its flagellum and surrounded itself with a somewhat gelatinous membrane. In this quiescent state the protoplast divides longitudinally into two and then into four portions, which soon become differentiated as complete individuals. These escape from the delicate membrane in an actively motile condition. At other times, especially when desiccation threatens, the individual comes to rest and surrounds itself with a thick membrane to form a *cyst* (Fig. 117). In the encysted condition the organism can withstand long periods of drought, and may be distributed by wind like particles of dust. In a favourable environment, *e.g.* an abundant supply of water, the protoplast within the cyst divides into two or four individuals which are liberated in an actively motile state when the wall of the cyst ruptures. There is no kind of sexual reproduction in *Euglena viridis*, and this is true of most of its nearest allies and of other Flagellata.

In so far as its mode of nutrition is that of an Alga, all the substances required for food being dissolved in the water, *Euglena viridis* is obviously plant-like, but in view of its power of movement, the absence of a cell-wall in the motile state, and of certain other structural features, it is also clearly a unicellular animal. Large numbers of the Flagellata possess a mixture of plant and animal characteristics, and this is probably an indication that both plants and animals have had a common origin in the very remote past.

If *Euglena viridis* is kept in water containing much organic food, its brilliant green colour tends to fade, and under such circumstances it feeds in a partly saprophytic manner. Other species of *Euglena* become entirely colourless and wholly saprophytic in their mode of nutrition if grown in darkness, in solutions of certain organic substances.

CHAPTER VIII

CONCLUSION

It will be useful, at the conclusion of our series of types, to sum up very briefly what we have learnt from them, with reference especially to the relationships of the groups which they represent. Throughout the book we have followed on the whole a descending order, proceeding from the more complex to the more simple, though there have been many exceptions to this rule, because it is impossible to arrange any set of plants in a single linear series, whether according to increasing or decreasing complexity.

In the present summary we will follow the reverse order, starting where we left off, with the lowest forms, following up the various lines of affinity, and concluding with the highest plants, with which we began in Part I. This is the natural succession, and the attempt to follow it will at any rate teach us how complicated the relationships are, and how difficult it is to arrange naturally even a few types such as ours.

The two groups near the end of the series, with which we therefore begin our summary, have no clear and evident affinities with any of the rest. As regards the Myxomycetes, it is doubtful whether any such relationship exists at all. These organisms have attained a fairly high stage of development (so far at least as their fruits are concerned) on lines of their own; they have typical nuclei which appear to undergo a sexual fusion. They are best regarded as forming by themselves a short but distinct line of descent, which may have arisen very far back, among organisms not yet characterised as either animals or plants. In the swarming and plasmodial stages the Myxomycetes would most naturally be regarded as animals, especially when we consider their mode of feeding. In the formation of their fruits and spores, however, they rather suggest plants of the nature of Fungi, but probably this is only a case of parallel development, not indicating a real blood-relationship to any undoubted members of the Vegetable Kingdom.

The Bacteria are still more difficult to place, for though in some ways we know so much about them, we still do not know what they are. Possibly several heterogeneous groups are included among them. The more typical Bacteria, however (such as *Bacillus subtilis* and its allies), which are often characterised by the production of endospores, are very unlike any other plants, and might be placed in a neutral group lying at the base of both the animal and vegetable kingdoms. Bacteria, however, offer so few points for morphological comparison with other groups, that nothing definite can be said at present as to their relationships.

Nostoc, representing the Cyanophyceæ, is another extremely simple type, so far at least as our present knowledge enables us to judge. It is probable, however, that further research may show the cell-structure to be more like that of the higher plants than it appears at present. It is doubtful whether the Cyanophyceæ should be classed with the Algæ or not; some botanists group them with the Bacteria in a class by themselves—the Schizophyta. As already pointed out, however, this relationship seems remote, except perhaps in the case of certain filamentous Bacteria. Until the cell-structure of the Cyanophyceæ and Bacteria is better understood, it may remain impossible to assign them to their true position. The absence of sexual reproduction and of ciliated swarm-cells tends to keep the Cyanophyceæ in an isolated position, though it now appears probable that they approach the Algæ in possessing bodies of the nature of incipient nuclei.

Leaving these dubious forms, we come to the Flagellata, a group of motile unicellular organisms, some of which are animals, while others are partly plant-like and partly animal-like. From the Flagellata both the Algæ and certain groups of the Protozoa may have evolved. *Euglena viridis* is an animal by virtue of its organisation, but a plant as regards its mode of nutrition.

Chlamydomonas is a remarkable unicellular Alga which betrays its relationship to simple animals of the group Flagellata by the fact that the cells display active movement throughout their ordinary vegetative life. In higher plants the power of locomotion becomes restricted to certain reproductive cells, and is ultimately lost altogether, its last appearance being in the spermatozoids of the Vascular Cryptogams and of such Gymnosperms as the Cycads. The sexual reproduction of *Chlamydomonas* ranges, in the different species, from conjugation, or the union of similar sexual cells, to fertilisation, in which the gametes are clearly differentiated.

From unicellular Green Algæ, or their ancestors among the flagellate Protozoa, quite a number of distinct lines of affinity branch out. In one direction we reach the Conjugatæ, some of which are themselves unicellular, while in others, such as *Spirogyra*, the cells are usually united in filaments. This group always remains at a low level anatomically, for no more complex thallus than a simple filament is ever produced. On the other hand, the histological structure shows a great advance, as indicated especially by the highly-differentiated chloroplasts, which not only assume strange and varied forms, but are much specialised internally, possessing protein bodies (pyrenoids) around which the formation of starch is partly localised. In many members of the group, namely among the Desmids, the external form of the individual cells is also very complex. The sexual process is well marked, but of the simplest kind, consisting in the union of the contents of similar vegetative cells, with, at most, only the slightest indication of any difference between the sexes. The Conjugatæ, so far as we know, do not lead on to any of the higher groups of plants.

In quite a different direction green unicellular organisms appear to have given rise to the remarkable group represented among our types by *Vaucheria*. Here the cell has become multinucleate, and grows out into a large and in some cases complex thallus, but without dividing, so that we find a highly organised plant with non-cellular structure. *Vaucheria* itself, though its thallus is simple compared with that of other Siphoneæ, has attained a very high level as regards reproduction, for its sexual organs are as sharply differentiated as in the higher multicellular plants.

Differences of sex appear quite independently in many diverse lines of affinity, among which that represented by *Vaucheria* is one of the most distinct.

Vaucheria is also of great interest because of its analogy with the lower Fungi. There can be no doubt of the close parallelism between such a Fungus as *Pythium* and such an Alga as *Vaucheria*. This parallelism, however, is no indication of affinity (see p. 223). Fungi and Algæ have probably been wholly distinct lines from a stage as simple as that of the unicellular Flagellata. Spermatozoids are not differentiated in *Pythium*, the male protoplasm being carried to the ovum by the fertilising tube. This condition has been compared to the change from fertilisation by spermatozoids to fertilisation by a pollen-tube, in passing from Cryptogams to Phanerogams. In both cases the disappearance of motile male cells is correlated with the loss of aquatic environment.[1] The gradual extinction of zoospores among the allies of *Pythium* is due to similar causes, and has been fully traced above (p. 218).

The Zygomycetes, as represented by *Pilobolus*, are Fungi which are adapted to a terrestrial mode of life. Though they still show signs of analogy with non-cellular Algæ, there can here be no question of affinity.

When we come to the higher Fungi, beginning with such forms as *Sphærotheca*, we find it impossible, in the present state of our knowledge, to determine their relation to the lower forms. The ascus has been sometimes regarded as corresponding to the sporangium of the Zygomycetes ; this, however, is probably untrue, for there is invariably a nuclear fusion in young asci and not in sporangia. In any case, the Ascomycetes have diverged very widely from the Phycomycete stock, as shown not only by their fructification but by their septate mycelium. This group reaches a very high development, the ascus-fruit having become a very complex structure in forms like *Xanthoria*. The Lichens, most of which are Ascomycetes, are the only Fungi which form a highly-organised aërial thallus. By their association with a green assimilating organism (the captive Alga) they have placed themselves on a level with the higher chlorophyll-containing plants.

Among the Ascomycetes the conidial form of fruit, though often important, is subordinate. In the remaining groups the conidia (the origin of which could already be traced in the Phycomycetes) and basidiospores have completely displaced the sporangia, and assume very

[1] See, however, p. 311.

various forms, constituting the great means of propagation. In the Uredineæ the æcidium is to be regarded as a sexually produced fruit, while the spermatia formed in the spermogonia play a very important part in the sexual process that leads to the development of æcidia. The uredospores, teleutospores, and sporidia, however, are all of an asexual nature.[1]

The Uredineæ are adapted to a strictly parasitic mode of life, and in habit differ greatly from other Basidiomycetes, especially if we consider a highly organised representative of the latter, such as the Mushroom. The fructifications of these highest of the Fungi are very complex, but the basidium itself appears to be comparable to a teleutospore germinating *in situ*, an homology which is strikingly confirmed by the occurrence of nuclear fusion, alike in the teleutospore and the basidium.

We see, then, that the Fungi form by themselves a highly-complex cycle of relationship, approaching the lower Algæ at one or two points, but otherwise distinct from the rest of the Vegetable Kingdom. Sexuality has proved to be far more general among this mass of saprophytic and parasitic organisms than was once supposed, though it often assumes a curiously reduced form comparable to the fusion of nuclei in apogamous Fern-prothalli.

We must now retrace our steps to the Algæ. The Red Seaweeds form a perfectly definite group by themselves, without clear connections either below or above. *Callithamnion* is a fair average type ; some forms are simpler, especially in the development of the fruit, but even the simplest of the undoubted Florideæ are highly organised

[1] It must be remembered, however, that the derivatives of the paired nuclei of the ascidiospores only fuse with one another in the young teleutospores (see p. 267).

plants, quite unlike any other group. Many are more complex than our type, but they are complex in their own peculiar way, and do not show any transition towards the higher groups of plants.

The sexual process in Florideæ is quite peculiar among Algæ, for no definite oöspore is ever formed as the result of fertilisation. The whole carpogonium when fertilised remains in complete continuity with the tissues of the thallus, and sends out branches, which ultimately produce numerous spores, usually after various subsidiary cell-fusions have taken place. This continuity of the spore-fruit with the thallus completely severs the Florideæ from the Bryophyta, with which there is otherwise a certain analogy, in so far as in both groups the result of fertilisation is a fruit. The Florideæ are also remarkable for the entire absence of motile ciliated cells, a point in which they differ from the great majority of the Algæ, though certain isolated groups, such as the Conjugatæ (which certainly have nothing to do with them), have the same peculiarity. The most striking result of recent work on the Florideæ has been to show that there is often a regular alternation between the sexual and asexual plants.[1]

The Phæophyceæ are also much isolated from other Algæ, but they have more in common with Chlorophyceæ than is the case with the Red Seaweeds. Ciliated cells are almost universal throughout the group, though in the highest Brown Algæ—the Fucales—they only appear as spermatozoids. Fertilisation usually, if not always, takes place *outside* the oögonium, a point in which these plants differ from the Green Algæ. The Fucales are on a much higher level than the rest of the group, but transitional forms are not altogether absent. On the whole, we may say that the Brown Algæ are a natural

[1] See also p. 208.

group, attaining very great complexity on their own lines, and not clearly related either to the lower or higher plants. In some of the Orders (*e.g.* the Laminariales and the Dictyotales) an alternation of sexual and asexual generations occurs.

We must now return to the Chlorophyceæ. We have in *Ulothrix* a form scarcely more complex than *Spirogyra*, but evidently on quite a different line of descent. Here the reproductive cells are all ciliated and active. It is between certain of these ciliated cells that conjugation takes place, and not between vegetative cells, as in *Spirogyra*. Evidently the origin of sexuality was quite distinct in these two groups. If we had been able to take a wider survey of the Vegetable Kingdom, we should have found evidence that this important step from asexual to sexual reproduction was made independently in many groups.

Ulothrix is also interesting from the great difference in subsequent development between the asexual and the sexual swarm-cells. The asexual zoospore merely reproduces the ordinary plant, whereas the zygospore gives rise to a dwarf individual quite distinct from the typical form. In fact we have here, coinciding with the first appearance of sexuality, some slight suggestion of regularly alternating generations. For this reason *Ulothrix* has been regarded as lying more or less in the direct line of descent of the archegoniate plants, in which regular alternation of generations is so striking a character. Of course this cannot be taken literally. No form now living can possibly be in the direct line of descent of any other form, any more than a man's cousin can be his ancestor! One cousin, it is true, may more than another inherit the characteristics of some remote ancestor, and this is all we mean in speaking of lines of descent with reference to recent plants. In the case of *Ulothrix*, however, there is probably nothing more than a remote analogy with the course of development of Bryophyta. A better parallel is really found in the alternation of sexual and asexual individuals in some of the higher Algæ.

Œdogonium makes in some respects a great step in advance. Here the sexual cells are perfectly differentiated ; instead of two similar conjugating motile cells, we find a small moving spermatozoid and a large resting ovum. The casual difference in size sometimes observed in *Ulothrix* has here become fixed, and other differences are added. Evidently there is a more perfect adjustment of function here ; for while both partners are still on equal terms as regards the union of their nuclei, it is the female cell alone which assumes the whole duty of accumulating food-supplies for the next generation. In order to do this most effectually, it remains at rest, in connection with the vegetative body of the plant. So far as the sexual division of labour is concerned, *Œdogonium* is almost as far advanced as any other plant.

The formation of dwarf male plants in some members of this group is without exact parallel in any other plants, and shows how far the specialisation of the sexes may go, even in simple organisms. This, however, is only a special case, for, as we have seen, there are some species of *Œdogonium* without dwarf males.

The division of the germinating oöspore into four swarm-spores is an interesting fact. These spores are just like the zoospores produced by the plant in its ordinary condition. The fact that they are always formed by the sexually produced resting-spore immediately on germination, indicates some approach to a regular alternation of sexual and asexual reproduction. In this respect, however, *Œdogonium* shows no advance on *Ulothrix*.

In another genus of green fresh-water Algæ, *Coleochæte*, the oöspore, while still enclosed in the oögonium, divides up into a group of cells, in each of which a zoospore is formed. This process shows a remote analogy with the formation of the simplest forms of sporogonium in the Liverworts, among which there is a genus (*Riccia*) in which the fruit consists of nothing but a mass of spore mother-cells surrounded by an epidermis. There is, however, no trace of affinity between the Alga and the Hepatic, in which the simplicity of the fruit may be the result of reduction.

The sporophyte of the higher plants, whatever its origin may have been, is specially adapted to the formation of aërial, as distinguished from aquatic, spores. The spores of the Archegoniatæ, from the lowest Bryophyta upwards, differ from those of any of the Algæ in being almost always suited for dissemination by the air. The sporophyte which bears them is essentially the *aërial* generation, while the gametophyte is dependent on water for the act of fertilisation. The difference is very well shown in *Pellia*, where the sexual generation is a low-growing thallus, keeping close to the damp ground, or even living under water, while the sporophyte grows up high into the air, exposing its spores as freely as possible, so that they may be dispersed by the wind. The result in other cases is attained in a different way, but the general rule holds good, that the function of the sporophyte—the dissemination of spores—requires exposure to dry air, while the most important function of the gametophyte—the act of fertilisation—requires the presence of water.

It is improbable that the aërial, asexual spores of the Bryophyta are homologous with the aquatic, asexual spores of the Green Algæ ; a better comparison is with the *tetraspores* of the Florideæ and the Dictyotales among the Brown Seaweeds. The origin of the Archegoniatæ must have taken place in enormously remote geological ages, when plants were first adapting themselves to terrestrial life, and we cannot be surprised that no transitional forms connecting them with the Algæ are known to us, unless it be among the remains of the older Devonian Flora. It is commonly assumed that the Bryophyta represent an earlier stage of evolution than the Pteridophyta, but palæontological evidence lends no support to this assumption.

So far as the vegetative structure of the thallus is concerned, *Pellia* is a very simple Liverwort; others are more complex, but in all alike the archegonium and antheridium are totally different from the sexual organs of any Thallophytes.

In the true Mosses both the sexual and asexual generations are more highly developed than in the Hepaticæ. Not only is the thallus replaced by a leafy stem (a change which is already accompanied in many Liverworts), but the anatomical structure is much more perfect, and a definite system of conducting tissue is differentiated. The sporophyte never develops into anything more than a fruit, yet it is anatomically the more elaborate of the two generations, as shown not only by the arrangements for dispersing the spores, but also by the vegetative tissues of the sporogonium, which have some resemblance to those of vascular plants, especially in the possession of true stomata. The mosses are highly organised plants in their own way, but appear to have no direct affinity with superior groups.

If we found a wide gap to cross in passing from Algæ to Bryophyta, this is equally the case when we pass on to the Vascular Cryptogams. The latter, as at present existing, form three quite distinct stocks,—Horsetails, Ferns, and Club Mosses,—but in none of the three is there the least evidence for any near relationship to the Bryophytes. The sexual generation presents little difficulty;

for instance, the thallus of a *Pellia* may well be compared with the prothallus of a Fern. It is the sporophyte which is so different in the two classes. Speaking broadly, the asexual generation in the Bryophyta is always a fruit, while in Vascular Cryptogams it is always a plant. Recently, however, among early Devonian fossils, organs with the structure of Bryophytic sporogonia have been found, borne on the stems of a simple form of Pteridophyte.

It is too soon to discuss the significance of these observations ; it may be that the further investigation of the oldest known land-plants will throw light on the origin of the two great divergent series of the Bryophyta and Pteridophyta. We have long had direct geological proof of the enormous antiquity of the Vascular Cryptogams which appear to form the bulk even of the older Devonian Flora. At that period, however, some of them seem to have been of a relatively primitive nature, differing from any of the existing races, and possibly combining their characters, while at the same time suggesting perhaps some lingering trace of old Algal affinities. The Devonian sporogonia, referred to above, are the first indications yet discovered of the high antiquity of a Bryophytic type of reproductive structure, though not of the Bryophyta themselves, as now understood.

The Pteridophyta are much more highly organised than any of the previous groups ; their advance is entirely confined to the asexual generation, for the gametophyte remains throughout at a very low level (below that of the simplest Bryophyta), and indeed degenerates as we reach the higher forms. No doubt the aquatic mode of fertilisation has kept back the gametophyte, which throughout the Pteridophyta is always a damp-loving organism, and never completely adapts itself to terrestrial life. The asexual plant, on the other hand, has attained the greatest complexity, rivalling that of the Flowering Plants.

In all the three main groups both homosporous and heterosporus forms have occurred, though among the Horsetails heterospory is now extinct.

Most Ferns, like our type, are homosporous, though even here rudimentary purely male prothalli are not uncommon. Heterospory in plants of the Fern alliance is now limited to two small families. The origin of heterospory may be compared to the origin of sexual differentiation (see above, p. 306). Just as nutritive functions came to be assigned specially to the ovum, so here they are assigned specially to the female prothallus, which has to feed the embryo during its development. The male prothallus can safely be treated by the plant on strictly economical principles, for it has nothing more to do than to produce a few minute spermatozoids. We see this change beginning both among Ferns and Horsetails ; in the latter the smaller, less vigorous prothalli, are exclusively male. Then the difference extends further back. The insignificant male prothallus only needs a small spore to grow from, while the female must be fed well from the first, and so a large spore, full of reserve food, is set apart for its formation. On the other hand, it is an advantage to have plenty of males in order to ensure fertilisation. Thus we get a large number of microspores forming small male prothalli, and a small number of megaspores forming comparatively large female prothalli.

In *Selaginella* we have an extreme case of heterospory, with an enormous difference between microspores and megaspores. There is, however, another change going on concurrently with that just sketched. Both prothalli have become the mere bearers of sexual organs, and are losing the character of distinct organisms. Hence the female prothallus, though so highly developed relatively to the male, is itself reduced, as compared with the prothallus of a homosporous Cryptogam.

The homologies between *Selaginella* and a gymnospermous Flowering Plant, such as *Picea*, are quite clear, and have been fully demonstrated above (p. 31). It is certain, however, that there is no near affinity between *Selaginella* and the Gymnosperms, which, as palæontological evidence indicates, probably sprang from a stock quite unlike any existing Cryptogam. In any case the homologies hold good, whatever the unknown heterosporous group or groups may have been from which the Gymnosperms sprang.

In the latter Class the female gametophyte has lost its independence altogether, and never leaves the megaspore, which itself remains shut up in the sporangium enclosed in an integument. Fertilisation leads to the formation of a seed—the characteristic structure of Flowering Plants, consisting of integument, sporangium, prothallus, and embryo, united to form one body, and fed entirely at the expense of the parent sporophyte.

Until the year 1896 there appeared to be a sharp distinction between Cryptogams and Phanerogams in the method of fertilisation, for active spermatozoids were supposed to be peculiar to the former, the generative cell in Phanerogams being carried passively to the ovum by the growth of the pollen-tube. The researches of the Japanese botanists Hirase and Ikeno, followed by those of Webber in America, first broke down this distinction, for they proved that in some Gymnosperms true motile spermatozoids are formed. The plants in which this important discovery was first made are the Maidenhair Tree (*Ginkgo biloba*), a tree of an ancient type, nearly related to the Coniferæ, and *Cycas revoluta* (see p. 34). In both these cases (to which others have now been added) a pollen-tube is formed, which enters the nucellus, but does not reach the archegonia, and serves chiefly to anchor the pollen-grain in the right position. Two generative cells are formed in the usual way, but each of these gives rise to a large, spirally-coiled spermatozoid, with numerous cilia (see Figs. 118 and 119, which have been sketched from the original preparations, kindly sent by Prof. Ikeno and Dr. Hirase; Fig. 119*). The spermatozoids break out from the pollen-tube, and by their own active movements swim to the necks of the archegonia through a cavity, filled with sap, which is formed in the upper part of the nucellus. These plants thus present a beautiful transition between the Cryptogamic and Phanerogamic methods of fertilisation. The male cells are conveyed for a short distance by the growth of the pollen-tube, but they have to com-

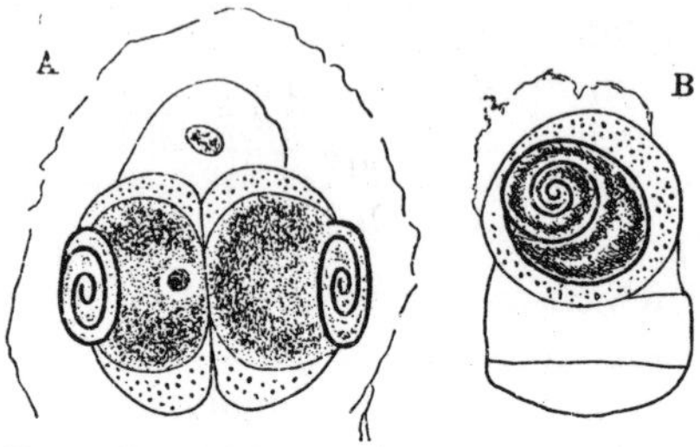

FIG. 118.—*Ginkgo biloba*. A, pair of generative cells in the pollen-tube. On the outer side of each cell a spiral coil is seen in connection with the nucleus. B, generative cell, showing the spiral spermatozoid in surface view. × 225. From a preparation by Dr. Hirase. (R. S.)

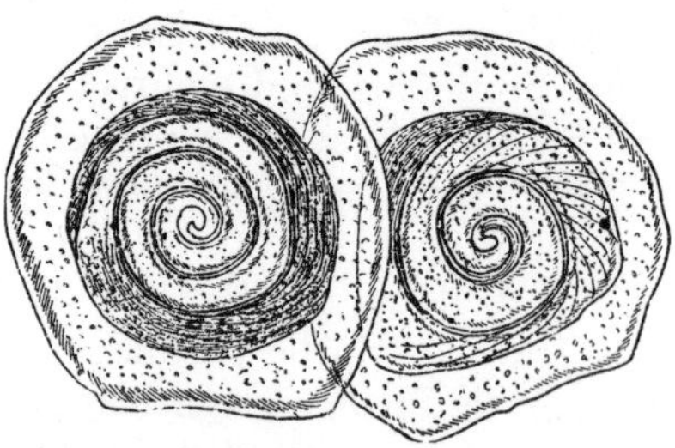

FIG. 119.—*Cycas revoluta*. Pair of generative cells from a pollen-tube, showing the spirally-coiled spermatozoids, surrounded by the cytoplasm of the cell. The fine striation overlying the spiral coil indicates the cilia. × 190. From a preparation by Prof. Ikeno. (R. S.)

plete the journey to the ovum by means of their own movements.

A condition still more cryptogamic has since been discovered by Caldwell in the Cuban Cycad *Microcycas calocoma*, in which each pollen-tube produces no less than eight pairs of generative cells, giving rise to sixteen spermatozoids. This multiplication of sperm-cells, probably unexampled among Seed-plants, is correlated with the immense number of archegonia (200 or more) produced in the embryo-sac of this strange plant.

These remarkable discoveries confirm, in the most striking way, the theoretical conclusions at which Hofmeister arrived forty years before.

FIG. 119*.—*Cycas revoluta*. Pollen-tube, branched above, and containing the two actively moving, ciliated spermatozoids. Magnified 44. (After Miyake.)

From Gymnosperms to Angiosperms is another great step, and here the gulf has not yet been bridged. In Angiosperms the female prothallus has almost disappeared, and even the archegonia are no longer recognisable. The embryo-sac (the equivalent of the megaspore) proceeds, after only a few preliminary divisions, to the formation of the ovum, and the development of the endosperm is dependent upon fertilisation, — obviously an expedient arrangement, for it is not formed at all unless it is wanted. The processes in the pollen-grain are also simplified. The great characteristic of Angiosperms is the high development of the flower and fruit. Not only does the megaspore remain enclosed in the integumented sporangium or ovule, but the latter is itself enclosed in the ovary, so that fertilisation has to take place through the mediation of the stigma and style. The remarkable development of the floral leaves, characteristic of most Angiosperms, is connected with the occurrence of pollination by insects, for which so many Angiosperms are specially adapted.

At present we are not in a position to determine either the relation of Angiosperms to Gymnosperms, nor that of Monocotyledons to Dicotyledons. The latter classes are mainly distinguished by vegetative characters, the reproductive phenomena being the same in both. On both these questions, however, we may hope for further light, especially from palæontological research, for the first appearance of Angiosperms falls within a geological period from which abundant fossil remains have come down to us. Dr. Wieland's discovery that the Bennettiteæ, Mesozoic plants allied to Cycads, produced elaborate hermaphrodite "flowers" somewhat comparable to those of the Angiosperms, has already thrown some light on these questions.

This brief summary has had one main object,—to indicate the complicated and difficult nature of all questions as to the affinities of plants. Naturalists in these days are agreed that the different forms of plants and animals arose from one another by descent. If this be so, a natural classification of the Vegetable Kingdom would take the form of a genealogical tree, just like the pedigree of a family. The genealogical tree of plants must have been complex beyond all power of conception, with boughs, branches, and twigs of every degree starting from each other at every possible point,—some long and some short, a few reaching on to our own day, while the immense majority came to an end in the long-past geologic ages.

If we attempted to construct such a tree, say for our

twenty-eight types, almost every branch would be marked with a query. If the reader has gained some idea of the difficulty and complexity of the profoundly interesting problems which the comparative study of plants presents to us, the object of this concluding chapter will have been attained.

GLOSSARY

ÆCIDIOSPORES.—The spores produced in chains in the cluster-cups (Æcidia) of Rust Fungi.

AMPHITHECIUM.—The outer layer of the young capsule in Bryophytes.

ANALOGOUS.—Corresponding in function.

ANDROSPORE.—The small zoöspore producing a dwarf male in some Œdogoniaceæ.

ANNULUS.—1. The row of cells, which causes a Fern sporangium to open.

2. The ring of abortive leaves at the base of an *Equisetum* cone.

3. The ring of cells in a Moss capsule, which splits when the lid comes off.

4. The ring underneath the pileus of a mushroom.

ANTHERIDIUM.—the organ in which the male cells are produced in Flowerless Plants.

APICAL CELL.—The large cell at the growing point, by the divisions of which, in many Flowerless Plants, all the new tissues and organs are ultimately produced.

APOGAMY.—The production of a sporophyte from a gametophyte without the intervention of the sexual organs.

APOPHYSIS.—The basal, sterile portion of a Moss capsule.

APOSPORY.—The production of a gametophyte from a sporophyte without the intervention of the spores.

APOTHECIUM.—the open, disc-shaped fruit of some Lichens and Ascomycetes.

ARCHEGONIUM.—The organ in which the female cell is produced in Gymnosperms, Vascular Cryptogams, and Bryophytes.

ARCHESPORIUM.—The cell or cells from which the spores are ultimately produced in a sporangium or sporogonium.

ASCOGONIUM.—the female organ of Ascomycetes, which gives rise to the ascus or ascogenous hyphæ.

ASCOSPORES.—Spores produced in an ascus.

ASCUS.—The characteristic reproductive cell of the Ascomycetes, producing a definite number of spores, usually 4 or 8, in its interior.

AUXILIARY CELLS.—The cells which fuse with the fertilised carpogonium in Red Seaweeds.

BASIDIOSPORES.—Spores produced on a basidium.

BASIDIUM.—The cell which gives rise externally to certain spores in the Basidiomycetes.

BIOLOGIC FORM.—A physiological variety of a species of parasitic fungus confined to a particular host or hosts.

BIPINNATE.—When the leaflets of a pinnate leaf are themselves pinnate.

CALYPTRA.—The envelope enclosing the young fruit in Bryophytes.

CAPILLITIUM.—A web of threads investing the spores: *e.g.* in the sporangium of a Myxomycete.

CAPSULE.—The spore-case in Bryophytes.

CARINAL.—Opposite a rib of the stem.

CARPOGONIUM.—The cell from which the fruit originates in Red Seaweeds.

CARPOSPORES.—The sexually produced spores of Red Seaweeds.

CHLAMYDOSPORE.—A thick-walled spore in Fungi usually formed in a hypha.

CILIA.—Protoplasmic threads, by means of which spermatozoids and other motile cells move.

CIRCINATE.—Coiled like a crozier or a snail.

COLUMELLA.—The central, sterile part of a Moss capsule and of the sporangium of certain Fungi.

COMMENSALISM.—The association of two different organisms for mutual benefit, as the Alga and Fungus in a Lichen.

CONCEPTACLE.—The cavity containing the sexual organs in Fucales.

CONIDIA.—Propagative cells of Fungi, asexually and externally produced.

CONIDIOPHORE.—The organ which bears the conidia.

CONJUGATION.—The union of two similar sexual cells.

CYST.—A thick-walled resting cell, *e.g.* the compartments of the resting plasmodium in Myxomycetes.

CYSTOCARP.—The sexually produced fruit of Red Seaweeds.

DICHOTOMY.—The equal division of the growing point to form two equivalent branches.

ELATERS.—1. The arm-like appendages attached to the spore in *Equisetum*.
2. The spirally thickened cells in the capsule of Liverworts, helping to disperse the spores.

ENDOSPORE.—A spore formed inside a cell in some Bacteria.

ENDOTHECIUM.—The inner layer of the young capsule in Bryophytes.

EPIBASAL.—Forming the upper part of the embryo.

EPITHECIUM.—The layer formed by the tips of the paraphyses above the asci in the fructification of a Lichen.

FERMENTATION.—The production of gases, especially CO_2, by certain micro-organisms under anærobic conditions.

FLAGELLA.—See CILIA.

FOOT.—The part of the embryo which absorbs food from the pro thallus in Vascular Cryptogams.

GAMETE.—A sexual cell.

GAMETOPHYTE.—The sexual generation.

GEMMÆ.—Cells or buds serving to propagate the plant.

GLYCOGEN.—A substance related to starch found in Fungi, etc.

GONIDIA.—The algal cells of Lichens.

HAUSTORIUM.—The sucker of a parasite, absorbing food from the host plant.

HERMAPHRODITE.—Containing both sexes, in the same flower or the same conceptacle.

HETEROCYSTS.—Large, thick-walled cells occurring in *Nostoc* and its allies.

HETERŒCIOUS.—Infesting different host plants, at different stages of its life-history (used of parasitic Fungi).

HETEROGAMY.—The sexual fusion of unlike gametes (♂ and ♀).

HETEROSPOROUS.—Producing two kinds of spores: microspores (male) and megaspores (female).

HETEROTHALLISM.—The occurrence of physiologically, but not morphologically, distinct sexual forms in certain Fungi.

HOMOLOGOUS.—Corresponding in origin.

HOMOSPOROUS.—Producing one kind of spore only, as Ferns and Horsetails.

HORMOGONIA.—Actively creeping filaments, in *Nostoc* and its allies.

HYALOPLASM.—Clear protoplasm, free from granules, as in a pseudopodium of a Myxomycete.

HYMENIUM.—The layer of cells producing the spores in a Fungus.

HYPHÆ.—The filaments of which Fungi, and sometimes Algæ, are composed.

HYPOBASAL.—Forming the lower part of the embryo.

HYPOTHECIUM.—The part of the fructification of a Lichen below the hymenium.

INDUSIUM.—A membranous envelope enclosing a group of sporangia, or sometimes a single sporangium, in Ferns.

INTERCALARY.—Lying between apex and base, used of growing points which are not terminal.

INVOLUCRE.—An envelope enclosing the reproductive organs; as the pocket round the archegonia in *Pellia*.

ISOGAMY.—The sexual fusion of similar gametes.

LAMELLÆ.—The "gills" of a Mushroom or Toadstool, which bear the spores.

LIGULE.—A membranous scale borne on the upper side of the leaf-base in *Selaginella*, *Isoëtes*, and their fossil allies.

MALIC ACID.—$C_4H_6O_5$: occurs in apples and most fruits.

MEGASPORANGIUM.—A spore-sac containing one or more megaspores.

MEGASPORE.—The large spore, which produces the female prothallus in heterosporous Cryptogams.

MICROCYST.—A swarm-spore, which has enclosed itself in a cell-wall, in Myxomycetes.

MICROSPORANGIUM.—A spore-sac containing microspores.

MICROSPORE.—A small spore, producing a male prothallus

MICROZOOSPORES.—The minute swarm-cells of certain Green Algæ.

MONOSTELIC.—Containing a single vascular cylinder.

MYCELIUM.—The vegetative part of a Fungus, as Mushroom spawn.

MYXAMŒBÆ.—The creeping swarm-spores of Myxomycetes.

OCTANTS.—The eight cells into which the fertilised ovum divides in some Vascular Cryptogams.

OÖGAMY.—The sexual fusion of a stationary female gamete with a male gamete.

OÖGONIUM.—The cell which contains the ovum or ova in many Algæ and Fungi.

OÖSPORE.—The fertilised ovum.

OPERCULUM.—The lid of a Moss capsule.

OVUM.—The female cell which, when fertilised, gives rise to the new generation.

PALEÆ.—Chaffy scales, as those of Ferns.

PARAMYLON.—A substance related to starch found in certain Flagellata.

PARAPHYSES.—Sterile hairs accompanying reproductive organs.

PARASITE.—A plant which grows on another plant and feeds on it.

PELTATE.—Shield-shaped.

PERIDIUM.—The wall of the fruit in certain Fungi, as in the Æcidium of the Rusts.

PERIPLASM.—The outer part of the protoplasm in an antheridium or oögonium, when not used in the formation of the sexual cells.

PERISTOME.—The ring of teeth round the mouth of a Moss capsule.

PERITHECIUM.—The envelope enclosing the ascus or asci in some Fungi, as in *Sphærotheca*.

PILEUS.—The upper, expanded part of the fruit in a Mushroom or Toadstool.

PINNATE.—Compound, with the leaflets arranged along each side of a common midrib.

PLANOGAMETE.—An actively moving sexual cell.

PLASMODIUM.—A naked protoplasmic body, containing many nuclei and capable of creeping movements.

POLYSTELIC.—Containing several vascular cylinders.

PROCARP.—The female reproductive branch in Red Seaweeds, before fertilisation.

PROTHALLUS.—The sexual generation in Vascular Cryptogams, and, in a modified form, in Flowering Plants.

PROTONEMA.—The filamentous growth developed from the spore (or other cells) of a Moss, and giving rise to the leafy plants.

PSEUDOPODIA.—The protrusions of living protoplasm put out, *e.g.*, by the moving plasmodium of Myxomycetes.

PUSTULE.—Dense aggregations of the spores of Fungi on the host.

PYCNIDIUM.—A flask-shaped fructification in certain Fungi which forms conidia (pycnospores).

PYRENOID.—A protein body, round which starch is formed, in the chloroplasts of many Algæ.

RAMENTA.—The chaffy scales or hairs of Ferns.

RECEPTACLE.—The end of a branch, on which the reproductive organs are borne.

RHIZOID.—A filamentous organ, discharging the function of a root, in the prothallus of Vascular Crytogams and in Bryophytes and Thallophytes.

RHIZOME.—An underground or prostrate stem, from which roots and herbaceous stems arise.

RHIZOPHORE.—A leafless branch, bearing the roots, as in *Selaginella*.

SAPROPHYTE.—A plant which lives on dead organic matter.

SCALARIFORM.—Like a ladder, as the pitted walls of the tracheids in Ferns.

SCLERENCHYMA.—Thick-walled, fibrous tissue, serving as a mechanical support.

SCLEROTIUM.—The thick-walled, resting condition of the plasmodium in Myxomycetes and the compact mycelium of certain Fungi.

SETA.—The stalk of the fruit in Liverworts and Mosses.

SOREDIA.—Detached groups of associated algal cells and fungal hyphæ, serving for the propagation of Lichens.

SORUS.—A group of sporangia.

SPECIALISED PARASITISM.—The occurrence of distinct physiological forms of a parasitic Fungus on particular hosts.

SPERMATIUM.—A male cell, which has no power of active movement, in some Algæ and Fungi.

SPERMATOZOID.—A male cell, which has the power of active movement by means of cilia.

SPERMOGONIUM.—The organ in which spermatia are produced in some Fungi.

SPORANGIOLE.—A small sporangium, containing one or a few spores, in Zygomycetes.

SPORANGIOPHORE.—An organ, in some Vascular Cryptogams, specially adapted for bearing sporangia; also used in Fungi.

SPORANGIUM.—A sac or case in which spores are produced.

SPORE.—The reproductive body in Cryptogams, usually consisting of a single cell.

SPORIDIUM.—A small spore, formed on the promycelium of Rust Fungi.

SPOROGONIUM.—the fruit, or asexual generation, in Liverworts and Mosses.

SPOROPHYLL.—A leaf, bearing sporangia.

SPOROPHYTE.—The asexual generation.

STELE.—A vascular cylinder, consisting of xylem and phloëm, usually associated with cellular tissue.

STERIGMA.—The little stalk on which a basidiospore or sporidium is borne.

STIPE.—The stem or stalk of a Mushroom or Toadstool.

SUSPENSOR.—1. A filament of cells, forming a temporary organ of attachment for the embryo: as in *Selaginella* and most Flowering Plants. 2, The hypha which bears a sexual cell in Zygomycetes.

SYMBIOSIS.—The association of two organisms living together for mutual benefit, as the Fungus and Alga in a Lichen.

TAPETUM.—The layer of cells immediately surrounding the spore-producing tissue (archesporium).

TELEUTOSPORE.—The spore of Rust Fungi which produces the promycelium on germination.

TETRASPORANGIUM.—The cell which divides up to form the tetraspores.

TETRASPORES.—The asexual spores, usually produced in fours, in Red Seaweeds.

THALLUS.—The vegetative body of the plant, when not differentiated into stem and leaf.

THECA.—The capsule of a Liverwort or a Moss.

TRABECULÆ.—Cell filaments connecting the stele with the cortex, as in *Selaginella*.

TRAMA.—The tissue underneath the hymenium in the higher Fungi.

TRICHOGYNE.—The hair-like organ which receives the male cells in Red Seaweeds.

TROPISM.—The bending movement of a plant organ in response to an external stimulus.

TUBER.—A short and thick underground branch, serving for propagation.

UREDOSPORE.—The unicellular summer spore of the Rust Fungi.

VALLECULAR.—Opposite a furrow of the stem.

VELUM.—The membrane which covers the gills in a young Mushroom or allied Fungus.

VENTER.—The lower part of an archegonium, containing the ovum.

WHORL.—A ring of leaves or other organs all inserted at the same level on the stem.

ZOOGLŒA.—The gelatinous colony formed by certain Bacteria in the resting condition.

ZOOSPORE.—An actively moving ciliated spore in Algæ and Fungi.

ZYGOSPORE.—The product of the conjugation of two similar sexual cells.

INDEX